Abelian Varieties with Complex Multiplication and Modular Functions

Princeton Mathematical Series

1. The Classical Groups *by Hermann Weyl*
3. An Introduction to Differential Geometry *by Luther Pfahler Eisenhart*
4. Dimension Theory *by W. Hurewicz and H. Wallman*
8. Theory of Lie Groups: I *by C. Chevalley*
9. Mathematical Methods of Statistics *by Harold Cramér*
10. Several Complex Variables *by S. Bochner and W. T. Martin*
11. Introduction to Topology *by S. Lefschetz*
12. Algebraic Geometry and Topology *edited by R. H. Fox, D. C. Spencer, and A. W. Tucker*
14. The Topology of Fibre Bundles *by Norman Steenrod*
15. Foundations of Algebraic Topology *by Samuel Eilenberg and Norman Steenrod*
16. Functionals of Finite Riemann Surfaces *by Menahem Schiffer and Donald C. Spencer*
17. Introduction to Mathematical Logic, Vol. I *by Alonzo Church*
19. Homological Algebra *by H. Cartan and S. Eilenberg*
20. The Convolution Transform *by I. I. Hirschman and D. V. Widder*
21. Geometric Integration Theory *by H. Whitney*
22. Qualitative Theory of Differential Equations *by V. V. Nemytskii and V. V. Stepanov*
23. Topological Analysis *by Gordon T. Whyburn* (revised 1964)
24. Analytic Functions *by Ahlfors, Behnke, Bers, Grauert et al.*
25. Continuous Geometry *by John von Neumann*
26. Riemann Surfaces *by L. Ahlfors and L. Sario*
27. Differential and Combinatorial Topology *edited by S. S. Cairns*
28. Convex Analysis *by R. T. Rockafellar*
29. Global Analysis *edited by D. C. Spencer and S. Iyanaga*
30. Singular Integrals and Differentiability Properties of Functions *by E. M. Stein*
31. Problems in Analysis *edited by R. C. Gunning*
32. Introduction to Fourier Analysis on Euclidean Spaces *by E. M. Stein and G. Weiss*
33. Étale Cohomology *by J. S. Milne*
34. Pseudodifferential Operators *by Michael E. Taylor*
36. Representation Theory of Semisimple Groups: An Overview Based on Examples *by Anthony W. Knapp*
37. Foundations of Algebraic Analysis *by Masaki Kashiwara, Takahiro Kawai, and Tatsuo Kimura. Translated by Goro Kato*
38. Spin Geometry *by H. Blaine Lawson, Jr., and Marie-Louise Michelsohn*
39. Topology of 4-Manifolds *by Michael H. Freedman and Frank Quinn*
40. Hypo-Analytic Structures: Local Theory *by François Treves*
41. The Global Nonlinear Stability of the Minkowski Space *by Demetrios Christodoulou and Sergiu Klainerman*
42. Essays on Fourier Analysis in Honor of Elias M. Stein *edited by C. Fefferman, R. Fefferman, and S. Wainger*
43. Harmonic Analysis: Real-Variable Methods, Orthogonality, and Oscillatory Integrals *by Elias M. Stein*
44. Topics in Ergodic Theory *by Ya. G. Sinai*
45. Cohomological Induction and Unitary Representations *by Anthony W. Knapp and David A. Vogan, Jr.*
46. Abelian Varieties with Complex Multiplication and Modular Functions *by Goro Shimura*

ABELIAN VARIETIES WITH COMPLEX MULTIPLICATION AND MODULAR FUNCTIONS

Goro Shimura

PRINCETON UNIVERSITY PRESS
PRINCETON, NEW JERSEY

Published by Princeton University Press, 41 William Street,
Princeton, New Jersey 08540
In the United Kingdom: Princeton University Press,
Chichester, West Sussex

Library of Congress Cataloging-in-Publication Data

Shimura, Gorō, 1930–
Abelian varieties with complex multiplication and modular
functions / Goro Shimura.
p. cm. — (Princeton mathematical series ; 46)
Includes bibliographical references and index.
ISBN 0-691-01656-9 (alk. paper)
1. Abelian varieties. 2. Modular functions. I. Title.
II. Series.
QA564.S458 1997
514.3—dc21 97-8673 CIP

This book has been composed in Times Roman

Princeton University Press books are printed on acid-free paper and meet the guidelines for permanence and durability of the Committee on Production Guidelines for Book Longevity of the Council on Library Resources

http://pup.princeton.edu

Printed in the United States of America

1 3 5 7 9 10 8 6 4 2

Contents

Preface

It was in 1961 that the book *Complex Multiplication of Abelian Varieties and Its Applications to Number Theory*, coauthored by Yutaka Taniyama and myself, appeared as No. 6 of Publications of the Mathematical Society of Japan. In its preface, which the reader will find after this one, I presented a short history of the subject, as well as a brief account of the contents of the book. In view of the progress made after its publication, it is natural to write a sequel to it. The present volume is my attempt, if modest, at the realization of this project. To make it self-contained, I included the essential contents of the first sixteen sections of the 1961 book. In that sense, this may naturally be viewed as an expanded edition, though its raison d'être is mainly in the newly written seventeen sections, and the last two sections of the old book are not included.

When I started to work on this sequel, I was well aware that it would have been more desirable to write a completely reorganized book incorporating both old and new material. Having found the task difficult, I chose a compromise plan as described above. However, I made considerable revisions throughout the old part, stating a few theorems in stronger forms and inserting explanatory sentences at many points.

The principal feature of the new part is as follows. We first reformulate the main theorem of complex multiplication in the adelic language, and next, in Chapter V, we show that the zeta function of an abelian variety A with complex multiplication over a number field k can be given as a product of certain Hecke L-functions. Such was already given in the previous book, but we now try to present more precise results. We first treat the case in which "imaginary endomorphisms" are rational over k, and next the case in which only "real endomorphisms" are rational over k. Eventually we determine the zeta function of A over any number field of definition under a certain condition which is satisfied if A is simple. We then investigate the problem of finding a model of A over a given field and a given Hecke character.

In Chapter VI we first consider families of abelian varieties parametrized by the points on certain hermitian symmetric spaces, and then relate them to modular forms and functions on such spaces. The classical theory of complex multiplication asserts that the values of elliptic modular functions at a point on the upper half-plane belonging to an imaginary quadratic field K generate abelian extensions of K. This is essentially a reformulation of our theorem specialized to the one-dimensional case, though naturally it requires some nontrivial facts on such modular functions. Now the higher-dimensional version of this fact can be formulated in terms of Hilbert or Siegel modular functions. We present here a formulation in a more general setting so that those functions become special cases.

In the historical part of the preface of the 1961 book, I referred to a "successful" work of Hecke on complex multiplication in two variables, but that was an overstatement, because the work was seriously flawed, though he stated basically correct assertions, which could have been proved if everything had been worked out. I also stated that this work of Hecke was included as a special case of the results in the book, but that was another overstatement, since we treated there only abelian functions instead of modular functions. However, the theorems in Section 26 of the present volume include what he expected as one of the easiest cases.

In Chapter VII we first recall the classical theory of theta functions in the sense of Jacobi-Riemann, and study the model of an arbitrary polarized abelian variety $\mathcal{P}$ that is the image of the projective embedding given by theta functions. Our interest is in the field of rationality of the model, which turns out to be an explicitly given finite algebraic extension k of the field of moduli of $\mathcal{P}$. We shall then determine the periods of the holomorphic 1-forms on $\mathcal{P}$ rational over k in terms of the values of certain modular forms.

In the case of complex multiplication, these periods lead to a collection of certain invariants which we call "the period symbol." The final two sections concern various algebraic relations among the periods on several abelian varieties which are not necessarily isogenous. We shall express the relations as certain linear properties of the period symbol. In a sense, this part may be viewed as the culmination of what has been developed in some of the preceding sections.

All these results in the new part were published in my papers listed in the Supplementary References at the end of the book. I have tried, however, to present better formulations than in those papers. I wish to acknowledge my gratitude to Alice Silverberg and Hiroyuki Yoshida, who kindly read the first draft of the entire new part and contributed numerous suggestions, which have been incorporated in the final version.

The 1961 book was published under the names of two authors. Though I believe that was the right thing to do, in fact, I was solely responsible for the whole text of the volume, as I indicated at the end of its preface. The reader who is interested in knowing about the life of Taniyama and the circumstances under which I wrote the book will find relevant pieces of information in my article "Yutaka Taniyama and His Time," published in the *Bulletin of the London Mathematical Society*, 21 (1989), pp. 186–196. It is my hope that the reader will long remember that to a large extent the main results up to §16 of the present volume have their source in our collaborative work until 1956.

Princeton, February 1996 G. S.

Preface to *Complex Multiplication of Abelian Varieties and Its Applications to Number Theory* (1961)

The history of complex multiplication began with the works of Gauss and Abel on elliptic functions. It would be right, however, to call Kronecker the initiator of number-theoretic investigation of the subject. The main theorem of Kronecker's theory asserts that the abelian extensions of every imaginary quadratic field are generated by the special values of certain elliptic or elliptic modular functions; as Kronecker left the work unfinished, the accomplishment needed the efforts of the late authors, Weber and Takagi. A similar and simpler result, which is also due to Kronecker, holds for the field of rational numbers: the abelian extensions of the field of rational numbers are generated by the roots of unity, the special values of the exponential functions. Hilbert conceived an idea to generalize these results, namely, to construct abelian extensions over any given algebraic number field by means of special values of analytic functions; he took this up as the 12th problem in his Paris Vortrag and emphasized its importance in number theory; it should be also mentioned that Kronecker had already thought of the problem. The first essential progress in this direction was made by Hecke, by following the idea of Hilbert. He succeeded in constructing unramified abelian extensions of certain biquadratic fields by means of singular values of Hilbert modular functions of two variables. This was the last work, as well as the first, till recent years, which attacked the problem successfully. On the other hand, a new development took place in the theory of complex multiplication of elliptic functions. First H. Hasse perceived the connection between complex multiplication and the Riemann hypothesis for congruence zeta functions, which was later proved by A. Weil in a fully general form. This observation led M. Deuring to establish a purely algebraic treatment of complex multiplication of elliptic curves. He could, moreover, along the same line of ideas, determine the zeta functions of elliptic curves with complex multiplication. The definition of zeta function of an algebraic curve defined over an algebraic number field is originally due to Hasse; and Weil is the first contributor to this subject.

Now the advancement in algebraic geometry of late years, especially in the abstract theory of abelian varieties, due to Weil, enabled us to approach the problem in a fairly general form, as was shown in three papers, published in the Proceedings of the International Symposium on algebraic number theory, Tokyo-Nikko, 1955, of Weil and the authors of the present monograph. It is the purpose of the monograph to provide a full exposition of the results announced in these memoirs.

Our chief object is the arithmetic of an abelian variety A of dimension n, whose endomorphism-ring is isomorphic to an order in an algebraic number field K of degree $2n$ over the field of rational numbers. The first task is to show that the field of moduli of A, whose definition must and can be given by virtue of the notion of polarization, and the fields generated by the coordinates of the points on A of finite order, are class fields over a certain algebraic number field K^*, corresponding to the ideal-groups determined by the arithmetical structure of A (Main Theorems 1, 2, 3 in Chapter IV). The number field K cannot be taken arbitrarily; it must be a totally imaginary quadratic extension of a totally real number field. K^* is the algebraic number field determined by K and the representation of K in the linear space of invariant differential forms on A. If $n = 1$, we have $K = K^*$, while if $n > 1$, both the cases $K = K^*$ and $K \neq K^*$ may occur. The abelian extensions of K^* thus obtained from A do not provide all the abelian extensions of K^* unless $n = 1$; at any rate, the classical results of Kronecker and Hecke are included in our main theorems as particular cases. It is noteworthy that the prime ideal decomposition of the $N(\mathfrak{p})$-th power endomorphism $\pi_{\mathfrak{p}}$ of $A(\mathfrak{p})$ is fundamental in our whole theory, where $A(\mathfrak{p})$ denotes the reduction of the variety A modulo a prime ideal $\mathfrak{p}$ of a field of definition k for A. The above result is in close connection with the investigation of the zeta function of the abelian variety A. In fact, a more precise analysis of $\pi_{\mathfrak{p}}$ shows that the correspondence $\mathfrak{p} \rightarrow \pi_{\mathfrak{p}}$ determines a Grössen-character of the field k. We are then led to the expression of the zeta function of A by the product of several Hecke L-series attached to Grössen-characters (Main Theorem 4 of Chapter IV); this is a generalization of the results of Weil and Deuring mentioned above.

We now give a summary of the contents. Chapter I is an exposition of more or less known results on abelian varieties, which are mostly given without proofs. The only exception is Section 2, where we have given a detailed (but elementary) treatment of invariant differential forms on abelian varieties. Section 3 deals with the analytic representation of abelian varieties, their homomorphisms and divisors by means of complex tori. In Section 4, first the notion of polarized varieties is introduced and then the definitions of field of moduli and Kummer variety are given. Chapter II is devoted to the algebraic part of the theory of complex multiplication. Sections 5 and 6 contain a necessary and sufficient condition that an algebraic number field K of degree $2n$ be realized as the endomorphism-algebra of an abelian variety of dimension n. Section 7 is the study of mutually isogenous abelian varieties in connection with the ideals of the endomorphism-rings; Section 8 concerns the phenomena which are essential only in the case of dimension $n > 1$ and related to the definition of the number field K^*. Chapter III contains the theory of reduction of algebraic varieties modulo a prime divisor of the basic field. We shall prove in Section 13 the fundamental theorem concerning the prime ideal-decomposition of $N(\mathfrak{p})$-th power homomorphism. Our final aims are achieved in Chapter IV. The first step (Section 14) is the investigation of the relations between abelian varieties, of the same type of complex multiplication, whose polarizations are also of the "same type." Then, in Section 15, we prove the first main theorem; an unramified class field is obtained by the field of moduli. A similar argument together with the analysis of the points of finite order gives us also class fields, whose characterization is the object of Section 16. These results are obtained assuming the endomorphism-ring to be the principal order of the number field. In Section 17, the case of nonprincipal order is completely investigated. The last section is devoted to the determination of the zeta-function of an abelian variety of the type described above.

A large part of the contents was prepared in collaboration of both authors during 1955–56 and published in 1957 in Japanese as the first six chapters of the book with the title "Kindai-teki Seisu-ron" (Modern number theory). The English version was then planned; but owing to the sudden death of the second named author in the autumn of 1958, the work had to be completed by the person left behind. The present volume is not a mere translation, however; we have written afresh from beginning to end, revising at many points, and adding new results such as Section 17 and several proofs of propositions which were previously omitted.

The present monograph owes much to the idea of Weil [54], though we have not necessarily indicated explicit references in the text. I take this opportunity to acknowledge my cordial gratitude to Professor André Weil for his constant advice, suggestions, and encouragement. I wish to acknowledge also my thanks to Mr. Taira Honda who read the manuscipt and contributed many useful suggestions.

University of Tokyo
February 1960 *Goro Shimura*

Notation and Terminology

We denote by **Z**, **Q**, **R**, and **C** the ring of rational integers, the fields of rational numbers, real numbers, and complex numbers, respectively, and by $\overline{\mathbf{Q}}$ the algebraic closure of **Q** in **C**. If ℓ is a rational prime, $\mathbf{Z}_\ell$ and $\mathbf{Q}_\ell$ denote the ring of ℓ-adic integers and the field of ℓ-adic numbers, respectively.

Given an associative ring R with identity element and an R-module S, we denote by $R^\times$ the group of all invertible elements of R, and by S_n^m the module of all $m \times n$-matrices with entries in S. We shall often put $S^m = S_1^m$. Thus $\mathbf{C}^n$ (resp. $\mathbf{R}^n$) is the vector space of n-dimensional complex (resp. real) column vectors. If V is a vector space over **C** of dimension n, a *lattice of* V means a discrete subgroup of V isomorhic to $\mathbf{Z}^{2n}$.

We denote the identity element of the ring R_n^n by 1_n, or simply by 1, and the transpose of a matrix X by ${}^tX$. We put $GL_n(R) = (R_n^n)^\times$, and

$$SL_n(R) = \{\, \alpha \in GL_n(R) \mid \det(\alpha) = 1 \,\}$$

if R is commutative. For a complex number or more generally for a complex matrix α we denote by $\mathrm{Re}(\alpha)$, $\mathrm{Im}(\alpha)$, and $\overline{\alpha}$ the real part, the imaginary part, and the complex conjugate of α. If $\alpha_1, \ldots, \alpha_r$ are square matrices, $\mathrm{diag}[\alpha_1, \ldots, \alpha_r]$ denotes the matrix with $\alpha_1, \ldots, \alpha_r$ in the diagonal blocks and 0 in all other blocks.

Terminology and basic notation concerning algebraic geometry are essentially the same as in Weil's trilogy [44], [45], and [46]. In particular, a variety or a subvariety always means an absolutely irreducible one. If V is a variety, we view V as the set of all its points rational over the universal domain. For a finite algebraic extension k' of a field k we denote by $[k' : k]_i$ and $[k' : k]_s$ the inseparable and separable factors of the degree of k' over k. If σ is an isomorphism of a field k_1 onto a field k_2, then for $z \in k_1$ we denote by z^σ the image of z under σ with the rule $z^{\sigma\tau} = (z^\sigma)^\tau$ for another isomorphism τ of k_2 onto a field. Furthermore, if Y is an algebro-geometric object defined with respect to k_1, we denote by Y^σ the image of Y under σ. If Y is defined by some polynomial equations, then Y^σ is defined by the transforms of the equations under σ. If k is a field and x is a point of an affine (resp. a projective) space, we denote by $k(x)$ the field generated over k by the coordinates (resp. the quotients of the coordinates) of the point x. We use also the notation $k(x)$ for the points on an abstract variety (cf. [44]).

Method of citation. There are two lists of references: Bibliography and Supplementary References. The former is the same as in the 1961 book and its articles are numbered from 1 through 57. Each item in the latter list contains a roman capital with or without more letters and numerals. Therefore the reader should be able to determine in which list the article in question appears.

Abelian Varieties with Complex Multiplication and Modular Functions

CHAPTER I

Preliminaries on Abelian Varieties

1. Homomorphisms and Divisors

The purpose of this section is to recall briefly some of the basic concepts on abelian varieties defined over arbitrary ground fields. For the general theory of abelian varieties, we refer the reader to Weil [46] and Lang [26].

1.1. By a *group variety* we understand an algebraic variety (affine, projective, or abstract) G with a group structure such that the map $(x, y) \mapsto xy$ of $G \times G$ into G and also the map $x \mapsto x^{-1}$ of G into G are both rational mappings defined everywhere. Such a G must be nonsingular. We say that a group variety G is defined over a field k if the variety G and these maps are defined over k.

A group variety is called an *abelian variety* if it is complete in the sense of [44], which is the case if it is a projective variety. In fact, every (abstract) abelian variety defined over k is biregularly isomorphic to a projective variety over k, and therefore we practically lose nothing by assuming it to be projective. It is a well-known fact that the group law of an abelian variety is commutative. Therefore we use the additive notation and denote the zero element by 0. Whenever we speak of *an abelian variety A defined over a field k*, we always assume that A as a group variety is defined over k in the above sense. If a subvariety of an abelian variety A is a subgroup, then it has a natural structure of abelian variety, and is called an *abelian subvariety* of A. An abelian variety is called *simple* if it has no abelian subvarieties other than $\{0\}$ and itself. (Some authors call it *absolutely simple*.)

Let A and B be two abelian varieties. By a *homomorphism* of A into B, or an *endomorphism* when $A = B$, we shall always understand a rational mapping λ of A into B satisfying $\lambda(x + y) = \lambda(x) + \lambda(y)$ generically. If that is so, then λ is defined everywhere and $\lambda(x + y) = \lambda(x) + \lambda(y)$ for every $x, y \in A$. We denote by $\mathrm{Ker}(\lambda)$ the kernel of λ. We shall often write λx for $\lambda(x)$. If such a λ is birational, then it must be biregular, and we call it an *isomorphism*, or an *automorphism* when $A = B$. We denote by $\mathrm{Hom}(A, B)$ the set of all homomorphisms of A into B, defined over any extension of a given field of definition for A and B, and put $\mathrm{End}(A) = \mathrm{Hom}(A, A)$. It is a basic fact that $\mathrm{Hom}(A, B)$ is a free $\mathbf{Z}$-module of finite rank; moreover if A and B are defined over k, then every element of $\mathrm{Hom}(A, B)$ is defined over a separably algebraic extension of k. We put also $\mathrm{Hom}_{\mathbf{Q}}(A, B) = \mathrm{Hom}(A, B) \otimes_{\mathbf{Z}} \mathbf{Q}$ and $\mathrm{End}_{\mathbf{Q}}(A) = \mathrm{End}(A) \otimes_{\mathbf{Z}} \mathbf{Q}$. Clearly $\mathrm{End}_{\mathbf{Q}}(A)$ has a structure of an algebra over $\mathbf{Q}$, and $\mathrm{End}(A)$ is an order of the algebra. We denote the identity element of $\mathrm{End}_{\mathbf{Q}}(A)$ by 1_A. More generally, if $\lambda \in \mathrm{Hom}_{\mathbf{Q}}(A, B)$ and $\mu \in \mathrm{Hom}_{\mathbf{Q}}(B, C)$, then we can define the product $\mu\lambda$ naturally as an element of $\mathrm{Hom}_{\mathbf{Q}}(A, C)$.

For two abelian varieties A and B of the same dimension, there exists a homomorphism of A onto B if and only if there exists a homomorphism of B onto A, in which case A and B are called *isogenous*, and any such homomorphism is called an *isogeny*. Given $\lambda \in \mathrm{Hom}(A, B)$ with A and B of the same dimension, take a field of definition k for A, B, and λ and take also a generic point x of A over k. If λ is an isogeny, we put

$$\nu(\lambda) = [k(x) : k(\lambda x)],$$

$$\nu_s(\lambda) = [k(x) : k(\lambda x)]_s, \qquad \nu_i(\lambda) = [k(x) : k(\lambda x)]_i,$$

and otherwise we put $\nu(\lambda) = \nu_s(\lambda) = \nu_i(\lambda) = 0$. These numbers do not depend on the choice of k and x. If λ is an isogeny, then $\nu_s(\lambda)$ is the order of $\mathrm{Ker}(\lambda)$. For every isogeny λ of A onto B there is a unique element λ' of $\mathrm{Hom}_{\mathbf{Q}}(B, A)$ such that $\lambda'\lambda = 1_A$ and $\lambda\lambda' = 1_B$. We write then $\lambda' = \lambda^{-1}$.

1.2. The ℓ-adic representation of homomorphisms. For an abelian variety A and a rational prime ℓ we put

$$\mathfrak{g}_\ell(A) = \bigcup_{\alpha=1}^{\infty} \mathrm{Ker}(\ell^\alpha 1_A).$$

If A is of dimension n and ℓ is different from the characteristic of a field of definition for A, then $\mathfrak{g}_\ell(A)$ is isomorphic to the direct sum $\mathfrak{M}$ of $2n$ copies of the additive group $\mathbf{Q}_\ell/\mathbf{Z}_\ell$. We call any one of the isomorphisms of $\mathfrak{g}_\ell(A)$ onto $\mathfrak{M}$ an *ℓ-adic coordinate-system of* $\mathfrak{g}_\ell(A)$. We consider every element of $\mathfrak{M}$ a column vector of dimension $2n$ with components in $\mathbf{Q}_\ell/\mathbf{Z}_\ell$. Let B be another abelian variety of dimension m and λ a homomorphism of A into B. Choose ℓ-adic coordinate systems $\mathfrak{v}$ of $\mathfrak{g}_\ell(A)$ and $\mathfrak{w}$ of $\mathfrak{g}_\ell(B)$. Then there exists an element M of $(\mathbf{Z}_\ell)_{2n}^{2m}$ such that $\mathfrak{w}(\lambda t) = M\mathfrak{v}(t)$ for every $t \in \mathfrak{g}_\ell(A)$. If we fix $\mathfrak{v}$ and $\mathfrak{w}$, the correspondence $\lambda \mapsto M$ can be uniquely extended to a $\mathbf{Q}$-linear map of $\mathrm{Hom}_{\mathbf{Q}}(A, B)$ into $(\mathbf{Q}_\ell)_{2n}^{2m}$, which we call the ℓ-adic representation of $\mathrm{Hom}_{\mathbf{Q}}(A, B)$ with respect to $\mathfrak{v}$ and $\mathfrak{w}$. In particular, if $A = B$ and $\mathfrak{v} = \mathfrak{w}$, this is a ring-homomorphism of $\mathrm{End}_{\mathbf{Q}}(A)$ into $(\mathbf{Q}_\ell)_{2n}^{2n}$.

Now let M_ℓ denote an ℓ-adic representation of $\mathrm{End}_{\mathbf{Q}}(A)$ with respect to a fixed $\mathfrak{v}$ as above. Given $\xi \in \mathrm{End}_{\mathbf{Q}}(A)$, let

$$P(x) = X^{2n} + a_1 X^{2n-1} + \cdots + a_{2n}$$

be the characteristic polynomial of $M_\ell(\xi)$. Then the following facts are known: *the a_i are rational numbers, and*

$$P(\xi) = \xi^{2n} + a_1\xi^{2n-1} + \cdots + a_{2n} = 0;$$

moreover, the polynomial P is determined by ξ independently of the choice of ℓ and ℓ-adic coordinate-system; furthermore, if $\xi \in \mathrm{End}(A)$, then $a_i \in \mathbf{Z}$ and

(1) $$\nu(\xi) = \det\bigl(M_\ell(\xi)\bigr).$$

We call P *the characteristic polynomial* of ξ and the roots of P *the characteristic roots of* ξ; we also put

(2) $$\mathrm{tr}(\xi) = \mathrm{tr}\bigl(M_\ell(\xi)\bigr).$$

1.3. The Picard variety of an abelian variety. Given an abelian variety A, let $\mathcal{G}_a(A)$ and $\mathcal{G}_l(A)$ denote, respectively, the set of divisors on A algebraically equivalent to 0 and the set of divisors on A linearly equivalent to 0. Then there exists an abelian variety A^* canonically isomorphic to $\mathcal{G}_a(A)/\mathcal{G}_l(A)$, which is called the *Picard variety* of A. Every divisor Y contained in $\mathcal{G}_a(A)$ defines a point of A^*, which we denote by $\mathrm{Cl}(Y)$. Let B be an abelian variety and B^* the Picard variety of B. For every homomorphism λ of A into B, we obtain a homomorphism λ^* of B^* into A^* such that

(3) $$\lambda^*(\mathrm{Cl}(Y)) = \mathrm{Cl}(\lambda^{-1}(Y))$$

whenever $\lambda^{-1}(Y)$ is defined. The mapping $\lambda \to \lambda^*$ is uniquely extended to an isomorphism of $\mathrm{Hom}_{\mathbf{Q}}(A, B)$ onto $\mathrm{Hom}_{\mathbf{Q}}(B^*, A^*)$; we denote by ${}^t\alpha$ the image of α by this isomorphism and call it the *transpose* of α. If $\alpha \in \mathrm{Hom}_{\mathbf{Q}}(A, B)$ and $\beta \in \mathrm{Hom}_{\mathbf{Q}}(B, C)$, we have ${}^t(\beta\alpha) = {}^t\alpha{}^t\beta$. Let X be a divisor on A; we shall denote by X_u the transform of X by the translation $x \to x + u$ on A. Now define the mapping φ_X of A into A^* by the relation

(4) $$\varphi_X(u) = \mathrm{Cl}(X_u - X)$$

for $u \in A$. Then φ_X is a homomorphism of A into A^*. The divisor X is said to be *non-degenerate* if φ_X is an isogeny. For any two divisors X, Y on A, we have $\varphi_X = \varphi_Y$ if and only if X and Y are algebraically equivalent (Barsotti [3], Serre [31]). Assuming X to be non-degenerate, put for every $\xi \in \mathrm{End}_{\mathbf{Q}}(A)$,

(5) $$\xi' = \varphi_X^{-1}{}^t\xi\varphi_X.$$

Then, it can be proved that $\xi \to \xi'$ is an involution of $\mathrm{End}_{\mathbf{Q}}(A)$ and if a suitable multiple of X is ample, we have, for every $\xi \neq 0$,

(6) $$\mathrm{tr}(\xi\xi') > 0.$$

We call this involution *the involution of* $\mathrm{End}_{\mathbf{Q}}(A)$ *determined by* X. Let λ be a homomorphism of A into B and Y a divisor on B; assume that $\lambda^{-1}(Y)$ is defined. Then, putting $X = \lambda^{-1}(Y)$, we have

$$\varphi_X = {}^t\lambda\varphi_Y\lambda. \tag{7}$$

1.4. l-adic representations of divisors. Let a be an integer and Y a divisor on A such that aY is linearly equivalent to 0. Then there exist two functions Φ and Ψ on A such that $(\Phi) = aY$, $\Phi(ax) = \Psi(x)^a$, where (Φ) denotes the divisor of Φ. For every point u on A such that $au = 0$, put

$$e_a(u, Y) = \Psi(x+u)\Psi(x)^{-1};$$

then $e_a(u, Y)$ is an a-th root of unity. Now let X be a divisor on A, and let u, v be two points on A such that $au = av = 0$. Since $a(X_v - X)$ is linearly equivalent to 0, we can consider $e_a(u, X_v - X)$. Put

$$e_{X,a}(u, v) = e_a(u, X_v - X).$$

Let k be a field of definition for A; and let l be a rational prime other than the characteristic of k. Let U_l denote the set of roots of unity, contained in the algebraic closure of k, whose orders are powers of l; then U_l is isomorphic to $\mathbf{Q}_l/\mathbf{Z}_l$. Take an isomorphism of U_l onto $\mathbf{Q}_l/\mathbf{Z}_l$ and denote it by lg; choose an l-adic coordinate-system $\mathfrak{v}$ of $\mathfrak{g}_l(A)$. Then there exists a matrix $E_l(X)$ with coefficients in $\mathbf{Z}_l$ satisfying

$$\lg e_{X,l^\nu}(s, t) \equiv l^\nu \cdot {}^t\mathfrak{v}(s)E_l(X)\mathfrak{v}(t) \qquad \mathrm{mod}\ \mathbf{Z}_l$$

for every point s, t on A such that $l^\nu s = l^\nu t = 0$. We call $E_l(X)$ the *l-adic representation* of X with respect to $\mathfrak{v}$. We have $E_l(X) = 0$ if and only if X is algebraically equivalent to 0.

1.5. q-th power homomorphisms. Let A be an abelian variety defined over a field k and σ an isomorphism of k onto a field k^σ. Then we obtain in a natural way an abelian variety A^σ, defined over k^σ, taking the transform 0^σ of the origin 0 of A as the origin of A^σ. If B is an abelian variety and λ is a homomorphism of A into B, both defined over k, we denote by λ^σ the homomorphism of A^σ into B^σ, whose graph is the transform by σ of the graph of λ. Now suppose that the characteristic p of the universal domain is not 0; let $q = p^f$ ($f > 0$) be a power of p. We shall denote by X^q the transform of any algebro-geometric object X by the automorphism $z \to z^q$ of the universal domain. We can define a homomorphism π of A onto A^q by

$$\pi x = x^q$$

for $x \in A$. We call π the *q-th power homomorphism* of A. If A is defined over a finite field with q elements, A^q coincides with A, and hence π is an endomorphism of A; we call then π the *q-th power endomorphism* of A. All the characteristic roots of the q-th power endomorphism have absolute value $q^{1/2}$; this is the so-called "Riemann hypothesis for congruence zeta-functions" proved by A. Weil. Let λ be a homomorphism of A into B; let π_A and π_B denote respectively the q-th power homomorphisms of A and B. We have then

$$\lambda^q \pi_A = \pi_B \lambda.$$

In particular, if A is defined over a finite field k with q elements, we have

$$\alpha\pi_A = \pi_A\alpha$$

for every endomorphism α of A, defined over k.

2. Differential Forms

2.1. Definitions. In this section, the varieties are all assumed to be defined over fields contained in a universal domain Ω which we fix once for all. Let V be a variety and k a field of definition for V. We shall denote by $k(V)$ the field of rational functions on V defined over k and by $\Omega(V)$ the field of all rational functions on V. If x is a generic point of V over k, the mapping $k(V) \ni f \to f(x)$ gives an isomorphism of $k(V)$ onto $k(x)$. We denote by $\mathcal{D}(V)$ and $\mathcal{D}(V;k)$ respectively the set of all derivations of $\Omega(V)$ over Ω and the set of all derivations of $k(V)$ over k. If V is of dimension n, $\mathcal{D}(V;k)$ is a vector space of dimension n over $k(V)$ and $\mathcal{D}(V)$ is a vector space obtained from $\mathcal{D}(V;k)$ by the scalar extension $\Omega(V)$ over $k(V)$. We shall denote by $\mathfrak{D}(V)$ the dual space of $\mathcal{D}(V)$ and by $\eta \cdot D$ the scalar product of $\eta \in \mathfrak{D}(V)$ and $D \in \mathcal{D}(V)$; then $(\eta, D) \to \eta \cdot D$ is a bilinear mapping of $\mathfrak{D}(V) \times \mathcal{D}(V)$ into $\Omega(V)$. Now, by a *differential form* of degree m on V, we shall understand a homogeneous element of degree m in the Grassmann algebra defined over $\mathfrak{D}(V)$. If f is a function on V, the mapping $\mathcal{D}(V) \ni D \to Df$ gives a linear mapping of $\mathcal{D}(V)$ into $\Omega(V)$, and hence defines an element of $\mathfrak{D}(V)$, a differential form of degree one on V; we denote it by df; then we have $df \cdot D = Df$. We see that $\mathfrak{D}(V)$ is generated over $\Omega(V)$ by the forms df for $f \in \Omega(V)$. If V is of dimension n, then there exists a set of n functions $\{g_1, \dots, g_n\}$ in $k(V)$ such that $k(V)$ is separably algebraic over $k(g_1, \dots, g_n)$. If $\{g_1, \dots, g_n\}$ is such a set, $dg_1, \dots, dg_n$ form a basis of $\mathfrak{D}(V)$ over $\Omega(V)$. By our definition, every differential form ω on V has an expression

$$\omega = \sum_{(i)} f_{(i)} dg_{i_1} \cdots dg_{i_r},$$

where the $f_{(i)}$ are elements of $\Omega(V)$. We shall say that a differential form ω on V is *defined over* k if ω can be written in the form

$$\omega = \sum_{(i)} \varphi_{(i)} d\psi_{i_1} \cdots d\psi_{i_r},$$

with the $\varphi_{(i)}$ and the ψ_i in $k(V)$. The set $\{g_1, \ldots, g_n\}$ being as above, a differential form

$$\sum_{i_1 < \cdots < i_r} f_{(i_1 \cdots i_r)} dg_{i_1} \cdots dg_{i_r}$$

is defined over k if and only if the $f_{(i)}$ are contained in $k(V)$.

Let V' be a simple subvariety of V. We shall say that a differential form ω on V is *finite along* (or *at*) V' if ω can be written in the form $\omega = \sum_{(i)} f_{(i)} dg_{i_1} \cdots dg_{i_r}$ where the $f_{(i)}$ and the g_i are functions on V which are all defined and finite along V'. If that is so, denoting by the $f'_{(i)}$ and the g'_i the functions on V' induced by the $f_{(i)}$ and the g_i, we obtain a differential form $\omega' = \sum_{(i)} f'_{(i)} dg'_{i_1} \cdots dg'_{i_r}$ on V' which is determined only by ω and V'; ω' does not depend upon the choice of the $f_{(i)}$ and the g_i. We call ω' the differential form on V' induced by ω.

2.2. Local parameters. Let K be a field and $u_1, \ldots, u_n$ be n independent variables over K. If K_1 is a separably algebraic extension of $K(u_1, \ldots, u_n)$, there exist n derivations $D_1, \ldots, D_n$ of K_1 over K such that

$$D_i u_i = 1, \quad D_i u_j = 0 \quad \text{for } i \neq j.$$

The D_i are uniquely determined by these relations; we shall denote D_i by $\partial/\partial u_i$ for each i.

Now let V be a variety of dimension n, defined over k, and y a simple point on V. We call a set of n functions $\{\tau_1, \ldots, \tau_n\}$ in $k(V)$ a *system of local parameters* for V at y defined over k, if the following conditions are satisfied.

(L1) *$k(V)$ is separably algebraic over $k(\tau_1, \ldots, \tau_n)$.*
(L2) *The τ_i are all defined and finite at y.*
(L3) *For every f in $k(V)$, defined and finite at y, the function $\partial f/\partial \tau_i$ is defined and finite at y for every i.*

Let x be a generic point of V over k; let V_α, y_α, x_α be affine representatives of V, y, x and S the ambient space for V_α; let N be the dimension of S. Then, by Koizumi [23], we know that n functions $\tau_1, \ldots, \tau_n$ in $k(V)$ form a system of local parameters for V at y, defined over k, if and only if the following conditions are satisfied.

(L′1) *The τ_1 are all defined and finite at y.*
(L′2) *There exists a set of N polynomials $F_i(X_1, \ldots, X_N, T_1, \ldots, T_n)$ with coefficients in k such that $F_i(x_\alpha, \tau(x)) = 0$ for $1 \leq i \leq N$ and $\det(\partial F_i/\partial X_j(y_\alpha, \tau(y))) \neq 0$.*

We shall prove this in §10.3 in a more general setting. If $\{\tau_1, \ldots, \tau_n\}$ is a system of local parameters at y, then $d\tau_1, \ldots, d\tau_n$ form a basis of $\mathfrak{D}(V)$; and, by (L3), every differential form

$$\omega = \sum_{i_1 < \cdots < i_r} f_{(i_1, \cdots i_r)} d\tau_{i_1} \cdots d\tau_{i_r}$$

is finite at y if and only if the $f_{(i)}$ are all defined and finite at y.

2.3. Let V and W be two varieties and T a rational mapping of V into W; we denote by the same notation T the graph of T, and by Y the projection of T on W. For every f in $\Omega(W)$, defined and finite along Y, we shall denote by $f \circ T$ the function on V defined by $(f \circ T)(x) = f(T(x))$ with respect to a field k of definition for V, W, T and f, where x is a generic point of V over k. Assuming that Y is simple on W, let ω be a differential form on W finite along Y. Then ω can be written in the form $\omega = \sum f_{(i)} dg_{i_1} \cdots dg_{i_r}$ where the $f_{(i)}$ and the g_i are elements of $\Omega(W)$ which are all defined and finite along Y. Put

$$\omega' = \sum (f_{(i)} \circ T) d(g_{i_1} \circ T) \cdots d(g_{i_r} \circ T).$$

Then ω' is a differential form on V which is determined only by ω and T, and does not depend on the choice of the $f_{(i)}$ and the g_i. We shall denote the form ω' by $\omega \circ T$. If V, W, T, ω are all defined over k, then $\omega \circ T$ is also defined over k. If T is defined at a simple point y on V and if $T(y)$ is simple on W, then for every differential form ω which is finite at $T(y)$, $\omega \circ T$ is finite at y.

The symbols V, W, T, ω being as above, let U be a variety and S a rational mapping of U into V; suppose that the image X of Y by S is simple on V and T is defined along X. Then we obtain a rational mapping $T \circ S$ of U into W, defined by $(T \circ S)(x) = T(S(x))$. Suppose further that the image Z of U by $T \circ S$ is simple on W and ω is finite along Z. We obtain then a differential form $\omega \circ (T \circ S)$ on U. It can be easily verified that the differential form $(\omega \circ T) \circ S$ is also defined and equal to $\omega \circ (T \circ S)$.

2.4. Differential forms of the first kind. Let V be a complete non-singular variety. We say that a differential form ω on V is of *the first kind* if ω is everywhere finite on V. Let W be another complete non-singular variety and T a rational mapping of V into W. Then, for every differential form ω of the first kind on W, $\omega \circ T$ is also a differential form of the first kind on V. This is easily proved by means of Proposition 5 of Koizumi [23].

PROPOSITION 1. *Let V_1 and V_2 be two complete non-singular varieties and p_i the projection from $V_1 \times V_2$ onto V_i for $i = 1, 2$. Then, for any differential form ω on $V_1 \times V_2$ of the first kind and of degree 1, there exist differential* *forms* ω_i *on* V_i $(i = 1, 2)$ *of the first kind and of degree 1 such that* $\omega = \omega_1 \circ p_1 + \omega_2 \circ p_2$.

This is a restatement of Theorem 3 of [23].

2.5. Let V be a variety defined over k and σ an isomorphism of k onto a field k^σ. Let f be a function in $k(V)$ and Γ the graph of f. We shall denote by f^σ the function on V^σ whose graph is Γ^σ; then, $f \to f^\sigma$ gives an isomorphism of $k(V)$ onto $k^\sigma(V^\sigma)$. Let ω be a differential form on V defined over k. Then ω has an expression $\omega = \sum_{(i)} f_{(i)} dg_{i_1} \cdots dg_{i_r}$ with the $f_{(i)}$ and the g_i in $k(V)$. It can be easily shown that the differential form $\omega' = \sum_{(i)} f_{(i)}^\sigma dg_{i_1}^\sigma \cdots dg_{i_r}^\sigma$ on V^σ is determined only by ω and σ, and does not depend upon the choice of the $f_{(i)}$ and the g_i. We shall denote the form ω' by ω^σ.

2.6. Invariant differential forms on group varieties. Let G be a group variety and t a point on G. We shall denote by T_t the left translation $x \to tx$ of G. T_t is obviously a birational correspondence of G onto itself. We call a differential form ω on G a *left invariant* differential form on G if $\omega \circ T_t = \omega$ for every t on G.

PROPOSITION 2. *Let G be a group variety and* ω *a left invariant differential form on G. Then* ω *is everywhere finite on G.*

PROOF. Take a field of definition k for G and ω, and a generic point x of G over k. Clearly ω is finite at x. Then, the relation $\omega = \omega \circ T_{xy}$ shows that ω is finite at any point y^{-1} on G.

LEMMA 1. *Let G be a group variety and* ω *a differential form on G; let k be a field of definition for G and* ω*, and t be a generic point of G over k. If the relation* $\omega \circ T_t = \omega$ *holds,* ω *is left invariant.*

PROOF. Let s be any generic point of G over k. Then there exists an isomorphism σ of $k(t)$ onto $k(s)$ over k such that $t^\sigma = s$. By the relation $\omega = \omega \circ T_t$, we have $\omega = \omega^\sigma = (\omega \circ T_t)^\sigma = \omega \circ T_s$. Now let x be a point of G. Take a generic point y of G over $k(x)$. Then, since xy and y^{-1} are generic on G over k, we have $\omega \circ T_{xy} = \omega \circ T_{y^{-1}} = \omega$, so that

$$\omega \circ T_x = \omega \circ (T_{xy} \circ T_{y^{-1}}) = (\omega \circ T_{xy}) \circ T_{y^{-1}} = \omega \circ T_{y^{-1}} = \omega.$$

This shows that ω is left invariant.

We shall denote by $\mathfrak{D}_0(G)$ the set of all left invariant differential forms on G of degree one and by $\mathfrak{D}_0(G; k)$ the set of elements of $\mathfrak{D}_0(G)$ which are defined over k.

PROPOSITION 3. *Let G be a group variety of dimension n, defined over k. Then $\mathfrak{D}_0(G;k)$ is a vector space of dimension n over k; and we have $\mathfrak{D}_0(G) = \mathfrak{D}_0(G;k) \otimes_k \Omega$ and $\mathfrak{D}(G) = \mathfrak{D}_0(G;k) \otimes_k \Omega(G)$.*

PROOF. It is obvious that $\mathfrak{D}_0(G;k)$ is a vector space over k. Denote by e the identity element of the group G. Let $\{\tau_1, \dots, \tau_n\}$ be a system of local parameters for G at e, defined over k. Then, every element ω of $\mathfrak{D}(G)$, defined over k, can be expressed in the form $\omega = \sum_i f_i d\tau_i$ with the f_i in $k(G)$. Let x and t be independent and generic on G over k; put $f_i' = f_i \circ T_{t^{-1}}$, $\tau_i' = \tau_i \circ T_{t^{-1}}$. We have then

$$\omega \circ T_{t^{-1}} = \sum_i f_i' d\tau_i', \quad f_i'(x) = f_i(t^{-1}x), \quad \tau_i'(x) = \tau_i(t^{-1}x).$$

Since $\{d\tau_1, \dots, d\tau_n\}$ is a basis of $\mathfrak{D}(G)$, there exist n^2 functions g_{ij}^t $(1 \le i \le n, 1 \le j \le n)$ in $\Omega(G)$ such that

$$d\tau_i' = \sum_j g_{ij}^t d\tau_j \qquad (1 \le i \le n).$$

The g_{ij}^t are contained in $k(t)(G)$, since the τ_i' and τ_i are contained in $k(t)(G)$. Let g_{ij} be the element of $k(G \times G)$ defined by $g_{ij}(t,x) = g_{ij}^t(x)$ for each i and j. If y is a point on G where the τ_i are local parameters and if the τ_i' are all defined and finite at y, then the g_{ij}^t are all defined and finite at y, so that the g_{ij} are all defined and finite at $t \times y$, and we have $g_{ij}(t,y) = g_{ij}^t(y)$. Applying this to the points t and e, we see that the g_{ij} are all defined and finite at $t \times t$ and $t \times e$, and we have

$$g_{ij}(t,t) = g_{ij}^t(t), \quad g_{ij}(t,e) = g_{ij}^t(e).$$

Now suppose that ω is left invariant. Then, the relation $\omega = \omega \circ T_t$ implies $\sum_i f_i d\tau_i = \sum_i f_i' d\tau_i' = \sum_{i,j} f_i' g_{ij}^t d\tau_j$; hence we have $f_j = \sum_i f_i' g_{ij}^t$, namely

$$f_j(x) = \sum_i f_i(t^{-1}x) g_{ij}(t,x) \qquad (1 \le j \le n). \tag{1}$$

Conversely, if n elements $f_1, \dots, f_n$ of $k(G)$ satisfy these relations, then $\omega = \sum_i f_i d\tau_i$ satisfies the relation $\omega \circ T_t = \omega$, and hence, by Lemma 1, ω is left invariant. Assume that ω is left invariant, namely that the f_j satisfy (1). The f_j are defined and finite at e since ω is finite at e. Hence we can specialize (t,x) to (t,t) in equation (1). We have then $f_j(t) = \sum_i f_i(e) g_{ij}(t,t)$ for $1 \le j \le n$. Let h_{ij} be the element of $k(G)$ defined by $h_{ij}(t) = g_{ij}(t,t)$ for each i and j; put $\omega_i = \sum_j h_{ij} d\tau_j$ for each i. Then we have $\omega = \sum_j f_j d\tau_j = \sum_i f_i(e)\omega_i$. This shows that every element of $\mathfrak{D}_0(G;k)$ is a linear combination of the ω_i with coefficients in k. Hence the dimension of the vector space $\mathfrak{D}_0(G;k)$ over

k is not greater than n. We shall now prove that the ω_q $(1 \leq q \leq n)$ are all left invariant. To prove this, it is sufficient to show that equality (1) holds if we substitute h_{qi}, h_{qj} for f_i, f_j, namely,

$$(2) \qquad h_{qj}(x) = \sum_i h_{qi}(t^{-1}x)g_{ij}(t, x) \qquad (1 \leq q \leq n, 1 \leq j \leq n),$$

or, as we have $h_{qi}(x) = g_{qi}(x, x)$,

$$(2') \quad g_{qj}(x, x) = \sum_i g_{qi}(t^{-1}x, t^{-1}x)g_{ij}(t, x) \quad (1 \leq q \leq n, 1 \leq j \leq n).$$

Let s be a generic point of G over $k(t, x)$; put

$$\tau_i'' = \tau_i \circ T_{s^{-1}}, \quad \tau_i''' = \tau_i' \circ T_{s^{-1}} = \tau_i \circ T_{(st)^{-1}} \qquad (1 \leq i \leq n).$$

Then we have

$$d\tau_i''' = (d\tau_i') \circ T_{s^{-1}} = \sum_j (g_{ij}^t \circ T_{s^{-1}}) d(\tau_j \circ T_{s^{-1}}) = \sum_j (g_{ij}^t \circ T_{s^{-1}}) d\tau_j''.$$

Define the functions g_{ij}^s and g_{ij}^{st} on G by

$$g_{ij}^s(x) = g_{ij}(s, x), \quad g_{ij}^{st}(x) = g_{ij}(st, x);$$

we have then $d\tau_i'' = \sum_j g_{ij}^s d\tau_j$, $d\tau_i''' = \sum_j g_{ij}^{st} d\tau_j$ for $1 \leq i \leq n$. These are obtained from the relation $d\tau_i' = \sum_j g_{ij}^t d\tau_j$ by isomorphisms between the fields $k(t)$, $k(s)$ and $k(st)$. It follows that

$$\sum_j g_{ij}^{st} d\tau_j = d\tau_i''' = \sum_j (g_{ij}^t \circ T_{s^{-1}}) d\tau_j'' = \sum_{j,l} (g_{ij}^t \circ T_{s^{-1}}) g_{jl}^s d\tau_l.$$

This implies $g_{il}^{st} = \sum_j (g_{ij}^t \circ T_{s^{-1}}) g_{jl}^s$ for $1 \leq i \leq n$, $1 \leq l \leq n$, or

$$(3) \quad g_{il}(st, x) = \sum_j g_{ij}(t, s^{-1}x) g_{jl}(s, x) \qquad (1 \leq i \leq n, 1 \leq l \leq n).$$

Specializing (s, t, x) to $(t, t^{-1}x, x)$, we obtain (2′); hence the ω_i are left invariant. Our theorem is completely proved if we show that the ω_i are linearly independent over $\Omega(G)$. To see this, it is sufficient to prove $\det(h_{ij}) \neq 0$, since $d\tau_1, \ldots, d\tau_n$ are linearly independent over $\Omega(G)$ and $\omega_i = \sum_j h_{ij} d\tau_j$. As t^{-1} is generic on G over k, the τ_i are local parameters at t^{-1} and consequently the τ_i' are local parameters at e. Hence, from the relation $d\tau_i' = \sum_j g_{ij}^t d\tau_j$ it follows that $\det(g_{ij}^t(e)) \neq 0$, namely, $\det(g_{ij}(t, e)) \neq 0$. Specializing (s, t, x) to (s, t, s) in relation (3), we get

$$g_{il}(st, s) = \sum_j g_{ij}(t, e) g_{jl}(s, s) \qquad (1 \leq i \leq n, 1 \leq l \leq n).$$

Hence we have $\det(g_{ij}(s, s)) \neq 0$; this completes the proof, because of the relation $h_{ij}(s) = g_{ij}(s, s)$.

Now specialize x to e in relation (2); we see then that the h_{ij} are defined at e and

(4) $$\det(h_{ij}(e)) \neq 0.$$

We need this fact later.

2.7. Invariant differential forms on abelian varieties. As an abelian variety is a commutative group, we call a left invariant differential form on an abelian variety simply an *invariant differential form.*

PROPOSITION 4. *Let A be an abelian variety. Then every differential form of degree 1 on A is of the first kind if and only if it is invariant.*

PROOF. By Proposition 2, we have only to prove the "only if" part. Let ω be a differential form on A of degree 1 and of the first kind. Taking A as G in Proposition 3, we use the letters τ_i, ω_i, h_{ij} in the same sense as in that proof. Then, by that proposition, there exist n functions a_i in $\Omega(A)$ such that $\omega = \sum_i a_i\omega_i$. As the ω_i are invariant, we have, for every x on A,

$$\omega \cdot T_x = \sum_i (a_i \circ T_x)\omega_i = \sum_{i,j} (a_i \circ T_x) h_{ij} d\tau_j.$$

Since ω is everywhere finite on A, $\omega \circ T_x$ is also everywhere finite on A; in particular, $\omega \circ T_x$ is finite at e. Therefore, recalling that the τ_j are local parameters at e, we see that the n functions $\sum_i (a_i \circ T_x) h_{ij} (1 \leq j \leq n)$ are finite at e, so that, by (4), the $a_i \circ T_x$ are finite at e. As x is an arbitrary point of A, this shows that the a_i are everywhere finite on A. Such functions must be constant, since A is a complete variety. Hence ω is a linear combination of the ω_i with constant coefficients; so ω is an invariant differential form.

2.8. Differentials of homomorphisms. Let A and B be two abelian varieties and λ a homomorphism of A into B. If ω is an element of $\mathfrak{D}_0(B)$, then $\omega \circ \lambda$ is defined and contained in $\mathfrak{D}_0(A)$. The mapping $\omega \to \omega \circ \lambda$ gives an Ω-linear mapping of $\mathfrak{D}_0(B)$ into $\mathfrak{D}_0(A)$. We shall denote this linear mapping by $\delta\lambda$, namely,

$$(\delta\lambda)\omega = \omega \circ \lambda$$

for $\omega \in \mathfrak{D}_0(B)$. If k is a field of definition for A, B and λ, then $\delta\lambda$ gives a k-linear mapping of $\mathfrak{D}_0(B; k)$ into $\mathfrak{D}_0(A; k)$. For every homomorphism μ of B into an abelian variety C, we have

$$\delta(\mu\lambda) = \delta\lambda\delta\mu.$$

PROPOSITION 5. *Let A and B be two abelian varieties; and let λ and μ be two homomorphisms of A into B. Then we have*

$$\delta(\lambda + \mu) = \delta\lambda + \delta\mu.$$

PROOF. Define homomorphisms α, β, γ, λ_0, μ_0 as follows:

$$A \xrightarrow{\alpha} A \times A \xrightarrow{\beta,\lambda_0,\mu_0} B \times B \xrightarrow{\gamma} B,$$

$$\alpha(x) = x \times x, \quad \beta(x \times y) = \lambda(x) \times \mu(y), \quad \gamma(z \times w) = z + w,$$

$$\lambda_0(x \times y) = \lambda(x) \times 0, \quad \mu_0(x \times y) = 0 \times \mu(y).$$

We have then $\lambda+\mu = \gamma\beta\alpha, \gamma\lambda_0\alpha = \lambda, \gamma\mu_0\alpha = \mu$, so that $\delta(\lambda+\mu) = \delta\alpha\delta\beta\delta\gamma$, $\delta\lambda = \delta\alpha\delta\lambda_0\delta\gamma$, $\delta\mu = \delta\alpha\delta\mu_0\delta\gamma$. Hence our proposition is proved if we show $\delta\beta = \delta\lambda_0 + \delta\mu_0$. Let p_1 and p_2 be respectively the projection from $B \times B$ onto the first and the second factors of $B \times B$. By Proposition 1, for every $\omega \in \mathfrak{D}_0(B \times B)$, there exist two elements ω_1, ω_2 in $\mathfrak{D}_0(B)$ such that $\omega = \omega_1 \circ p_1 + \omega_2 \circ p_2$. It follows that $\delta\beta\omega = \delta\beta(\omega_1 \circ p_1) + \delta\beta(\omega_2 \circ p_2)$. We see easily

$$p_1 \circ \beta = p_1 \circ \lambda_0, \quad p_2 \circ 0 = p_2 \circ \lambda_0, \quad p_2 \circ \beta = p_2 \circ \mu_0, \quad p_1 \circ 0 = p_1 \circ \mu_0,$$

so that we have

$$\begin{aligned}
\delta\beta(\omega_1 \circ p_1) &= \omega_1 \circ p_1 \circ \beta &&= \omega_1 \circ p_1 \circ \lambda_0 &&= \delta\lambda_0(\omega_1 \circ p_1), \\
\delta\beta(\omega_2 \circ p_2) &= \omega_2 \circ p_2 \circ \beta &&= \omega_2 \circ p_2 \circ \mu_0 &&= \delta\mu_0(\omega_2 \circ p_2), \\
\delta\lambda_0(\omega_2 \circ p_2) &= \omega_2 \circ p_2 \circ \lambda_0 &&= \omega_2 \circ p_2 \circ 0 &&= 0, \\
\delta\mu_0(\omega_1 \circ p_1) &= \omega_1 \circ p_1 \circ \mu_0 &&= \omega_1 \circ p_1 \circ 0 &&= 0.
\end{aligned}$$

It follows that $\delta\beta\omega = \delta\lambda_0\omega + \delta\mu_0\omega$; this completes the proof.

LEMMA 2. *Let $k(x)$ be an extension of a field k and s the smallest number of quantities u_i in $k(x)$ such that $k(x)$ is separably algebraic over $k(u)$. Then there exist s and no more than s linearly independent derivations of $k(x)$ over k. Moreover, if the characteristic p of k is not 0, we have $[k(x) : k(x^p)] = p^s$ and $[k(x) : k(x^q)] \leq q^s$ for any power $q = p^f$ with $f > 0$.*

PROOF. Let $(u) = (u_1, \dots, u_s)$ be a set of quantities in $k(x)$ such that $k(x)$ is separably algebraic over $k(u)$. By our definition of s, for every i, $k(x)$ is not separably algebraic over $k(u_1, \dots, u_{i-1}, u_{i+1}, \dots, u_s)$. Hence there exists, for each i, a derivation D_i of $k(x)$ such that $D_i u_i = 1$, $D_i u_j = 0$ $(i \neq j)$; the derivations D_i are obviously linearly independent. Let D be a derivation of $k(x)$ over k; put $Du_i = y_i$ and $D' = D - \sum_i y_i D_i$; then D' is a derivation of $k(x)$ over $k(u)$. As $k(x)$ is separably algebraic over $k(u)$, we have $D' = 0$; this proves the first assertion. Now suppose that k is of characteristic $p \neq 0$. As $k(x)$ is purely inseparable over $k(x^q)$ for any power $q = p^f$ with $f > 0$, $k(x)$ is purely inseparable and separable over $k(x^q, u)$; so we have $k(x) = k(x^q, u)$. It follows that we have $[k(x) : k(x^q)] \leq q^s$, since the u_i^q are contained in $k(x^q)$; in particular, we have $[k(x) : k(x^p)] \leq p^s$. Suppose that we have $[k(x) : k(x^p)] = p^r < p^s$; then there exist r quantities v_j in $k(x)$ such that $k(x) = k(x^p, v)$. By the first assertion of our lemma, the number of linearly independent derivations of $k(x)$ over $k(x^p)$ is not greater than r. This is a contradiction since every derivation of $k(x)$ over k gives a derivation of $k(x)$ over $k(x^p)$. Hence we must have $[k(x) : k(x^p)] = p^s$. This completes the proof.

THEOREM 1. *Let A and B be two abelian varieties and λ a homomorphism of A into B; let k be a field of definition for A, B and λ, and x a generic point of A over k. If the linear mapping $\delta\lambda$ of $\mathfrak{D}_0(B)$ into $\mathfrak{D}_0(A)$ is of rank r, then the following assertions hold:*

(i) *$k(x)$ is separably generated over $k(\lambda x)$ if and only if* $\dim_k(\lambda x) = r$.
(ii) *Assuming that A and B have the same dimension n, we have $\nu_i(\lambda) = 1$ if and only if $n = r$.*
(iii) *For n as in* (ii), *if k is of characteristic $p \neq 0$ and if $k(\lambda x) \supset k(x^q)$ for a power $q = p^e$ $(e > 0)$ of p, then we have $\nu(\lambda) = \nu_i(\lambda) \leq q^{n-r}$.*

PROOF. Let n and m be respectively the dimensions of A and B. Let F denote the subfield $\{f \circ \lambda \mid f \in k(B)\}$ of $k(A)$. Let D be a derivation of $k(A)$ over k. We shall prove that $(\delta\lambda\omega) \cdot D = 0$ for all $\omega \in \mathfrak{D}_0(B)$ if and only if $DF = 0$. Take a basis $\{\omega_1, \ldots, \omega_m\}$ of $\mathfrak{D}_0(B; k)$ over k; then, for every $f \in k(B)$, there exist, by Proposition 3, m functions g_i in $k(B)$ such that $df = \sum_i g_i\omega_i$. If $(\delta\lambda\omega) \cdot D = 0$ for all $\omega \in \mathfrak{D}_0(B)$, we have

$$D(f \circ \lambda) = d(f \circ \lambda) \cdot D = \sum_i (g_i \circ \lambda)(\delta\lambda\omega_i) \cdot D = 0;$$

this shows $DF = 0$. Conversely, suppose that $DF = 0$. Every ω in $\mathfrak{D}_0(B)$ can be expressed in the form $\omega = \sum_i f_i dh_i$ with $f_i \in \Omega(B)$, $h_i \in k(B)$. We have hence

$$(\delta\lambda\omega) \cdot D = \sum_i (f_i \circ \lambda) d(h_i \circ \lambda) \cdot D = \sum_i (f_i \circ \lambda) D(h_i \circ \lambda) = 0.$$

Thus we have proved that $(\delta\lambda\omega) \cdot D = 0$ for all $\omega \in \mathfrak{D}_0(B)$ if and only if $DF = 0$. Now, by our assumption, $\delta\lambda[\mathfrak{D}_0(B)]$ is a vector subspace of $\mathfrak{D}_0(A)$ of dimension r. By Proposition 3, any linearly independent elements of $\mathfrak{D}_0(A)$ over Ω are linearly independent over $\Omega(A)$. Hence there exist exactly

$n-r$ linearly independent derivations of $k(A)$ over F. If x is generic on A over k, the mapping $f \to f(x)$ gives an isomorphism of $k(A)$ onto $k(x)$, and F corresponds to $k(\lambda x)$ by this isomorphism; so there exist exactly $n-r$ linearly independent derivations of $k(x)$ over $k(\lambda x)$. By Lemma 2, we can find $n-r$ elements $u_1, \ldots, u_{n-r}$ in $k(x)$ such that $k(x)$ is separably algebraic over $k(\lambda x, u_1, \ldots, u_{n-r})$. If $\dim_k(\lambda x) = r$, the u_i must be independent variables over $k(\lambda x)$, so that $k(x)$ is separably generated over $k(\lambda x)$. Conversely, if $k(x)$ is separably generated over $k(\lambda x)$, then by Proposition 16 of [44, Chapter I], the dimension of $k(x)$ over $k(\lambda x)$ is $n-r$, so that we have $\dim_k(\lambda x) = r$. This proves assertion (i). If $m = n$ and $\nu_i(\lambda) = 1$, $k(x)$ is separably algebraic over $k(\lambda x)$; consequently, by what we have just proved, we have $n = \dim_k(\lambda x) = r$. Conversely, if rank $\delta\lambda = n$, there is no derivation other than 0 in $k(x)$ over $k(\lambda x)$, so that $k(x)$ is separably algebraic over $k(\lambda x)$, namely, $\nu_i(\lambda) = 1$; this implies (ii). Suppose now that k is of characteristic $p \neq 0$ and $k(\lambda x) \supset k(x^q)$ for $q = p^e$ with $e > 0$. Then, by Lemma 2, we have

$$[k(x) : k(\lambda x)] = [(k(x) : k(\lambda x, x^q)] \leq q^{n-r};$$

this proves (iii) of our theorem.

COROLLARY. *Let B be an abelian variety and A an abelian subvariety of B. If α denotes the injection of A into B, we have*

$$\delta\alpha(\mathfrak{D}_0(B)) = \mathfrak{D}_0(A).$$

This is an easy consequence of (i) of Theorem 1.

PROPOSITION 6. *Let A, B, λ, k, x be the same as in Theorem 1. Suppose that the characteristic p of k is not 0. Then:*

(i) *We have $\delta\lambda = 0$ if and only if $k(\lambda x) \subset k(x^p)$.*
(ii) *Assume that A and B are of the same dimension; if $\nu_i(\lambda) = 1$, we have $k(x) = k(x^q, \lambda x)$ for every power $q = p^e$ with $e > 0$; conversely, if $k(x) = k(x^q, \lambda x)$ for some $q = p^e$ with $e > 0$, we have $\nu_i(\lambda) = 1$.*

PROOF. The proof of Theorem 1 implies that $\delta\lambda = 0$ if and only if $DF = 0$ for all derivations D of $k(A)$ over k, the notation being as there. Hence we have $\delta\lambda = 0$ if and only if $F \subset k(A)^p \cdot k$; the latter condition is equivalent to $k(\lambda x) \subset k(x^p)$; this proves (i). If $\nu_i(\lambda) = 1$, $k(x)$ is separably algebraic over $k(\lambda x)$; as $k(x)$ is purely inseparable over $k(x^q)$ for every power $q = p^e$ with $e > 0$, we have $k(x) = k(x^q, \lambda x)$. Conversely, suppose that $k(x) = k(x^q, \lambda x)$ for some power $q = p^e$ with $e > 0$. Then there is no derivation of $k(x)$ over $k(\lambda x)$ other than 0, so that $k(x)$ is separably algebraic over $k(\lambda x)$, namely $\nu_i(\lambda) = 1$.

PROPOSITION 7. *Let A be an abelian variety of dimension n, defined over a field of characteristic $p \neq 0$. Then, $\nu_i(p1_A)$ is a multiple of p^n and the order of* Ker$(p1_A)$ *is a divisor of p^n.*

PROOF. By Proposition 5, we have $\delta(p1_A) = \delta(\overbrace{1_A + \cdots + 1_A}^{p}) = p\delta 1_A = 0$. Hence, by (i) of Proposition 6, we have $k(px) \subset k(x^p)$, where k is a field of definition for A and x a generic point of A over k. It follows that $\nu_i(p1_A) = [k(x) : k(px)]_i \geq [k(x) : k(x^p)] = p^n$. Since the order of Ker$(p1_A)$ is equal to $\nu_s(p1_A)$ and $\nu_s(p1_A)\nu_i(p1_A) = \nu(p1_A) = p^{2n}$, we obtain our proposition.

Let A be an abelian variety of dimension n and k a field of definition for A. Denote by End$(A; k)$ the set of all elements in End(A) defined over k and by End$_{\mathbf{Q}}(A; k)$ the subset End$(A; k) \otimes \mathbf{Q}$ of End$_{\mathbf{Q}}(A)$. For every $\lambda \in$ End$(A; k)$, $\delta\lambda$ gives a linear transformation of $\mathfrak{D}_0(A; k)$. We have seen above that the relations $\delta(\lambda + \mu) = \delta\lambda + \delta\mu$, $\delta(\lambda\mu) = \delta\mu\delta\lambda$ hold, so that the mapping $\lambda \to \delta\lambda$ gives an anti-representation of End$(A; k)$. As $\mathfrak{D}_0(A; k)$ is a vector space of dimension n over k, we obtain, with respect to a basis of $\mathfrak{D}_0(A; k)$ over k, an anti-representation of End$(A; k)$ by matrices of degree n *with coefficients in k*. If k is of characteristic $p \neq 0$, we have $\delta(p1_A) = 0$; so our representation is not faithful. If k is of characteristic 0, we get a faithful representation. In fact, if $\delta\lambda = 0$, the rank of $\delta\lambda$ is 0, so that by (i) of Theorem 1, we have $\dim_k(\lambda x) = 0$; this implies $\lambda = 0$. In case of characteristic 0, we can extend uniquely the representation to a representation of End$_{\mathbf{Q}}(A; k)$. We shall call this anti-representation a *representation of* End$_{\mathbf{Q}}(A; k)$ *by invariant differential forms*.

2.9. Differential forms on a curve and its Jacobian variety. In the sequel, we denote by $\mathfrak{D}_0(V)$ the set of all differential forms on an algebraic variety V of degree 1 of the first kind.

PROPOSITION 8. *Let C be a complete curve without singular point and J a Jacobian variety of C and φ a canonical mapping of C into J. Then, $\omega \to \omega\circ\varphi$ gives an isomorphism of $\mathfrak{D}_0(J)$ onto $\mathfrak{D}_0(C)$.*

PROOF. Let g be the genus of C; denote by C_g and J_g the product $C \times \cdots \times C$ of g copies of C and the product $J \times \cdots \times J$ of g copies of J, respectively. Let k be a field of definition for C, J, and φ, and $x_1 \times \cdots \times x_g$ a generic point of C_g over k; define a rational mapping Ψ of C_g into J by $\Psi(x_1, \ldots, x_g) = \sum_{i=1}^{g} \varphi(x_i)$. Then Ψ is everywhere defined on C_g. Putting $z = \Psi(x_1, \ldots, x_g)$, we see that $k(x_1, \ldots, x_g)$ is separably algebraic over $k(z)$. Let σ and τ denote respectively the rational mappings of J_g into J and of C_g into J_g defined by

$$\sigma(z_1, \ldots, z_g) = z_1 + \cdots + z_g, \qquad \tau(x_1, \ldots, x_g) = \varphi(x_1) \times \cdots \times \varphi(x_g).$$

Denote further by p_i the projection of C_g onto the i-th factor and by q_i the projection of J_g onto the i-th factor. We have then $\Psi = \sigma \circ \tau$, $q_i \circ \tau = \varphi \circ p_i$ and $\delta\sigma = \delta q_1 + \cdots + \delta q_g$ by virtue of Proposition 5, so that we have, for every $\omega \in \mathfrak{D}_0(J)$,

$$\begin{aligned}\omega\circ\Psi &= \omega\circ\sigma\circ\tau = (\omega\circ q_1 + \cdots + \omega\circ q_g)\circ\tau \\ &= \omega\circ q_1\circ\tau + \cdots + \omega\circ q_g\circ\tau = \omega\circ\varphi\circ p_1 + \cdots + \omega\circ\varphi\circ p_g.\end{aligned}$$

Hence, if $\omega\circ\varphi = 0$, we have $\omega\circ\Psi = 0$; as the mapping Ψ is separably algebraic, $\omega\circ\Psi = 0$ implies $\omega = 0$. This shows that the mapping $\omega \to \omega\circ\varphi$ of $\mathfrak{D}_0(J)$ into $\mathfrak{D}_0(C)$ is one-to-one. Our proposition is thereby proved, since $\mathfrak{D}_0(J)$ and $\mathfrak{D}_0(C)$ are of the same dimension g.

The symbols C, J and φ being as above, let C' be another complete non-singular curve, J' its Jacobian variety, and φ' a canonical mapping of C' into J'; and let k be a field of definition for C, J, φ, C', J', φ'. Let X be a positive divisor of $C \times C'$ rational over k. X determines a homomorphism of J into J' as follows (cf. Weil [45, 46]). Take a generic point x of C over k and put

$$X \cdot (x \times C') = \sum_{\nu=1}^{n} x \times y_\nu;$$

then, there exists a homomorphism λ of J into J', defined over k, and a point b on J', rational over k, such that

$$\lambda[\varphi(x)] + b = \sum_{\nu=1}^{n} \varphi'(y_\nu);$$

λ and b do not depend on the choice of k and x.

Proposition 9. *The notation being as above, let C_0 be a complete non-singular curve with a generic point z over the algebraic closure k_1 of k such that $k(z) = k(x, y_1, \ldots, y_n)$; let p and the q_ν be the rational mappings of C_0 into C and into C' defined by $p(z) = x$, $q_\nu(z) = y_\nu$ with respect to k_1. Then, for every $\omega \in \mathfrak{D}_0(J')$, we have*

$$\omega\circ\lambda\circ\varphi\circ p = \sum_{\nu=1}^{n} \omega\circ\varphi'\circ q_\nu.$$

Proof. Define the rational mappings α, β, γ as follows:

$$C_0 \xrightarrow{\alpha} C' \times \cdots \times C' \xrightarrow{\beta} J' \times \cdots \times J' \xrightarrow{\gamma} J',$$

$$H_0(u, v) = H_0(v, u),$$
$$H(d_1, d_2) \equiv H_0(d_1, d_2) \mod \mathbf{Z} \quad \text{for } d_1, d_2 \in D.$$

Putting

$$E(x, y) = H(x, y) - H(y, x),$$

we call E the *alternating form defined by* f. If f is holomorphic, $E(x, y)$ is a Riemann form on $\mathbf{C}^n/D$; we call then E the *Riemann form defined by* f. A theta function f is said to be *normalized* if H is skew-hermitian and b is real valued; if that is so, we have

(2) $$H(x, y) = \frac{1}{2}\left[E(x, y) - \sqrt{-1}E(x, \sqrt{-1}y)\right].$$

Conversely, let $E(x, y)$ be a Riemann form on $\mathbf{C}^n/D$. Then there exists a holomorphic theta function f on $\mathbf{C}^n/D$ such that E is the Riemann form defined by f.

If f is a theta function on $\mathbf{C}^n/D$, then the divisor (f) of f is defined on $\mathbf{C}^n/D$, which is an analytic divisor of $\mathbf{C}^n/D$. Conversely, if X is an analytic divisor of $\mathbf{C}^n/D$, there exists a theta function f on $\mathbf{C}^n/D$ such that $(f) = X$. We can prove that the alternating form E defined by f is determined only by X and independent of the choice of f; so we call E the alternating (or Riemann) form defined by X, and denote it by $E(X)$. It is a well-known basic fact that *a complex torus $\mathbf{C}^n/D$ has a structure of abelian variety if and only if there exists a non-degenerate Riemann form on $\mathbf{C}^n/D$.*

Let A be an abelian variety defined over $\mathbf{C}$. Then we can find a complex torus $\mathbf{C}^n/D$ and an analytic isomorphism θ of A onto $\mathbf{C}^n/D$. We call the pair $(\mathbf{C}^n/D, \theta)$ or simply the isomorphism θ an *analytic coordinate-system* of A. If X is a divisor on A, then $\theta(X)$ is an analytic divisor of $\mathbf{C}^n/D$, and vice versa; we write $E(X) = E(\theta(X))$ and call it the *alternating* (or *Riemann*) *form defined by* X with respect to θ. We have $E(X) = E(Y)$ if and only if X and Y are algebraically equivalent.

3.2. Analytic and rational representations of homomorphisms. Let A_1 and A_2 be two abelian varieties defined over $\mathbf{C}$; let $(\mathbf{C}^n/D_1, \theta_1)$ and $(\mathbf{C}^m/D, \theta_2)$ be analytic coordinate-systems of A_1 and A_2, respectively. Consider now a homomorphism λ of A_1 into A_2. There exists a linear mapping Λ of $\mathbf{C}^n$ into $\mathbf{C}^m$ such that

$$\theta_2 \circ \lambda = \Lambda \circ \theta_1;$$

Λ must satisfy $\Lambda(D_1) \subset D_2$. Conversely, every linear mapping of $\mathbf{C}^n$ into $\mathbf{C}^m$ satisfying this condition corresponds to a homomorphism of A_1 into A_2. With respect to the coordinate-systems (z_i) in $\mathbf{C}^n$ and (w_i) in $\mathbf{C}^m$, Λ is represented where the numbers of the factors in the products are both equal to n, and

$$\alpha(z) = y_1 \times \cdots \times y_n, \qquad \beta(u_1 \times \cdots \times u_n) = \varphi'(u_1) \times \cdots \times \varphi'(u_n),$$
$$\gamma(v_1 \times \cdots \times v_n) = v_1 + \cdots + v_n.$$

Put $\lambda_1 = \lambda + b$; we have then

$$(\gamma\circ\beta\circ\alpha)(z) = \sum_\nu \varphi'(y_\nu) = (\lambda\circ\varphi)(x) + b = (\lambda_1\circ\varphi\circ p)(z).$$

Let ω be an element of $\mathfrak{D}_0(J')$; as ω is an invariant form, we have $\omega\circ\lambda_1 = \omega\circ\lambda$. Denote by γ_ν the projection of $J' \times \cdots \times J'$ onto the ν-th factor. Then we have $\gamma = \gamma_1 + \cdots + \gamma_n$, so that $\omega\circ\gamma = \omega\circ\gamma_1 + \cdots + \omega\circ\gamma_n$ by Proposition 5. As we have $(\gamma_\nu\circ\beta\circ\alpha)(z) = \varphi'(y_\nu) = (\varphi'\circ q_\nu)(z)$, we get $\omega\circ\gamma_\nu\circ\beta\circ\alpha = \omega\circ\varphi'\circ q_\nu$. Hence we have

$$\omega\circ\lambda\circ\varphi\circ p = \omega\circ\lambda_1\circ\varphi\circ p = \omega\circ\gamma\circ\beta\circ\alpha = \sum_{\nu=1}^{n} \omega\circ\gamma_\nu\circ\beta\circ\alpha = \sum_{\nu=1}^{n} \omega\circ\varphi'\circ q_\nu.$$

3. Analytic Theory of Abelian Varieties

In this section, we shall recall some of known results from the classical theory of abelian varieties; a modern treatment for this subject can be found in Weil [57].

3.1. Theta functions and Riemann forms. Let D be a discrete subgroup of $\mathbf{C}^n$ of rank $2n$; then $\mathbf{C}^n/D$ is a complex torus. An $\mathbf{R}$-bilinear form $E(x, y)$ on $\mathbf{C}^n$ with values in $\mathbf{R}$ is called a *Riemann form* on $\mathbf{C}^n/D$ if it satisfies the following conditions.

(R1) *The value* $E(x, y)$ *is an integer for every* $x \in D$, $y \in D$.
(R2) $E(x, y) = -E(y, x)$.
(R3) *The form* $(x, y) \mapsto E(x, \sqrt{-1}y)$ *is a positive (not necessarily non-degenerate) symmetric form.*

A meromorphic function f on $\mathbf{C}^n$ is called a *theta function* on $\mathbf{C}^n/D$ if we have

$$f(x + d) = f(x)\exp[l_d(x) + c_d]$$

for every $d \in D$, where $l_d(x)$ is a $\mathbf{C}$-linear form on $\mathbf{C}^n$ and c_d is a complex number, both depending on d. Then we can show that there exist two $\mathbf{R}$-bilinear forms H, H_0 and an $\mathbf{R}$-linear form b, with values in $\mathbf{C}$, such that

$$(1)\quad f(x + d) = f(x)\exp\left\{2\pi\sqrt{-1}\left[H(d, x) + \frac{1}{2}H_0(d, d) + b(d)\right]\right\} \qquad \text{for } d \in D,$$

by an $m \times n$ matrix $S = (s_{ij})$ with complex coefficients as follows: regarding the z_j and the w_i as functions on $\mathbf{C}^n$ and $\mathbf{C}^m$, we have

$$w_i \circ \Lambda = \sum_{j=1}^{n} s_{ij} z_j \quad (1 \leq i \leq m).$$

The mapping $\lambda \to \Lambda$ (or $\lambda \to S$) can be uniquely extended to a representation of $\mathrm{Hom}_{\mathbf{Q}}(A_1, A_2)$, which we call the *analytic representation of* $\mathrm{Hom}_{\mathbf{Q}}(A_1, A_2)$, with respect to the analytic coordinate-systems θ_1 and θ_2.

Put now

$$\omega_j = dz_j \circ \theta_1, \quad \eta_i = dq_i \circ \theta_2.$$

Then, we see easily that $\{\omega_1, \ldots, \omega_n\}$ is a basis of $\mathfrak{D}_0(A_1)$ and $\{\eta_1, \ldots, \eta_m\}$ is a basis of $\mathfrak{D}_0(A_2)$; we have obviously,

$$\delta\lambda(\eta_i) = \sum_{j=1}^{n} s_{ij} \omega_j \quad (1 \leq i \leq m).$$

This shows that $S = (s_{ij})$ is the transpose of the representation of $\delta\lambda$ with respect to the bases $\{\omega_j\}$ and $\{\eta_i\}$.

Let $\{u_1, \ldots, u_{2n}\}$ and $\{v_1, \ldots, v_{2m}\}$ be respectively bases of D_1 and D_2 over $\mathbf{Z}$. Since Λ maps D_1 into D_2, there exists a $2m \times 2n$ matrix $M = (r_{ij})$ with coefficients in $\mathbf{Z}$ such that

$$\Lambda(u_j) = \sum_{i=1}^{2m} r_{ij} v_i \quad (1 \leq j \leq 2n).$$

The correspondence $\lambda \to M$ is uniquely extended to a representation of $\mathrm{Hom}_{\mathbf{Q}}(A_1, A_2)$, which we call the *rational representation of* $\mathrm{Hom}_{\mathbf{Q}}(A_1, A_2)$ with respect to $\{u_i\}$ and $\{v_i\}$. We can easily verify that the rational representation of $\mathrm{End}_{\mathbf{Q}}(A)$ is equivalent to l-adic representations defined in §1.2. Let U denote the $n \times 2n$ matrix whose column-vectors are $u_1, \ldots, u_{2n}$ and V the $m \times 2m$ matrix whose column-vectors are $v_1, \ldots, v_{2m}$. We have then

$$SU = VM,$$

so that

$$\begin{pmatrix} S & 0 \\ 0 & \bar{S} \end{pmatrix} \begin{pmatrix} U \\ \bar{U} \end{pmatrix} = \begin{pmatrix} V \\ \bar{V} \end{pmatrix} M,$$

where bars denote complex conjugation. Now suppose that $A_1 = A_2$, $n = m$, $D_1 = D_2$, $\theta_1 = \theta_2$, and $U = V$. Since the matrix $\binom{U}{\bar{U}}$ is invertible, we see that *the rational representation M is equivalent to the direct sum of the analytic representation S and its complex conjugate $\bar{S}$.*

3.3. Dual abelian varieties. Let $(\mathbf{C}^n/D, \theta)$ be an analytic representation of an abelian variety A. We shall now define the dual of $\mathbf{C}^n/D$. Let $x = (x_\nu)$ and $y = (y_\nu)$ be two vectors in $\mathbf{C}^n$ with the components x_ν and y_ν. Put

$$\langle x, y\rangle = \sum_{\nu=1}^{n}(x_\nu \bar{y}_\nu + \bar{x}_\nu y_\nu). \tag{3}$$

Then, we see that $\langle x, y\rangle$ is a non-degenerate symmetric $\mathbf{R}$-bilinear form on $\mathbf{C}^n$ with values in $\mathbf{R}$; hence, $\mathbf{C}^n$ is considered as the dual vector space over $\mathbf{R}$ of itself, with respect to this inner product $\langle x, y\rangle$. We have furthermore

$$\langle\sqrt{-1}x, y\rangle = \langle x, -\sqrt{-1}y\rangle. \tag{4}$$

Denote by D^* the set of all vectors $y \in \mathbf{C}^n$ such that $\langle x, y\rangle$ is an integer for every $x \in D$; then D^* is a discrete subgroup of $\mathbf{C}^n$, so that $\mathbf{C}^n/D^*$ is a complex torus of dimension n, which we call the *dual* of $\mathbf{C}^n/D$. We can show that $\mathbf{C}^n/D^*$ has a structure of abelian variety, and consider $\mathbf{C}^n/D^*$ as an analytic representation of the Picard variety A^* of A in the following manner. Let T be the set of all homomorphisms of D into the group of complex numbers of absolute value 1. We see easily that the mapping

$$\mathbf{C}^n \ni y \to \exp[2\pi\sqrt{-1}\langle y, \quad\rangle]$$

gives an isomorphism of $\mathbf{C}^n/D^*$ onto T; so we identify T with $\mathbf{C}^n/D^*$ by this isomorphism. Let X be a divisor of A. Take a normalized theta function f on $\mathbf{C}^n/D$ such that $(f) = \theta(X)$. Suppose that X is algebraically equivalent to 0; we have then $E(X) = 0$. Since f is normalized, we can easily verify that f satisfies the formula

$$f(x + d) = \mu(d)f(x) \quad \text{for } d \in D,$$

where μ is an element of T; it can be seen that μ is determined only by X. Let $\psi(X)$ denote the point on $\mathbf{C}^n/D^*$ corresponding to μ. Then we can establish an isomorphism θ^* of A^* onto $\mathbf{C}^n/D^*$ by the relation

$$\theta^*(\mathrm{Cl}(X)) = \psi(X).$$

This implies, as we have said, that $\mathbf{C}^n/D^*$ is considered as an analytic representation of A^*; we call $(\mathbf{C}^n/D^*, \theta^*)$ the *dual* of $(\mathbf{C}^n/D, \theta)$.

Now we shall consider the mapping φ_Y of A into A^*, defined by (4) of §1.3. Let Y be a divisor on A and f a normalized theta function on $\mathbf{C}^n/D$ such that $(f) = \theta(Y)$. Put $E = E(Y)$; then f satisfies formula (1) of §3.1 with the form H given by (2). Put, for every $u \in \mathbf{C}^n$,

$$\Phi_u(x) = f(x)^{-1}f(x - u)\exp[2\pi\sqrt{-1}H(u, x)].$$

Let t be the point on A corresponding to u by θ. We observe that

(5) $$(\Phi_u) = \theta(Y_t - Y),$$

(6) $$\Phi_u(x+d) = \Phi_u(x)\exp[2\pi\sqrt{-1}E(u,d)] \quad \text{for } d \in D.$$

On the other hand, we obtain an **R**-linear mapping $\mathfrak{E}$ of $\mathbf{C}^n$ into itself by the relation

$$\langle \mathfrak{E}(u), v\rangle = E(u,v).$$

By properties (R1-3) of Riemann forms and the relation (4), we see that $\mathfrak{E}$ is **C**-linear and maps D into D^*, so that $\mathfrak{E}$ gives a homomorphism of $\mathbf{C}^n/D$ into $\mathbf{C}^n/D^*$. Then relations (5) and (6) show that $\mathfrak{E}(u)$ represents the point $\psi(Y_t - Y)$ on $\mathbf{C}^n/D^*$; namely, we have

$$\mathfrak{E}(\theta(t)) = \theta^*(\mathrm{Cl}(Y_t - Y)).$$

In other words, $\mathfrak{E}$ is the analytic representation of φ_Y with respect to θ and θ^*.

Let $\{u_1, \ldots, u_{2n}\}$ be a basis of D over **Z**; we can find $2n$ elements u_i^* of $\mathbf{C}^n$ such that

$$\langle u_i, u_j^*\rangle = \delta_{ij},$$

where the δ_{ij} denote Kronecker's delta; the u_j^* form a basis of D^* over **Z**. Let $E_0 = (e_{ij})$ be the matrix of degree $2n$ which represents the form E with respect to the basis $\{u_i\}$; this means

$$E\left(\sum_{i=1}^{2n} a_i u_i, \sum_{i=1}^{2n} b_i u_i\right) = {}^t b E_0 a = \sum_{i,j} e_{ij} b_i a_j,$$

where a and b denote respectively the vectors of $\mathbf{R}^{2n}$ with the components a_i and b_i. We have then

$$\mathfrak{E}(u_j) = \sum_{i=1}^{2n} e_{ij} u_i^*;$$

this shows that E_0 is the rational representation of φ_Y with respect to the bases $\{u_i\}$ and $\{u_i^*\}$.

Let N be a positive integer and u, v two vectors in $\mathbf{C}^n$ such that $Nu \in D$, $Nv \in D$; let s and t be the points on A corresponding to u and v by θ. The functions f and Φ_u being defined for the divisor Y as above, put

$$g(x) = \Phi_u(Nx), \quad h(x) = \Phi_u(x)^N.$$

We observe that g and h are considered as functions on $\mathbf{C}^n/D$ and satisfy

$$(h) = \theta(N(Y_s - Y)), \quad h(Nx) = g(x)^N,$$

$$g(x+v) = g(x)\exp[2\pi\sqrt{-1}E(u, Nv)].$$

Using the notation of §1.4, we have

$$e_{Y,N}(t,s) = \exp[2\pi\sqrt{-1}NE(u,v)]. \tag{7}$$

It follows that, for every rational prime l, E_0 coincides with the l-adic representation of Y, defined in §1.4, for a suitable choice of l-adic coordinate-system.

Let A_1 be another abelian variety defined over $\mathbf{C}$ and $(\mathbf{C}^m/D_1, \theta_1)$ an analytic representation of A_1; let A_1^* be the Picard variety of A_1 and $(\mathbf{C}^m/D_1^*, \theta_1^*)$ the dual of $(\mathbf{C}^m/D_1, \theta_1)$. We shall now consider the transpose ${}^t\lambda$ of a homomorphism λ of A into A_1. Take bases $\{u_i\}$ of D, $\{u_i^*\}$ of D^*, $\{v_i\}$ of D_1, $\{v_i^*\}$ of D_1^*, over $\mathbf{Z}$, such that

$$\langle u_i, u_j^*\rangle = \delta_{ij}, \quad \langle v_i, v_j^*\rangle = \delta_{ij}.$$

The homomorphism λ has an analytic representation Λ which is a linear mapping of $\mathbf{C}^n$ into $\mathbf{C}^m$ and a rational representation $M = (c_{ij})$, for which we have

$$\Lambda(u_j) = \sum_i c_{ij}v_i.$$

We can obtain a $\mathbf{C}$-linear mapping ${}^t\Lambda$ of $\mathbf{C}^m$ into $\mathbf{C}^n$ by the relation

$$\langle x, {}^t\Lambda(y)\rangle = \langle \Lambda(x), y\rangle$$

for $x \in \mathbf{C}^n$, $y \in \mathbf{C}^m$. We shall now show that ${}^t\Lambda$ is the analytic representation of ${}^t\lambda$ with respect to θ^* and θ_1^*, and ${}^tM$ is the rational representation of ${}^t\lambda$ with respect to $\{v_i^*\}$ and $\{u_1^*\}$. To prove this, take a divisor X on A_1 which is algebraically equivalent to 0 and a normalized theta function f on $\mathbf{C}^m/D_1$ such that $(f) = \theta_1(X)$; as is seen above, f satisfies

$$f(y+d_1) = f(y)\exp[2\pi\sqrt{-1}\langle d_1, y^*\rangle] \quad \text{for} \quad d_i \in D_1,$$

where y^* is an element of $\mathbf{C}^m$ which represents $\mathrm{Cl}(X)$. Put $g = f\ \Lambda$; we have then,

$$(g) = \theta(\lambda^{-1}(X)),$$

$$g(x+d) = g(x)\exp[2\pi\sqrt{-1}\langle d, {}^t\Lambda(y^*)\rangle] \quad \text{for} \quad d \in D.$$

These relations show that ${}^t\Lambda(y^*)$ is a point on $\mathbf{C}^n$ corresponding to the point ${}^t\lambda(\mathrm{Cl}(X)) = \mathrm{Cl}(\lambda^{-1}(X))$ on A^*. Thus we have proved that ${}^t\Lambda$ is the analytic representation of ${}^t\lambda$ with respect to θ^* and θ_1^*; it follows that ${}^tM$ is the rational representation of ${}^t\lambda$ with respect to $\{v_i^*\}$ and $\{u_i^*\}$.

A, A^*, $(\mathbf{C}^n/D, \theta)$ and $(\mathbf{C}^n/D^*, \theta^*)$ being as above, let Y be a non-degenerate divisor on A, and $E(x, y)$ the alternating form defined by Y. Let λ be an element of $\mathrm{End}_{\mathbf{Q}}(A)$; let $\mathfrak{E}$ and Λ be respectively the analytic representation of φ_Y and λ with respect to θ and θ^*. We have then

$$E(\mathfrak{E}^{-1\,t}\Lambda\mathfrak{E}(x), y) = \langle {}^t\Lambda\mathfrak{E}(x), y\rangle = \langle\mathfrak{E}(x), \Lambda(y)\rangle = E(x, \Lambda(y));$$

in other words, $\varphi_Y^{-1\,t}\lambda\varphi_Y$ is the adjoint of λ with respect to the form $E(x, y)$. The involution $\lambda \to \lambda' = \varphi_Y^{-1\,t}\lambda\varphi_Y$ of $\mathrm{End}_{\mathbf{Q}}(A)$ is thus described by means of the alternating form.

4. Fields of Moduli and Kummer Varieties

4.1. Polarization of a variety. Let V be a complete variety, non-singular in co-dimension 1, defined over a field k, and X a divisor on V which is rational over k. We denote by $L(X; k)$ the set of functions f on V, defined over k, such that $\mathrm{div}(f) \succ -X$, where $\mathrm{div}(f)$ denotes the divisor of the function f. If k' is an extension of k, we have $L(X; k') = L(X; k) \otimes_k k'$, so that the dimension of the vector space $L(X; k)$ over k (which is always finite) is independent of k; we denote this dimension by $l(X)$. Let $\{f_0, \ldots, f_r\}$ be a basis of $L(X; k)$ over k and x a generic point of V over k. Consider $(f_0(x), \ldots, f_r(x))$ as a point of the projective space P^r of dimension r and denote by U the locus of $(f_0(x), \ldots, f_r(x))$ over k. Then we obtain a rational mapping Φ of V onto U defined by

$$\Phi(x) = (f_0(x), \ldots, f_r(x))$$

with respect to k. We say that the divisor X (or the complete linear system defined by X) is *ample* if the mapping Φ is biregular and $\{\mathrm{div}(f) + X \mid f \in L(X; k)\}$ has no fixed component; this definition does not depend upon the choice of k and $\{f_i\}$.

V, X being as above, we denote by $\mathcal{C}(X)$ the set of all divisors X' on V for which there exist two positive integers m, m' such that mX is algebraically equivalent to $m'X'$. The set $\mathcal{C}(X)$ is called a *polarization* of V if it contains an ample divisor. A variety V is said to be *polarizable* if there exists a polarization of V. We understand by a *polarized variety* a couple $(V, \mathcal{C})$ formed by a variety V and a polarization $\mathcal{C}$ of V; every divisor X in $\mathcal{C}$ is called a *polar divisor* of $(V, \mathcal{C})$. Given a field of definition k for V, we say that $(V, \mathcal{C})$ or $\mathcal{C}$ is defined over k if $\mathcal{C}$ contains a divisor rational over k. If V has a structure of abelian variety, we call $(V, \mathcal{C})$ with this structure a *polarized abelian variety*; in this case, $(V, \mathcal{C})$ is said to be defined over k only when the structure of abelian variety is also defined over k. Let σ be an isomorphism of k into a field k'; $(V, \mathcal{C})$ being defined over k, we denote by $\mathcal{C}^\sigma$ the polarization $\mathcal{C}(X^\sigma)$ of V^σ, where X is a divisor in $\mathcal{C}$ rational over k.

PROPOSITION 10. *If V is defined over k, every polarized variety $(V, \mathcal{C})$ is defined over a finite algebraic extension of k.*

PROOF. Take a divisor X in $\mathcal{C}$. We can find a specialization X' of X over the algebraic closure of k, which is rational over a finite algebraic extension of k. Then, X' is algebraically equivalent to X, so that X' is contained in $\mathcal{C}$. This proves the proposition.

PROPOSITION 11. *If V is defined over k and polarizable, then V has a polarization defined over k.*

PROOF. By Proposition 10, every polarization contains a divisor X which is rational over a finite algebraic extension k' of k. Put $Y = p^m \sum X^\sigma$, where the sum is taken over all the isomorphisms σ of k' into the algebraic closure of k, and p denotes the characteristic or 1 according as the characteristic is a prime or 0. Then, for a suitable m, Y is rational over k; and we can easily see that Y determines a polarization of V; our proposition is thereby proved.

We shall now confine ourselves to abelian varieties. A theory for a more general case can be found in Matsusaka [28]. It may be noted, however, that the definition of the field of moduli in this paper is different from ours.

4.2. Fields of moduli of polarized abelian varieties. The following two propositions are due to Weill [56].

PROPOSITION 12. *Every abelian variety is polarizable.*

PROPOSITION 13. *Let X be a divisor on an abelian variety. If there exists an integer $n > 0$ such that nX is ample, then X is non-degenerate. Conversely, if X is a positive non-degenerate divisor, there exists an integer $n_0 > 0$ such that nX is ample for $n \geq n_0$.*

Let $(A, \mathcal{C})$ and $(A', \mathcal{C}')$ be two polarized abelian varieties, of the same dimension. A homomorphism (resp. an isomorphism) λ of A onto A' is called a *homomorphism* (resp. an *isomorphism*) of $(A, \mathcal{C})$ onto $(A', \mathcal{C}')$ if there exists a divisor X' in $\mathcal{C}'$ such that $\lambda^{-1}(X')$ is contained in $\mathcal{C}$; if that is so, for every Y in $\mathcal{C}'$, $\lambda^{-1}(Y)$ is contained in $\mathcal{C}$.

We now give an important theorem which is basic for the definition of field of moduli.

THEOREM 2. *Let A be an abelian variety and $\mathcal{C}$ a polarization of A. Then, there exists a field k_0 with the following property:*

(M) *k and σ being respectively a field of definition for $(A, \mathcal{C})$ containing k_0 and an isomorphism of k into a field, $(A, \mathcal{C})$ is isomorphic to $(A^\sigma, \mathcal{C}^\sigma)$ if and only if σ is the identity on k_0.*

A proof is given in Shimura [36]. For our later use, we need only the case where A is defined over an algebraic number field of finite degree. In this case the field k_0 is easily given by Galois theory. Before showing this, we give an easy consequence of Theorem 2.

PROPOSITION 14. *If the characteristic of the universal domain is 0, the field k_0 with property (M) of Theorem 2 is uniquely determined by $(A, \mathcal{C})$ and is contained in every field of definition for $(A, \mathcal{C})$.*

PROOF. Let k_0 and k_0' be two fields with property (M). Take a field of definition k for $(A, \mathcal{C})$ containing k_0 and k_0'. Then for every isomorphism σ of k into a field, σ is the identity on k_0 if and only if σ is so on k_0'; this implies $k_0 = k_0'$. Let k_1 be a field of definition for $(A, \mathcal{C})$ and τ an isomorphism of k_0k_1 into a field such that τ is the identity on k_1. Then we have $(A, \mathcal{C}) = (A^\tau, \mathcal{C}^\tau)$, so that by property (M), τ is the identity on k_0; this proves $k_0 \subset k_1$.

Now let us consider the case where A is defined over an algebraic number field k_1 of finite degree. By Proposition 10, $(A, \mathcal{C})$ is defined over a finite algebraic extension k' of k_1. Take a Galois extension k'' of $\mathbf{Q}$ containing k' and call G the Galois group of k'' over $\mathbf{Q}$. Let H be the subgroup of G composed of the elements $\sigma \in G$ such that $(A^\sigma, \mathcal{C}^\sigma)$ is isomorphic to $(A, \mathcal{C})$, and k_0 the subfield of k'' corresponding to H. Then, it is easy to see that k_0 has property (M).

We call the field k_0 with property (M), which is uniquely determined by $(A, \mathcal{C})$ if the characteristic is 0, as we have seen in Proposition 14, the *field of moduli* of $(A, \mathcal{C})$. Obviously, two polarized abelian varieties, isomorphic to each other, have the same field of moduli. We can define the field of moduli also in the case of positive characteristics; for details we refer the reader to [36].

If the characteristic is 0, $(A, \mathcal{C})$ determines a point z on the Siegel's space of degree n, where n is the dimension of A. In fact, the field of moduli of $(A, \mathcal{C})$ is generated over $\mathbf{Q}$ by the values of certain Siegel's modular functions, as will be explained in §§24 and 25.

PROPOSITION 15. *Let $\mathcal{C}$ be a polarization of an abelian variety. Then, there exists a divisor Y in $\mathcal{C}$ such that every divisor in $\mathcal{C}$ is algebraically equivalent to a multiple mY with a positive integer m.*

PROOF. Consider, for each $X \in \mathcal{C}$, the homomorphism φ_X of A into its Picard variety A^*. By Proposition 13, every X in $\mathcal{C}$ is non-degenerate, so that we have $\nu(\varphi_X) > 0$. Since the $\nu(\varphi_X)$ are integers, there exists a divisor Y in $\mathcal{C}$ such that $\nu(\varphi_Y) = \mathrm{Min}\{\nu(\varphi_X) \mid X \in \mathcal{C}\}$. We shall prove that Y has the property of our proposition. For every X in $\mathcal{C}$, by our definition, there exist two positive integers a and b such that aX is algebraically equivalent to bY; we have then $a\varphi_X = b\varphi_Y$. We can find two integers q and r such that $b = aq + r, 0 \leq r < a$. Assume that $r > 0$; then putting $Z = X - qY$, $\varphi_{aZ} = a\varphi_Z = a\varphi_X - aq\varphi_Y = r\varphi_Y = \varphi_{rY}$. This implies that aZ is algebraically equivalent to rY, and hence Z is contained in $\mathcal{C}$. By the relations $a\varphi_Z = r\varphi_Y$ and $r < a$, we get $\nu(\varphi_Z) < \nu(\varphi_Y)$; this is a contradiction; so r must be 0. We have then $b = aq$, and hence $\varphi_X = q\varphi_Y$. It follows that X is algebraically equivalent to qY; this completes the proof.

We call a divisor Y with the property of Proposition 15 a *basic polar divisor* of $\mathcal{C}$.

4.3. Quotients of an abelian variety. We want to prove

PROPOSITION 16. *Let A be an abelian variety and G a finite group of automorphisms of A; let k be a field of definition for A and the elements of G. Then, there exist a projective variety W and a rational mapping F of A onto W, both defined over k, satisfying the following conditions:*

(K1) *F is everywhere defined on A.*

(K2) *$F(u) = F(v)$ if and only if there exists an element $\gamma \in G$ such that $u = \gamma(v)$.*

(K3) *if F' is a rational mapping of A into a variety W' satisfying $F' = F' \ \gamma$ for every $\gamma \in G$, then there exists a rational mapping Φ of W into W' such that $F' = \Phi \ F$ and Φ is defined at $F(a)$ whenever F' is defined at a.*

PROOF. Here we borrow an idea from Serre [32], where quotients of a variety, which is not necessarily abelian, are treated in a little different form. We assume that A is a projective variety. This is possible by virtue of Propositions 11 and 12. For every point a of A, we can find a homogeneous polynomial $H(X)$ with coefficients in k, other than the constants, such that $H(\gamma(a)) \neq 0$ for every $\gamma \in G$. Put

$$B' = \{x \in A \mid H(x) \neq 0\}, \qquad B = \bigcap_{\gamma \in G} \gamma(B').$$

Then, B is an open set of A, in the sense of Zariski-topology, which is biregularly equivalent over k to an affine variety. By our choice of H, B contains the given point a; moreover, we have $\gamma(B) = B$ for every $\gamma \in G$. We obtain in this way a finite open covering $\{B_i\}$ of A such that each B_i is biregularly equivalent over k to an affine variety V_i and $\gamma(B_i) = B_i$ for every $\gamma \in G$. Let k' be an arbitrary extension of k, x a generic point of A over k', and ξ_i the point on V_i corresponding to x for each i. Now we fix our attention to one of the V_i, say V_1. Since A is non-singular, so is V_1; hence $k'[\xi_1]$ is integrally closed. As we have $\gamma(B_1) = B_1$, we can define the operation of the elements of G on the variety V_1. Every element γ of G determines an automorphism of $k'(x) = k'(\xi_1)$. We denote by u^γ the image of $u \in k'(x)$ by this automorphism. Moreover, since γ acts on V_1 and $k'[\xi_1]$ is integrally closed, this automorphism induces an automorphism of $k'[\xi_1]$. Denote by K, K', R, R' respectively the set of G-invariant elements of $k(x)$, $k'(x)$, $k[\xi_1]$, $k'[\xi_1]$. Then, it is easy to see that K and K' are respectively the quotient fields of R and R', and we have $R' = R \otimes_k k'$. Now let $(\xi_{1\lambda})$ be the coordinates of ξ_1, and η_μ the coefficients of the polynomials $P_\lambda(X) = \prod_\gamma (X - \xi_{1\lambda}^\gamma)$. Then we see that $k[\xi_1] \supset R \supset k[\eta]$ and $k[\xi_1]$ is integral over $k[\eta]$. As $k[\xi_1]$ is finitely generated as a $k[\eta]$-module, the submodule R is also finitely generated. Hence there exists a finite set of elements (ζ_1) such that $R = k[\zeta_1]$; we have then $R' = k'[\zeta_1]$. Since $k'[\xi_1]$ is integrally closed and $R' = K' \cap k'[\xi_1]$, we see that $k'[\zeta_1]$ is integrally closed. We obtain in the same manner, for each i, a finite set (ζ_i) such that

$$K = k(\zeta_i), \quad k[\zeta_i] = K \cap k[\xi_i], \quad k'[\zeta_i] = K' \cap k'[\xi_i].$$

Let a be a point of B_1 and α the point on V_1 corresponding to a. We have obviously

$$[x \to a; k'] = [\xi_1 \to \alpha; k'],$$

where $[u \to v; k']$ denotes the specialization-ring of v in $k'(u)$ (cf. §9.2). Let β be a specialization of ζ_1 over $\xi_1 \to \alpha$ ref. k'. We shall now prove

$$(1) \qquad K' \cap [\xi_1 \to \alpha; k'] = [\zeta_1 \to \beta; k'].$$

First we observe that for every $\gamma \in G$,

$$(2) \qquad K' \cap [\xi_1 \to \alpha; k'] = K' \cap [\xi_1 \to \gamma(\alpha); k'].$$

In fact, if u is an element of $K' \cap [\xi_1 \to \alpha; k']$, we have an expression $u = f(\xi_1)/g(\xi_1)$, where f and g are polynomials with coefficients in k' such that $g(\alpha) \neq 0$. As u is contained in K', we have $u = f(\xi_1^\gamma)/g(\xi_1^\gamma)$. Since the $\xi_{1\lambda}^\gamma$ are contained in $k[\xi_1]$, there exist polynomials h_λ with coefficients in k such that $\xi_{1\lambda}^\gamma = h_\lambda(\xi_1)$. Put

$$p(X) = f(\ldots, h_\lambda(X), \ldots), \qquad q(X) = g(\ldots, h_\lambda(X), \ldots).$$

Then we have $u = p(\xi_1)/q(\xi_1)$ and $q(\gamma^{-1}(\alpha)) = g(\alpha) \neq 0$. Hence u is contained in $K' \cap [\xi_1 \to \gamma^{-1}(\alpha); k']$; so we have

$$K' \cap [\xi_1 \to \alpha; k'] \subset K' \cap [\xi_1 \to \gamma^{-1}(\alpha); k'].$$

The inverse inclusion is similarly proved; so we obtain equality (2). Now u being an element of $K' \cap [\xi_1 \to \alpha; k']$, let d be a specialization of u over $\zeta_1 \to \beta$ ref. k'. Extend this specialization to a specialization

$$(\xi_1, \zeta_1, u) \to (\alpha', \beta, d) \quad \text{ref. } k'.$$

As $k'[\xi_1]$ is integral over $k'[\zeta_1]$, α' must be finite. Consider the polynomial

$$M(T) = \prod_\gamma \left(\sum_\lambda \xi_{1\lambda}^\gamma T_\lambda \right).$$

We observe that the coefficients of $M(T)$ are contained in $k[\zeta_1]$, so that the specializations

$$(\xi_1, \zeta_1) \to (\alpha, \beta) \quad \text{ref. } k', \qquad (\xi_1, \zeta_1) \to (\alpha', \beta) \quad \text{ref. } k'$$

lead to the equality

$$\prod_\gamma \left(\sum_\lambda \gamma(\alpha)_\lambda T_\lambda \right) = \prod_\gamma \left(\sum_\lambda \gamma(\alpha')_\lambda T_\lambda \right),$$

where $\gamma(\alpha)_\gamma$ and $\gamma(\alpha')_\lambda$ denote respectively the λ-th coordinates of the points $\gamma(\alpha)$ and $\gamma(\alpha')$. This shows that there exists an element γ of G such that $\gamma(\alpha) = \alpha'$. By relation (2), u is contained in $K' \cap [\xi_1 \to \alpha'; k']$; so its specialization d must be finite. This implies that every element of $K' \cap [\xi_1 \to \alpha; k']$ is integral over $[\zeta_1 \to \beta; k']$. On the other hand, as $k'[\zeta_1]$ is integrally closed, $[\zeta_1 \to \beta; k']$ is integrally closed; this proves equality (1). Now let W_i be the locus of ζ_i over k. Consider a specialization

$$(\zeta_1, \cdots, \zeta_h) \to (\beta_1, \ldots, \beta_h) \text{ ref. } k'.$$

Extend this to a specialization

$$(x, \xi_1, \ldots, \xi_h, \zeta_1, \ldots, \zeta_h) \to (a, \alpha_1, \ldots, \alpha_h, \beta_1, \ldots, \beta_h) \text{ ref. } k'.$$

Since $k'[\xi_i]$ is integral over $k'[\zeta_i]$ and $k'[\xi_i] \supset k'[\zeta_i]$, we see that, for every i, α_i is finite if and only if β_i is finite. As $\{B_i\}$ is a covering of A, at least one of the α_i is finite; and for such an i, we have

$$[x \to a; k'] = [\xi_i \to \alpha_i; k'],$$

$$K' \cap [\xi_i \to \alpha_i; k'] = [\zeta_i \to \beta_i; k'],$$

and hence

(3) $$K' \cap [x \to a; k'] = [\zeta_i \to \beta_i; k'].$$

Therefore the specialization-ring $[\zeta_i \to \beta_i; k]$ does not depend on the choice of i, provided β_i is finite. It follows that the affine varieties W_i determine an abstract variety W if we regard the ζ_i as corresponding generic points with respect to k. Denote by z the point on W whose representatives are the ζ_i, and by F the rational mapping of A onto W defined by $F(x) = z$ with respect to k. Equality (3) is then written in the form

$$K' \cap [x \to a; k'] = [z \to F(a); k']. \tag{4}$$

By our construction, F is everywhere defined on A and $F \circ \gamma = F$ for every $\gamma \in G$. We shall now prove that (W, F) satisfies condition (K3). A pair (W', F') being as in (K3), let k' be a field of definition for W' and F', containing k; let x be a generic point on A over k'. Put $F(x) = z$, $F'(x) = z'$. Since $F' \circ \gamma = F'$ for every $\gamma \in G$, $k'(z')$ is contained in $k'(z)$. Hence we obtain a rational mapping Φ of W into W' defined by $\Phi(z) = z'$ with respect to k'; we have then $F' = \Phi \circ F$. Now suppose that F' is defined at a point $a \in A$. Then we have

$$[x \to a; k'] \supset [z' \to F'(a); k'].$$

By equality (4), we have

$$[z \to F(a); k'] \supset [z' \to F'(a); k'];$$

this proves that Φ is defined at $F(a)$. Therefore (W, F) satisfies (K3). We have thus constructed an abstract variety W and a rational mapping F, both defined over k, satisfying conditions (K1), (K3) and the "if" part of (K2). It remains to prove the "only if" part of (K2) and realize W as a projective variety. For this purpose, take a generic point x on A over k and consider the Chow point y of the 0-dimensional cycle $\sum_{\gamma \in G}(\gamma(x))$ on A. Let W_1 be the locus of y over k and W_0 a projective normalization of W_1 with respect to k. Then we obtain a rational mapping f_1 of A onto W_1 defined by $f_1(x) = y$ with respect to k and a birational mapping f of W_0 onto W_1 defined by the normalization. Put $f_0 = f^{-1} \circ f_1$, $t = f_0(x)$. We have obviously $k(t) = K = k(z)$. Hence there exists a birational mapping Ψ of W onto W_0, defined over k, such that $\Psi(z) = t$. Let (a, b) be a specialization of (x, t) over k. As W_0 is a normalization of W_1 with respect to k, $[t \to b; k]$ is integrally closed. Using this fact, we can prove the equality

$$K \cap [x \to a; k] = [t \to b; k],$$

in the same way as in the proof of (1). Hence we have

$$[t \to b; k] = [z \to F(a); k].$$

It follows that W_0 and W are biregularly equivalent over k under Ψ. Consequently f_0 and f_1 are everywhere defined. Hence $f_1(a)$ is the Chow point of the cycle $\sum_{\gamma \in G} \gamma(a)$ for every $a \in A$. If $F(a) = F(b)$, then $f_1(a) = f_1(b)$, which means that $\sum_{\gamma \in G} \gamma(a) = \sum_{\gamma \in G} \gamma(b)$, and hence $a = \gamma(b)$ with some $\gamma \in G$. Thus F satisfies (K2). This completes our proof.

The couple (W, F) is uniquely determined by conditions (K1–3) up to biregular birational mappings. We call (W, F) a *quotient of A by G, defined over k*.

REMARK 1. The notation and assumptions being as in (K3), let k' be a field of definition for W' and F', containing k. Then the rational mapping Φ is defined over k'. This is included in the above proof.

4.4. Kummer varieties. By Weil [54], Matsusaka [28] we know

PROPOSITION 17. *Every polarized abelian variety has only a finite number of automorphisms.*

Here we reproduce the proof of [54]. Given a polarization $\mathcal{C}$ of an abelian variety A, take a divisor X in $\mathcal{C}$ and consider the involution $\alpha \to \alpha' = \varphi_X^{-1}{}^t\alpha\varphi_X$ of $\mathrm{End}_{\mathbf{Q}}(A)$. For every automorphism α of $(A, \mathcal{C})$, there exist two positive integers m and m' such that

$$m^t\alpha\varphi_X\alpha = m'\varphi_X$$

by virtue of relation (7) of §1.3. Taking the degree of both sides, we get $m = m'$, so that $\alpha\alpha' = 1_A$. Thus we obtain $\mathrm{tr}(\alpha\alpha') = 2n$ for every automorphism α of $(A, \mathcal{C})$, where n is the dimension of A. Since $\mathrm{tr}(\alpha\alpha')$ is a positive non-degenerate quadratic form and $\mathrm{End}(A)$ is finitely generated over $\mathbf{Z}$, only a finite number of such α can exist.

Given a polarized abelian variety $(A, \mathcal{C})$, let G be the group of automorphisms of $(A, \mathcal{C})$; Proposition 17 asserts that G is finite. By a *Kummer variety* of $(A, \mathcal{C})$, we understand a quotient of A by G.

THEOREM 3. *Let $(A, \mathcal{C})$ be a polarized abelian variety and k_0 the field of moduli of $(A, \mathcal{C})$. Then there exists a Kummer variety (W, F) of $(A, \mathcal{C})$ satisfying the following conditions:*

(N1) *W is defined over k_0.*
(N2) *F is defined over every field of definition for $(A, \mathcal{C})$ containing k_0.*
(N3) *Given a field of definition k for $(A, \mathcal{C})$ containing k_0, if σ is an isomorphism of k into a field and if η is an isomorphism of $(A, \mathcal{C})$ onto $(A^\sigma, \mathcal{C}^\sigma)$, then we have $F = F^\sigma \circ \eta$.*

We note that if $(A, \mathcal{C})$ is isomorphic to $(A^\sigma, \mathcal{C}^\sigma)$, then σ is the identity on k_0 by virtue of property (M) of Theorem 2, so that $W^\sigma = W$ by property (N1).

We shall call a Kummer variety (W, F) of $(A, \mathcal{C})$ satisfying conditions (N1–3) a *normalized Kummer variety* of $(A, \mathcal{C})$, which is uniquely determined for $(A, \mathcal{C})$ up to biregular birational mappings defined over k_0, if A is defined over a separably generated extension of k_0.

We give a proof of Theorem 3 only in the case where $(A, \mathcal{C})$ is defined over an algebraic number field; the same method is applicable to the general case; as for this, see Remark 2 at the end of the proof.

Let G be the group of automorphisms of $(A, \mathcal{C})$ and k a field of definition for $(A, \mathcal{C})$ and the elements of G; we assume that k is a finite Galois extension of $\mathbf{Q}$; this is possible by our restriction. Then, as seen above, there exists a Kummer variety (W, F) defined over k. Let $\mathfrak{G}$ denote the Galois group of k over k_0. For every $\sigma \in \mathfrak{G}$, by virtue of property (M) of k_0, there exists an isomorphism α_σ of $(A, \mathcal{C})$ onto $(A^\sigma, \mathcal{C}^\sigma)$. We see that $G^\sigma = \{\gamma^\sigma \mid \gamma \in G\}$ is the group of automorphisms of $(A^\sigma, \mathcal{C}^\sigma)$ and (W^σ, F^σ) is a Kummer variety of $(A^\sigma, \mathcal{C}^\sigma)$. Then $F^\sigma \circ \alpha_\sigma$ is a rational mapping of A onto W^σ which is everywhere defined on A; and for every $\gamma \in G$, we have $(F^\sigma \circ \alpha_\sigma) \circ \gamma = F^\sigma \circ \alpha_\sigma$; in fact, as α_σ is an isomorphism of $(A, \mathcal{C})$ onto $(A^\sigma, \mathcal{C}^\sigma)$, there exists an element $\gamma' \in G^\sigma$ such that $\alpha_\sigma \circ \gamma = \gamma' \circ \alpha_\sigma$, so that $F^\sigma \circ \alpha_\sigma \circ \gamma = F^\sigma \circ \gamma' \circ \alpha_\sigma = F^\sigma \circ \alpha_\sigma$. Hence, by (K3), there exists a rational mapping β_σ of W onto W^σ such that $F^\sigma \circ \alpha_\sigma = \beta_\sigma \circ F$; and β_σ is everywhere defined on W. We obtain similarly a rational mapping β'_σ of W^σ onto W, everywhere defined on W^σ, such that $\beta'_\sigma \circ F^\sigma = F \circ \alpha_\sigma^{-1}$. It is easy to see that β_σ is birational and biregular, and $\beta'_\sigma = \beta_\sigma^{-1}$. The birational mapping β_σ is uniquely determined by σ and does not depend upon the choice of α_σ; in fact, for any other isomorphism α'_σ of $(A, \mathcal{C})$ onto $(A^\sigma, \mathcal{C}^\sigma)$, there exists an element $\gamma \in G$ such that $\alpha'_\sigma = \alpha_\sigma \circ \gamma$; it follows that $F^\sigma \circ \alpha'_\sigma = F^\sigma \circ \alpha_\sigma$; this shows the uniqueness of β_σ. We can easily verify that β_σ is defined over k. Now σ and τ being two elements of $\mathfrak{G}$, we have $\beta_\sigma^\tau \circ F^\tau = F^{\sigma\tau} \circ \alpha_\sigma^\tau$ and $\beta_\tau \circ F = F^\tau \circ \alpha_\tau$ so that $\beta_\sigma^\tau \circ \beta_\tau \circ F = F^{\sigma\tau} \circ \alpha_\sigma^\tau \circ \alpha_\tau$. By the uniqueness, we have $\beta_\sigma^\tau \circ \beta_\tau = \beta_{\sigma\tau}$. Put $f_{\tau,\sigma} = \beta_\tau \circ \beta_\sigma^{-1}$ for every $\sigma \in \mathfrak{G}$, $\tau \in \mathfrak{G}$. Then we observe that the relations

$$f_{\tau,\rho} = f_{\tau,\sigma} \circ f_{\sigma,\rho}, \qquad f_{\tau\rho,\sigma\rho} = (f_{\tau,\sigma})^\rho$$

hold for every ρ, σ, $\tau \in \mathfrak{G}$. Hence, applying the results of Weil [55] to the present case, we obtain a variety W_0 and a birational biregular mapping φ of W_0 onto W such that: (i) W_0 is defined over k_0; (ii) φ is defined over k; (iii) $f_{\tau,\sigma} = \varphi^\tau \circ (\varphi^\sigma)^{-1}$ for every $\sigma, \tau \in \mathfrak{G}$. Put $F_0 = \varphi^{-1} \circ F$; then (W_0, F_0) is clearly a Kummer variety of $(A, \mathcal{C})$ and F_0 is defined over k. We shall now show that (W_0, F_0) satisfies conditions (N1–3). (N1) is just property (i). Let k' be a field of definition for $(A, \mathcal{C})$; then k' contains k_0 by virtue of Proposition 14. Let σ be an isomorphism of kk' onto a field which fixes every element of k_0;

denote by the same letter σ the element of $\mathfrak{G}$ induced by σ. Then, for every isomorphism α_σ of $(A, \mathcal{C})$ onto $(A^\sigma, \mathcal{C}^\sigma)$, we have

$$F_0^\sigma \circ \alpha_\sigma = (\varphi^\sigma)^{-1} \circ F^\sigma \circ \alpha_\sigma = (\varphi^\sigma)^{-1} \circ \beta_\sigma \circ F = \varphi^{-1} \circ F = F_0.$$

This shows that F_0 satisfies (N3). In particular, if σ is the identity on k', we have $(A^\sigma, \mathcal{C}^\sigma) = (A, \mathcal{C})$, so that we can take $\alpha_\sigma = 1_A$; we have hence $F_0^\sigma = F_0$ for every isomorphism σ of kk' over k'. It follows that F_0 is defined over k'; so F_0 satisfies (N2).

REMARK 2. We can prove Theorem 3 without any condition on the fields of definition, using also the theory of [55]. In order to perform this, it is necessary to avoid inseparable field-extensions, since [55] deals only with separable or regular extensions. This is certainly possible by virtue of Proposition 4 of [36]. For details, see [S61, §2.2].

REMARK 3. We can take the variety W in Theorem 3 to be a projective variety; this is proved by means of Proposition 16 and [55].

CHAPTER II

Abelian Varieties with Complex Multiplication

5. Structure of Endomorphism Algebras

Some of the results of §§5–6 have their source in Lefschetz [27]. The reader will also find a more general structure theory of $\mathrm{End}_{\mathbf{Q}}(A)$ in Albert [1, 2] and [S63].

5.1. Let $\mathfrak{R}$ be a simple algebra over $\mathbf{Q}$ and $\mathfrak{Z}$ the center of $\mathfrak{R}$; put $[\mathfrak{R} : \mathfrak{Z}] = f^2$, $[\mathfrak{Z} : \mathbf{Q}] = d$. Then, $\mathfrak{R}$ has d inequivalent irreducible representations in the algebraic closure of $\mathbf{Q}$, which are all of degree f. We call a representation S of $\mathfrak{R}$ in an extension of $\mathbf{Q}$ a *reduced representation* of $\mathfrak{R}$ if S is equivalent to the direct sum of those d irreducible representations. S being a reduced representation of $\mathfrak{R}$, the characteristic polynomial of $S(\alpha)$ has rational coefficients for every $\alpha \in \mathfrak{R}$. Put

$$N(\alpha) = \det S(\alpha), \qquad \mathrm{Tr}(\alpha) = \mathrm{tr} S(\alpha)$$

for $\alpha \in \mathfrak{R}$. We call $N(\alpha)$ and $\mathrm{Tr}(\alpha)$ the *reduced norm* and the *reduced trace* of α; these are independent of the choice of S.

LEMMA 1. *Let $\mathfrak{R}$ be a simple algebra over $\mathbf{Q}$ and S a representation of $\mathfrak{R}$ in an extension of $\mathbf{Q}$. Suppose that for every $\alpha \in \mathfrak{R}$, the characteristic polynomial of $S(\alpha)$ has rational coefficients. Then S is equivalent to the sum of a multiple of a reduced representation of $\mathfrak{R}$ and a 0-representation.*

PROOF. Let the S_i, for $1 \leq i \leq d$, denote the inequivalent irreducible representations of $\mathfrak{R}$ in the algebraic closure L of $\mathbf{Q}$. Then S is equivalent to the direct sum of representations $m_i S_i$ and a 0-representation, where the m_i denote the multiplicities. Let the σ_i, for $1 \leq i \leq d$, be all the isomorphisms of the center $\mathfrak{Z}$ of $\mathfrak{R}$ into L. Then we observe, after reordering, that $S_i(\alpha)$ is the diagonal matrix $\alpha^{\sigma_i} 1_f$ for $\alpha \in \mathfrak{Z}$, for every i. Hence, for every $\alpha \in \mathfrak{Z}$, the characteristic polynomial of $S(\alpha)$ is of the form $X^h \prod_i (X - \alpha^{\sigma_i})^{m_i f}$. Our assumption implies that the m_i are the same; this proves our lemma.

PROPOSITION 1. *Let A be an abelian variety of dimension n and $\mathfrak{S}$ a commutative semi-simple subalgebra of* $\mathrm{End}_{\mathbf{Q}}(A)$. *Then we have*

$$[\mathfrak{S} : \mathbf{Q}] \leq 2n.$$

If $[\mathfrak{S} : \mathbf{Q}] = 2n$, *then the commutor of $\mathfrak{S}$ in* $\mathrm{End}_{\mathbf{Q}}(A)$ *coincides with $\mathfrak{S}$.*

PROOF. Let the K_i denote the simple components of $\mathfrak{S}$. As $\mathfrak{S}$ is commutative, the K_i are fields. Put $[k_i : \mathbf{Q}] = d_i$. Let S_i be a reduced representation of K_i. Take a prime l other than the characteristic of the fields of definition for A, and consider an l-adic representation M_l of $\mathrm{End}_{\mathbf{Q}}(A)$. By Lemma 1, the restriction of M_l to K_i is equivalent to the direct sum of a multiple $m_i S_i$ of S_i and a 0-representation. Considering M_l on $\mathfrak{S}$, we see that M_l is equivalent to the direct sum of the $m_i S_i$ and a 0-representation. As M_l is faithful, every m_i must be positive. Hence we have $2n \geq \sum_i m_i d_i \geq \sum_i d_i = [\mathfrak{S} : \mathbf{Q}]$. This proves the first assertion. Now suppose that $2n = [\mathfrak{S} : \mathbf{Q}]$. Let $\mathfrak{S}'$ be the commutor of $\mathfrak{S}$ in $\mathrm{End}_{\mathbf{Q}}(A)$. We can find a matrix P with coefficients in the algebraic closure of $\mathbf{Q}_l$ such that $PM_l(\xi)P^{-1}$ is a diagonal matrix for every $\xi \in \mathfrak{S}$. As we have $[\mathfrak{S} : \mathbf{Q}] = 2n$, there exists an element α of $\mathfrak{S}$ such that the diagonal elements of $PM_l(\alpha)P^{-1}$ are distinct. For every $\eta \in \mathfrak{S}'$, $PM_l(\eta)P^{-1}$ commutes with $PM_l(\alpha)P^{-1}$, so that $PM_l(\eta)P^{-1}$ is a diagonal matrix for every $\eta \in \mathfrak{S}'$. This shows that $\mathfrak{S}'$ is a commutative semi-simple algebra. Then, applying to $\mathfrak{S}'$ what we have just proved for $\mathfrak{S}$, we get $[\mathfrak{S}' : \mathbf{Q}] \leq 2n$; so we must have $\mathfrak{S} = \mathfrak{S}'$; this completes the proof.

PROPOSITION 2. *Let A be an abelian variety of dimension n; let $\mathfrak{R}$ be a simple subalgebra of* $\mathrm{End}_{\mathbf{Q}}(A)$ *and $\mathfrak{Z}$ the center of $\mathfrak{R}$; put*

$$[\mathfrak{R} : \mathfrak{Z}] = f^2, \qquad [\mathfrak{Z} : \mathbf{Q}] = d.$$

Suppose that $\mathfrak{R}$ contains the identity element of $\mathrm{End}_{\mathbf{Q}}(A)$. *Then fd divides $2n$; and putting $2n = fdm$, we have, for every $\alpha \in \mathfrak{R}$,*

$$\nu(\alpha) = N(\alpha)^m, \qquad \mathrm{tr}(\alpha) = m\mathrm{Tr}(\alpha),$$

where $N(\alpha)$ and $\mathrm{Tr}(\alpha)$ *denote the reduced norm and trace of $\alpha \in \mathfrak{R}$.*

PROOF. Let S be a reduced representation of $\mathfrak{R}$. Take a prime l other than the characteristic and an l-adic representation M_l of $\mathrm{End}_{\mathbf{Q}}(A)$. By Lemma 1, the restriction of M_l to $\mathfrak{R}$ is equivalent to a multiple mS of S; so we have $2n = fdm$; and the characteristic polynomial of $M_l(\alpha)$ is the m-th power of that of $S(\alpha)$. This proves the proposition.

PROPOSITION 3. *Let A be an abelian variety of dimension n. If* $\mathrm{End}_{\mathbf{Q}}(A)$ *contains a field F of degree $2n$ over $\mathbf{Q}$, then A is isogenous to a product*

$B \times \cdots \times B$ with a simple abelian variety B; the commutor of F in $\mathrm{End}_{\mathbf{Q}}(A)$ *coincides with F; and, for every $\alpha \in F$, we have*

$$\nu(\alpha) = N_{F/\mathbf{Q}}(\alpha), \qquad \mathrm{tr}(\alpha) = \mathrm{Tr}_{F/\mathbf{Q}}(\alpha).$$

PROOF. By the results of n°55–6 of Weil [46], there exist simple abelian varieties $A_1, \ldots, A_s$ such that A is isogenous to the product

$$(A_1 \times \cdots \times A_1) \times \cdots \times (A_s \times \cdots \times A_s)$$

and the A_i are not isogenous to each other. Let h_i be the number of the factor A_i occurring in the product, for each i. Then, denoting by $\mathfrak{M}_i$ the total matrix ring of degree h_i over $\mathrm{End}_{\mathbf{Q}}(A_i)$, $\mathrm{End}_{\mathbf{Q}}(A_i \times \cdots \times A_i)$ is identified with $\mathfrak{M}_i$, and $\mathrm{End}_{\mathbf{Q}}(A)$ is identified with the direct sum of the $\mathfrak{M}_i$. Let ε_i denote the identity element of $\mathfrak{M}_i$ for each i. $F\varepsilon_i$ is not $\{0\}$ for at least one of the ε_i, say ε_1. Then, $F\varepsilon_1$ is isomorphic to F, since F is a field. We see that $F\varepsilon_1$ is a semi-simple commutative subalgebra of $\mathrm{End}_{\mathbf{Q}}(A_1 \times \cdots \times A_1)$; hence, by Proposition 1, we have $[F : \mathbf{Q}] \leq 2h_1 \cdot \dim(A_1)$. By the assumption $[F : \mathbf{Q}] = 2n$, we must have $s = 1$ and A is isogenous to $A_1 \times \cdots \times A_1$. This proves the first assertion of our proposition. The second assertion follows from Proposition 1 and the last one from Proposition 2.

PROPOSITION 4. *The symbols A, B and F being as in Proposition 3, let m be the dimension of B and h the number of the factor B in the product $B \times \cdots \times B$ which is isogenous to A. Let K be the center of* $\mathrm{End}_{\mathbf{Q}}(B)$. *Then, K is a subfield of F; and, if we put $[K : \mathbf{Q}] = f$, $[\mathrm{End}_{\mathbf{Q}}(B) : K] = g^2$, we have $2n = fgh$, $2m = fg$.*

PROOF. We first note that $\mathrm{End}_{\mathbf{Q}}(B)$ is a division algebra, since B is simple. $\mathrm{End}_{\mathbf{Q}}(A)$ is identified with the total matrix ring of degree h over $\mathrm{End}_{\mathbf{Q}}(B)$; hence K is the center of $\mathrm{End}_{\mathbf{Q}}(A)$ and is contained in F by virtue of Proposition 3. We observe that $\mathrm{End}_{\mathbf{Q}}(A)$ is a central simple algebra over K and $[\mathrm{End}_{\mathbf{Q}}(A) : K] = g^2h^2$. By a well-known theorem of ring theory, $[F : K] = 2n/f$ divides gh; and if we put $gh = [F : K]q$, the commutor of F in $\mathrm{End}_{\mathbf{Q}}(A)$ is an algebra of degree q^2 over F. By Proposition 3, we must have $q = 1$, so that $[F : K] = gh = 2n/f$ and hence $2m = fg$.

LEMMA 2. *Let K be an algebraic number field of finite degree; let ρ be an automorphism of K such that $\rho^2 = 1$ and K_0 the subfield of K consisting of all the elements of K fixed by ρ. Suppose that*

$$\mathrm{Tr}_{K/\mathbf{Q}}(\xi\xi^{\rho}) > 0$$

for every element $\xi \neq 0$ in K. Then K_0 is a totally real field. If ρ is not the identity on K, K is a totally imaginary field, and, for every isomorphism τ of K into $\mathbf{C}$, $\xi^{\rho\tau}$ is the complex conjugate of ξ^τ.

PROOF. We have $\mathrm{Tr}_{K_0/\mathbf{Q}}(\xi^2) > 0$ for every $\xi \neq 0$ in K_0. Let $\tau_1, \ldots, \tau_m$ denote all the isomorphisms of K_0 into $\mathbf{C}$. Assume that τ_1 is not real; then the complex conjugate of τ_1 coincides with one of the isomorphisms $\tau_2, \ldots, \tau_m$, say τ_2. We see that $K_0^{\tau_1}$ is dense in $\mathbf{C}$, so that there exists an element η in K_0 such that $\mathrm{Re}((\eta^2)^{\tau_1}) < -1$. By the approximation theorem, for any small positive number ε, we can find an element ξ in K_0 such that

$$|\xi^{\tau_1} - \eta^{\tau_1}| = |\xi^{\tau_2} - \eta^{\tau_2}| < \varepsilon, |\xi^{\tau_i}| < \varepsilon \quad (2 < i \leq m).$$

If we take a sufficiently small ε, we have

$$\begin{aligned} \mathrm{Tr}_{K_0/\mathbf{Q}}(\xi^2) &= \mathrm{Re}(\mathrm{Tr}_{K_0/\mathbf{Q}}(\xi^2)) \\ &= \mathrm{Re}((\xi^2)^{\tau_1}) + \mathrm{Re}((\xi^2)^{\tau_2}) + \cdots + \mathrm{Re}((\xi^2)^{\tau_m}) < -2. \end{aligned}$$

This is a contradiction; hence K_0 must be totally real. Now suppose that ρ is not the identity. Then we have $[K : K_0] = 2$, so that there exists an element ζ of K such that $K = K_0(\zeta)$, $\zeta^2 \in K_0$, $\zeta^\rho = -\zeta$. For every element $\alpha \neq 0$ of K_0, we have

$$2\mathrm{Tr}_{K_0/\mathbf{Q}}(\alpha^2\zeta^2) = -\mathrm{Tr}_{K/\mathbf{Q}}((\alpha\zeta)(\alpha\zeta)^\rho) < 0.$$

Again by the approximation theorem, we can find, for each i, an element α of K_0 such that

$$|\alpha^{\tau_i}| > 1, |\alpha^{\tau_j}| < \varepsilon \quad (j \neq i)$$

for a sufficiently small positive number ε; we have then $(\alpha^2\zeta^2)^{\tau_i} < 0$; this shows that $-(\zeta^2)^{\tau_i}$ is positive. Hence $-\zeta^2$ is totally positive, so that $K = K_0(\zeta)$ is totally imaginary; and for every isomorphism τ of K into $\mathbf{C}$, ζ^τ is a purely imaginary number. Therefore, we have, for every $\alpha, \beta \in K_0$,

$$(\alpha + \beta\zeta)^{\rho\tau} = (\alpha - \beta\zeta)^\tau = \alpha^\tau - \beta^\tau\zeta^\tau = \overline{\alpha^\tau + \beta^\tau\zeta^\tau} = \overline{(\alpha + \beta\zeta)^\tau},$$

where $\bar{z}$ denotes the complex conjugate of z; this completes our proof.

PROPOSITION 5. *Let B be a simple abelian variety and K the center of* $\mathrm{End}_{\mathbf{Q}}(B)$. *Then K is a totally real number field or a totally imaginary quadratic extension of a totally real number field.*

PROOF. In §1.3, we have noted that $\mathrm{End}_{\mathbf{Q}}(B)$ has an involution $\xi \mapsto \xi'$ with the property $\mathrm{tr}(\xi\xi') > 0$ for every $\xi \neq 0$ in $\mathrm{End}_{\mathbf{Q}}(B)$. As K is the center of $\mathrm{End}_{\mathbf{Q}}(B)$, the involution maps K onto itself; and by Proposition 2, we have $\mathrm{tr}(\xi) = m\mathrm{Tr}_{K/\mathbf{Q}}(\xi)$ for $\xi \in K$, for a suitable positive integer m. Hence we have $\mathrm{Tr}_{K/\mathbf{Q}}(\xi\xi') > 0$ for every $\xi \neq 0$ in K. Our proposition is then an immediate consequence of Lemma 2.

PROPOSITION 6. *The notation being as in Proposition 4, suppose that the characteristic of the fields of definition for A is zero. Then, we have $g = 1$ and $\mathrm{End}_{\mathbf{Q}}(B) = K$.*

PROOF. If the characteristic is 0, we can consider a rational representation of $\mathrm{End}_{\mathbf{Q}}(B)$, defined in §3.2, which is of degree $2m$, m being the dimension of B. As $\mathrm{End}_{\mathbf{Q}}(B)$ is a division algebra, the degree of any rational representation of $\mathrm{End}_{\mathbf{Q}}(B)$ is divisible by $[\mathrm{End}_{\mathbf{Q}}(B) : \mathbf{Q}] = fg^2$. Hence, by the equality $2m = fg$, we must have $g = 1$; this implies $\mathrm{End}_{\mathbf{Q}}(B) = K$.

5.2. CM-types. Let $\mathfrak{R}$ be an algebra over $\mathbf{Q}$, with an identity element 1. We understand by an *abelian variety of type* ($\mathfrak{R}$) a couple (A, ι) formed by an abelian variety A and an isomorphism ι of $\mathfrak{R}$ into $\mathrm{End}_{\mathbf{Q}}(A)$ such that $\iota(1) = 1_A$. When there is no fear of confusion, we write (A, ι) simply by A and identify an element α of $\mathfrak{R}$ with $\iota(\alpha)$.

Let F be an algebraic number field, (A, ι) an abelian variety of type (F) and n the dimension of A. By Proposition 2, $[F : \mathbf{Q}]$ divides $2n$. We shall now investigate the structure of (A, ι) for which $[F : \mathbf{Q}] = 2n$ holds, assuming that the characteristic is 0. If the characteristic is 0, A is isomorphic to a complex torus, and we obtain a rational representation M and an analytic representation S of $\mathrm{End}_{\mathbf{Q}}(A)$, with respect to an analytic coordinate-system; M is of degree $2n$, and S is of degree n; M is equivalent to the direct sum of S and the complex conjugate $\bar{S}$ of S (cf. §3.2). Let $\varphi_1, \ldots, \varphi_{2n}$ be all the isomorphisms of F into $\mathbf{C}$. Then, by Lemma 1, the representation M restricted to F is equivalent to the direct sum of the φ_i; hence S is equivalent to the direct sum of half of $2n$ isomorphisms φ_i, say $\varphi_1, \ldots, \varphi_n$. Then $\bar{S}$ is equivalent to the direct sum of $\varphi_{n+1}, \ldots, \varphi_{2n}$, which is the direct sum of $\bar{\varphi}_1, \ldots, \bar{\varphi}_n$. Therefore, we observe that there are no two isomorphisms among $\varphi_1, \ldots, \varphi_n$ which are complex conjugate of each other. Moreover, we see that F must be totally imaginary. The set $\{\varphi_1, \ldots, \varphi_n\}$ being thus determined, we say that (A, ι) is of type $(F; \{\varphi_1, \ldots, \varphi_n\})$. Recalling that S is equivalent to the representation of $\mathrm{End}_{\mathbf{Q}}(A)$ by invariant differential forms, we can find n invariant differential forms $\omega_1, \ldots, \omega_n$ of degree 1 on A such that, for every $\alpha \in F$,

$$\delta\iota(\alpha)\omega_i = \alpha^{\varphi_i}\omega_i \quad (1 \leq i \leq n).$$

Conversely, if there exist such ω_i, (A, ι) is of type $(F; \{\varphi_i\})$; and the ω_i form a basis of $\mathfrak{D}_0(A)$. We shall often use these facts afterwards.

Now let us consider the center K of $\mathrm{End}_{\mathbf{Q}}(A)$, which is also the center of $\mathrm{End}_{\mathbf{Q}}(B)$, where B is a simple abelian variety determined as in Proposition 3. Proposition 5 asserts that K is totally real or a totally imaginary quadratic extension of a totally real field. By Propositions 4 and 6, we have $[K : \mathbf{Q}] = 2 \cdot \dim(B)$; so we can apply to B and K what we have proved for A and F; then we see that K must be totally imaginary. Let S' be an analytic representation of $\mathrm{End}_{\mathbf{Q}}(B)$. Then, as A is isogenous to the product of h copies of B, the restriction of S to K is equivalent to h times of S'. Hence the restriction of $\varphi_1, \ldots, \varphi_n$ to K yields exactly $f/2$ isomorphisms $\psi_1, \ldots, \psi_{f/2}$ of K into $\mathbf{C}$, each repeated h times, where $f = [K : \mathbf{Q}]$; and S' is equivalent to the direct sum of the ψ_j; there are no two isomorphisms among ψ_j which are complex conjugate of each other.

In general, for an algebraic number field F of degree $2n$ and n distinct isomorphisms φ_i of F into $\mathbf{C}$, we say that $(F; \{\varphi_1, \ldots, \varphi_n\})$ is a *CM-type* if there exists an abelian variety of dimension n of type $(F; \{\varphi_i\})$. The above discussion gives us a necessary condition for a CM-type.

THEOREM 1. *In order that $(F; \{\varphi_i\})$ be a CM-type, it is necessary and sufficient that F contains two subfields K and K_0 satisfying the following conditions:*

(CM1) *K_0 is totally real and K is a totally imaginary quadratic extension of K_0.*

(CM2) *There are no two isomorphisms among the φ_i which are complex conjugate of each other on K.*

The sufficiency will be proved in the following section (Theorem 3).

6. Construction of Abelian Varieties with Complex Multiplication

6.1. Analytic structure of an abelian variety of type $(F; \{\varphi_i\})$**.** Given (A, ι) of type $(F; \{\varphi_i\})$, put

$$\mathfrak{r} = \iota^{-1}[\mathrm{End}(A) \cap \iota(F)];$$

then $\mathfrak{r}$ is a subring of F which is finitely generated over $\mathbf{Z}$ and $F = \mathbf{Q}\mathfrak{r}$; hence $\mathfrak{r}$ is a free $\mathbf{Z}$-module of rank $2n$, where $n = \dim A$. Take an analytic coordinate-system $(\mathbf{C}^n/D, \theta)$ of A and denote by S the analytic representation of $\mathrm{End}_{\mathbf{Q}}(A)$ with respect to θ. We see that $S(\iota(\alpha))$ is non-singular for every $\alpha \neq 0$ in F. As we have $S(\iota(\alpha))D \subset D$ for every α in $\mathfrak{r}$, D is considered as an $\mathfrak{r}$-module. Choose a vector $x_0 \neq 0$ in D and put $D' = S(\iota(\mathfrak{r}))x_0$. Then the mapping $\alpha \to S(\iota(\alpha))x_0$ is an $\mathfrak{r}$-isomorphism of $\mathfrak{r}$ onto D'. Hence D' is of rank $2n$, so that there exists a positive integer g such that $gD \subset D'$. Fixing such a number

g, we obtain an $\mathfrak{r}$-isomorphism $x \to \mu$ of D into $\mathfrak{r}$ by means of the relation

$$gx = S(\iota(\mu))x_0.$$

Let $\mathfrak{m}$ denote the image of this isomorphism; $\mathfrak{m}$ is then an ideal of $\mathfrak{r}$. Put $x_1 = g^{-1}x_0$; we have then

$$D = S(\iota(\mathfrak{m}))x_1.$$

We observe that $S(\iota(\alpha))D \subset D$ if and only if $\alpha\mathfrak{m} \subset \mathfrak{m}$; hence $\mathfrak{r}$ consists of all the elements $\alpha \in F$ such that $\alpha\mathfrak{m} \subset \mathfrak{m}$, namely $\mathfrak{r}$ is the "order" of the module $\mathfrak{m}$. Now, for a suitable choice of coordinate-system, $S(\iota(\alpha))$ is the diagonal matrix with the diagonal elements $\alpha^{\varphi_1}, \dots, \alpha^{\varphi_n}$. Let $(b_1, \dots, b_n)$ be the components of the vector x_1 with respect to this coordinate-system. Then D is the set of all vectors with the components $(\mu^{\varphi_1}b_1, \dots, \mu^{\varphi_n}b_n)$ for $\mu \in \mathfrak{m}$. Assume that we have $b_i = 0$ for some i; then D is contained in a proper subspace of $\mathbf{C}^n$; this is a contradiction, since D is a discrete subgroup of $\mathbf{C}^n$ of rank $2n$. Therefore, we must have $b_i \neq 0$ for every i. Change the coordinate-system $(z_1, \dots, z_n)$ for $(b_1^{-1}z_1, \dots, b_n^{-1}z_n)$; we see that $S(\iota(\alpha))$ is expressed again by the same diagonal matrix as before with respect to this new system; and D is the set of all vectors with the components $(\mu^{\varphi_1}, \dots, \mu^{\varphi_n})$ for $\mu \in \mathfrak{m}$. We have thus proved

THEOREM 2. *Let F be an algebraic number field of degree $2n$ and the φ_i, for $1 \leq i \leq n$, n distinct isomorphisms of F into $\mathbf{C}$; denote by $u(\alpha)$ the vector in $\mathbf{C}^n$ with the components $(\alpha^{\varphi_1}, \dots, \alpha^{\varphi_n})$ for $\alpha \in F$, by $D(\mathfrak{m})$ the set of all vectors $u(\alpha)$ for α in a free $\mathbf{Z}$-module $\mathfrak{m}$ in F of rank $2n$, and by $S(\alpha)$ the diagonal matrix with the diagonal elements $\alpha^{\varphi_1}, \dots, \alpha^{\varphi_n}$. If (A, ι) is an abelian variety of type $(F; \{\varphi_i\})$, then there exist a module $\mathfrak{m}$ and an isomorphism θ of A onto $\mathbf{C}^n/D(\mathfrak{m})$ by which $\iota(\alpha)$ corresponds to the linear transformation of $\mathbf{C}^n$ given by $S(\alpha)$ for every $\alpha \in F$; and if we denote by $\mathfrak{r}$ the set of all elements α in F such that $\alpha\mathfrak{m} \subset \mathfrak{m}$, we have*

$$\iota(\mathfrak{r}) = \iota(F) \cap \mathrm{End}(A).$$

COROLLARY. *Any two abelian varieties of the same CM-type are isogenous to each other.*

PROOF. If $\mathfrak{m}$, $\mathfrak{m}'$ are two free $\mathbf{Z}$-submodules of F of rank $2n$, then there exists a positive integer g such that $g\mathfrak{m} \subset \mathfrak{m}'$; we have then $gD(m) \subset D(\mathfrak{m}')$. Hence $x \to gx$ gives a homomorphism of $\mathbf{C}^n/D(\mathfrak{m})$ onto $\mathbf{C}^n/D(\mathfrak{m}')$; this proves the assertion.

REMARK. The homomorphism $x \to gx$ commutes obviously with the operation of F.

6.2. Construction. We shall now prove that the existence of the fields K and K_0 satisfying (CM1, 2) is sufficient for $(F; \{\varphi_i\})$ to be a CM-type.

THEOREM 3. *The symbols F, φ_i, $D(\mathfrak{m})$, $S(\alpha)$ being the same as in Theorem 2, suppose that F contains two subfields K, K_0 satisfying conditions* (CM1, 2) *of Theorem 1. Then, for every free* **Z**-*submodule $\mathfrak{m}$ of F of rank $2n$, $\mathbf{C}^n/D(\mathfrak{m})$ is isomorphic to an abelian variety A, and, for every $\alpha \in F$, the linear transformation of $\mathbf{C}^n$ given by $S(\alpha)$ corresponds to an element of* $\mathrm{End}_{\mathbf{Q}}(A)$*; if we denote this element by $\iota(\alpha)$, then (A, ι) is of type $(F; \{\varphi_i\})$. Moreover, if F does not coincide with K, A is not simple.*

PROOF. Take a basis $\{\alpha_1, \dots, \alpha_{2n}\}$ of $\mathfrak{m}$ over **Z**; then $2n$ vectors $u(\alpha_1), \dots, u(\alpha_{2n})$ form a basis of $D(\mathfrak{m})$ over **Z**. Observe that $D(\mathfrak{m})$ is discrete in $\mathbf{C}^n$ if and only if the $u(\alpha_i)$ are linearly independent over **R**, which is so if and only if the matrix of degree $2n$ with the columns $(\alpha_i^{\varphi_1}, \dots, \alpha_i^{\varphi_n}, \overline{\alpha_i^{\varphi_1}}, \dots, \overline{\alpha_i^{\varphi_n}})$, for $1 \leq i \leq 2n$, is non-singular. The latter is the case, since the α_i form a basis of F over **Q** and $2n$ isomorphisms φ_i, $\bar{\varphi}_i$ give all the isomorphisms of F into **C**. Hence $\mathbf{C}^n/D(\mathfrak{m})$ is a complex torus. We shall now prove that $\mathbf{C}^n/D(\mathfrak{m})$ has a structure of abelian variety. It is sufficient to show this for a certain $\mathfrak{m}$, because, for every two free **Z**-submodules $\mathfrak{m}$ and $\mathfrak{m}'$ of F, $D(\mathfrak{m})$ and $D(\mathfrak{m}')$ are commensurable to each other. Put $[K_0 : \mathbf{Q}] = m$, $[F : K] = h$; we have then $n = mh$, $[K : \mathbf{Q}] = 2m$. Let $\mathfrak{n}$ be a free **Z**-submodule of K of rank $2m$ and $\{\gamma_1, \dots, \gamma_h\}$ a basis of F over K; put

$$\mathfrak{m} = \mathfrak{n}\gamma_1 + \cdots + \mathfrak{n}\gamma_h,$$

$$\Delta_\lambda = \{u(\beta) \mid \beta \in \mathfrak{n}\gamma_\lambda\}.$$

Then $D(\mathfrak{m})$ is the sum of $\Delta_1, \dots, \Delta_h$, and each Δ_λ has $2m$ generators which are linearly independent over **R**. Denote by V_λ the subspace of $\mathbf{C}^n$ generated over **R** by these $2m$ vectors. Then it is easy to see that $\mathbf{C}^n/D(\mathfrak{m})$ is isomorphic to the direct product of the V_λ/Δ_λ as a real analytic manifold. We shall now show that the V_λ are complex vector subspaces of $\mathbf{C}^n$. Let $\psi_1, \dots, \psi_k$ be the distinct isomorphisms of K into **C** induced by the φ_i. By virtue of conditions (CM1, 2), we see that $k = m$ and the φ_i are all the isomorphisms of F into **C** inducing the ψ_j on K. Denote by $v(\beta)$, for $\beta \in K$, the vector of $\mathbf{C}^m$ whose components are $\beta^{\psi_1}, \dots, \beta^{\psi_m}$, and by Δ the set of all vectors $v(\beta)$ for $\beta \in \mathfrak{n}$. Then, applying to Δ what we have proved above, we see that Δ is a discrete subgroup of $\mathbf{C}^m$ and $\mathbf{C}^m/\Delta$ is a complex torus. If $\{\beta_1, \dots, \beta_{2m}\}$ is a basis of $\mathfrak{n}$ over **Z**, $2m$ vectors $v(\beta_t)$ give a basis of $\mathbf{C}^m$ over **R**. The linear mapping $x \to \sqrt{-1}x$ of $\mathbf{C}^m$ onto itself, regarded as an **R**-linear mapping, determines a matrix (c_{st}) of degree $2m$ with real coefficients with respect to the basis $\{v(\beta_t)\}$:

$$(1) \qquad \sqrt{-1}\beta_t^{\psi_j} = \sum_{s=1}^{2m} c_{st}\beta_s^{\psi_j} \qquad (1 \leq j \leq m, 1 \leq t \leq 2m).$$

Now we see that the vectors $(\gamma_\lambda^{\varphi_1}\beta_t^{\varphi_1}, \dots, \gamma_\lambda^{\varphi_n}\beta_t^{\varphi_n})$, for $1 \leq t \leq 2m$, form a basis of V_λ over $\mathbf{R}$. As every φ_i induces one of the ψ_j on K, we find, multiplying (1) by $\gamma_\lambda^{\varphi_i}$,

$$(2) \qquad \sqrt{-1}\gamma_\lambda^{\varphi_i}\beta_t^{\varphi_i} = \sum_{s=1}^{2m} c_{st}\gamma_\lambda^{\varphi_i}\beta_s^{\varphi_i} \qquad (1 \leq i \leq n, 1 \leq t \leq 2m).$$

This implies $\sqrt{-1}V_\lambda \subset V_\lambda$ for every λ, so that V_λ is a complex subspace of $\mathbf{C}^n$ of dimension m; and it can be easily verified that $\mathbf{C}^n/D(\mathfrak{m})$ is isomorphic, as a complex manifold, to the direct product of the complex tori V_λ/Δ_λ. Moreover, if we let $v(\beta_t)$ correspond to $u(\gamma_\lambda\beta_t)$, we obtain an $\mathbf{R}$-isomorphism η_λ of $\mathbf{C}^m$ onto V_λ; we see easily $\eta_\lambda(\Delta) = \Delta_\lambda$. By relations (1) and (2), we have $\sqrt{-1}\eta_\lambda = \eta_\lambda\sqrt{-1}$, so that η_λ gives a complex analytic isomorphism of $\mathbf{C}^m/\Delta$ onto V_λ/Δ_λ. It follows that $\mathbf{C}^n/D(\mathfrak{m})$ is complex analytically isomorphic to the direct product of h copies of $\mathbf{C}^m/\Delta$. Hence, if we show that $\mathbf{C}^m/\Delta$ has a structure of abelian variety, $\mathbf{C}^n/D(\mathfrak{m})$ has also a structure of abelian variety. We prove this by constructing a non-degenerate Riemann form on $\mathbf{C}^m/\Delta$. By condition (CM1), there exists an element ζ in K such that $K = K_0(\zeta)$ and $\zeta^2 \in K_0$; as K is totally imaginary, $-\zeta^2$ must be totally positive. We can take ζ in such a way that

$$\operatorname{Im}(\zeta^{\psi_j}) > 0 \qquad (1 \leq j \leq m).$$

In fact, if this is not so, we choose an element α of K_0 such that $\alpha^{\psi_j}\operatorname{Im}(\zeta^{\psi_j}) > 0$ for $1 \leq j \leq m$, and adopt $\alpha\zeta$ in place of ζ. Now, z, w being two vectors of $\mathbf{C}^m$ with the components $(z_1, \dots, z_m)$ and $(w_1, \dots, w_m)$, we define an $\mathbf{R}$-bilinear form $E(z, w)$ on $\mathbf{C}^m$ by

$$E(z, w) = \sum_{j=1}^{m} \zeta^{\psi_j}(z_j\bar{w}_j - \bar{z}_j w_j).$$

We see easily $E(z, w) = -E(w, z)$, and

$$E(z, \sqrt{-1}w) = -\sqrt{-1}\sum_{j=1}^{m} \zeta^{\psi_j}(z_j\bar{w}_j + \bar{z}_j w_j).$$

Hence $E(z, \sqrt{-1}w)$ is a symmetric form and is positive non-degenerate since the ζ^{ψ_j} are purely imaginary and we have $\operatorname{Im}(\zeta^{\psi_j}) > 0$. Denoting by ρ the automorphism of K over K_0, other than the identity, we have $\zeta^\rho = -\zeta$ and $\xi^{\rho\psi_i} = \overline{\xi^{\psi_i}}$ for every ξ in L (cf. Lemma 2 of §5.1 and its proof). By means of these relations, we have, for every α, β in K,

$$E(v(\alpha), v(\beta)) = \mathrm{Tr}_{K/\mathbf{Q}}(\zeta\alpha\beta^\rho).$$

We can find a positive integer g such that all elements of $g\zeta\mathfrak{n}\mathfrak{n}^\rho$ are algebraic integers; then the values of $gE(z, w)$ on $\Delta \times \Delta$ are integers. Thus we obtain a non-degenerate Riemann form $gE(z, w)$ on $\mathbf{C}^m/\Delta$. This proves that $\mathbf{C}^n/D(\mathfrak{m})$ has a structure of abelian variety. The rest of our theorem is obvious.

The notation being as above, denote by $T(\xi)$ for $\xi \in K$ the diagonal matrix with the diagonal elements $\xi^{\psi_1}, \ldots, \xi^{\psi_m}$. Then, by virtue of the relation $\xi^{\rho\psi_i} = \overline{\xi^{\psi_i}}$, we have

$$E(z, T(\xi)w) = E(T(\xi^\rho)z, w)$$

for every $\xi \in K$. Thus we have proved the first part of the following theorem.

THEOREM 4. *Let K_0 be a totally real field of degree m, K a totally imaginary quadratic extension of K_0, and ρ the automorphism of K over K_0 other than the identity; let $(K; \{\psi_i\})$ be a CM-type and $\mathfrak{n}$ a free $\mathbf{Z}$-submodule of K of rank $2m$. Denote by $v(\beta)$ for $\beta \in K$ the vector of $\mathbf{C}^m$ with the components $\beta^{\psi_1}, \ldots, \beta^{\psi_m}$, and by $D(\mathfrak{n})$ the set of all vectors $v(\beta)$ for $\beta \in \mathfrak{n}$. Let ζ be a number of K such that $-\zeta^2$ is a totally positive element of K_0 and $\mathrm{Im}(\zeta^{\psi_i}) > 0$ for every i. Put, for two vectors $z = (z_1, \ldots, z_m)$ and $w = (w_1, \ldots, w_m)$ of $\mathbf{C}^m$,*

$$E(z, w) = \sum_{i=1}^{m} \zeta^{\psi_i}(z_i\bar{w}_i - \bar{z}_i w_i).$$

Then, for a suitable positive integer g, the form gE is a non-degenerate Riemann form on $\mathbf{C}^m/D(\mathfrak{n})$; and we have

$$E(z, T(\xi)w) = E(T(\xi^\rho)z, w) \tag{3}$$

for every ξ in K. Conversely, every non-degenerate Riemann form on $\mathbf{C}^m/D(\mathfrak{n})$ satisfying (3) is obtained from an element ζ of K in this manner. If $\mathbf{C}^m/D(\mathfrak{n})$ is simple, every Riemann form on $\mathbf{C}^m/D(\mathfrak{n})$, other than 0, is non-degenerate and satisfies relation (3).

PROOF. We have only to prove the second and the last assertions; we first prove the last. As is remarked in §1.3 and §3.3, for every non-degenerate Riemann form $E(z, w)$, we obtain an involution $\Lambda \to \Lambda'$ of $\mathrm{End}_{\mathbf{Q}}(C^m/D(\mathfrak{n}))$ by the relation

$$E(z, \Lambda w) = E(\Lambda' z, w).$$

If $\mathbf{C}^m/D(\mathfrak{n})$ is simple, every Riemann form on it, other than 0, is non-degenerate, and $\mathrm{End}_{\mathbf{Q}}(C^m/D(\mathfrak{n}))$ coincides with $T(K)$ by virtue of Proposition 6; so the involution corresponds to an automorphism τ of K; we have namely $T(\xi)' = T(\xi^\tau)$, and $\mathrm{Tr}_{K/\mathbf{Q}}(\xi\xi^\tau) > 0$ for every $\xi \neq 0$ in K. By Lemma 2, if we denote by K_1 the subfield of K consisting of the elements fixed by τ, K_1 is totally real and $[K : K_1] = 2$. As K_0 and K_1 are totally real, we must have $K_0 = K_1$, and hence $\tau = \rho$; so E satisfies (3). Now we prove the second assertion. If E is a Riemann form, the mapping $\xi \to E(v(\xi), v(1))$ is a $\mathbf{Q}$-linear mapping of K into $\mathbf{Q}$, so that there exists an element ζ of K such that $E(v(\xi), v(1)) = \mathrm{Tr}_{K/\mathbf{Q}}(\zeta\xi)$ for every $\xi \in K$. Suppose that E satisfies (3). We have then

$$\begin{aligned} E(v(\xi), v(\eta)) &= E(v(\xi), T(\eta)v(1)) = E(T(\eta^\rho)v(\xi),\ v(1)) \\ &= E(v(\eta^\rho\xi), v(1)) = \mathrm{Tr}_{K/\mathbf{Q}}(\zeta\xi\eta^\rho). \end{aligned}$$

Since E is alternating, we have

$$\mathrm{Tr}_{K/\mathbf{Q}}(\zeta\xi\eta^\rho) = -\mathrm{Tr}_{K/\mathbf{Q}}(\zeta\xi^\rho\eta) = -\mathrm{Tr}_{K/\mathbf{Q}}(\zeta^\rho\xi\eta^\rho);$$

this implies $\zeta = -\zeta^\rho$. Hence $-\zeta^2 = \zeta\zeta^\rho$ is contained in K_0 and $K = K_0(\zeta)$. As K is totally imaginary, $-\zeta^2$ must be totally positive. By the same argument as in the proof of Lemma 2, we find $\xi^{\rho\psi_i} = \overline{\xi^{\psi_i}}$ for every $\xi \in K$, so that

$$E(v(\xi), v(\eta))) = \mathrm{Tr}_{K/\mathbf{Q}}(\zeta\xi\eta^\rho) = \sum_{i=1}^{m} \zeta^{\psi_i}(\xi^{\psi_i}\overline{\eta^{\psi_i}} - \overline{\xi^{\psi_i}}\eta^{\psi_i}).$$

Since the vectors $v(\xi)$ for $\xi \in K$ form a dense subset of $\mathbf{C}^m$, we have

$$E(z, w) = \sum_{i=1}^{m} \zeta^{\psi_i}(z_i\bar{w}_i - \bar{z}_i w_i)$$

on $\mathbf{C}^m \times \mathbf{C}^m$. The inequality $\mathrm{Im}(\zeta^{\psi_i}) > 0$ follows from the fact that $E(z, \sqrt{-1}z) = -2\sqrt{-1}\sum_{i=1}^{m}\zeta^{\psi_i}|z_i|^2$ is a positive form. This completes our proof.

We can give another expression for the form E on $D(\mathfrak{n}) \times D(\mathfrak{n})$. Let $\{\gamma_1, \gamma_2\}$ be a basis of K over K_0; put

$$\eta = \zeta(\gamma_1\gamma_2^\rho - \gamma_1^\rho\gamma_2);$$

then we have $\eta^\rho = \eta$, so that η is an element of K_0. For $\alpha_1, \alpha_2, \beta_1, \beta_2 \in K_0$, we have

$$\mathrm{Tr}_{K/K_0}(\zeta(\alpha_1\gamma_1 + \alpha_2\gamma_2)(\beta_1\gamma_1 + \beta_2\gamma_2)^\rho) = \eta(\alpha_1\beta_2 - \alpha_2\beta_1).$$

Hence we obtain

$$E(v(\alpha_1\gamma_1 + \alpha_2\gamma_2), v(\beta_1\gamma_1 + \beta_2\gamma_2)) = \mathrm{Tr}_{K_0/\mathbf{Q}}(\eta(\alpha_1\beta_2 - \alpha_2\beta_1)).$$

6.3. The Picard variety. The notation being as in Theorem 4, let A be an abelian variety isomorphic to $\mathbf{C}^m/D(\mathfrak{n})$; we fix an isomorphism θ of A onto $\mathbf{C}^m/D(\mathfrak{n})$ and denote by $\iota(\alpha)$ for $\alpha \in K$ the element of $\mathrm{End}_{\mathbf{Q}}(A)$ corresponding to $T(\alpha)$; then (A, ι) is of type $(K; \{\psi_\nu\})$. We shall now consider the Picard variety A^* of A. Recall the form $\langle z, w\rangle$ on $\mathbf{C}^m$ introduced in §3.3. We have

$$\langle z, w\rangle = \sum_{\nu=1}^{m}(z_\nu \bar{w}_\nu + \bar{z}_\nu w_\nu)$$

for any two vectors z, w with the coordinates z_ν, w_ν, so that for every $\xi, \eta \in K$,

$$\langle v(\xi), v(\eta)\rangle = \mathrm{Tr}_{K/\mathbf{Q}}(\xi\eta^\rho). \tag{4}$$

Let $(\mathbf{C}^m/D^*, \theta^*)$ be the dual of $(\mathbf{C}^m/D(\mathfrak{n}), \theta)$, defined in §3.3, which is an analytic representation of A^*; D^* is the set of vectors z such that $\langle z, w\rangle \in \mathbf{Z}$ for every $w \in D(\mathfrak{n})$; we see then easily that every vector in D^* is of the form $v(\xi)$ for $\xi \in K$. Hence if we denote by $\mathfrak{n}^*$ the set of elements $\xi \in K$ such that

$$\mathrm{Tr}_{K/\mathbf{Q}}(\xi\mathfrak{n}^\rho) \in \mathbf{Z},$$

we find $D^* = D(\mathfrak{n}^*)$ in view of (4). Thus the dual of $\mathbf{C}^m/D(\mathfrak{n})$ is given by $\mathbf{C}^m/D(\mathfrak{n}^*)$. By relation (4), we see

$$\langle T(\alpha)v(\xi), v(\eta)\rangle = \langle v(\xi), T(\alpha^\rho)v(\eta)\rangle.$$

Since the vectors $v(\xi)$ for $\xi \in K$ form a dense subset in $\mathbf{C}^m$, we have

$$\langle T(\alpha)z, w\rangle = \langle z, T(\alpha^\rho)w\rangle$$

for every $\alpha \in K$. This shows that $T(\alpha^\rho)$ is the analytic representation of the element ${}^t\iota(\alpha)$ of $\mathrm{End}_{\mathbf{Q}}(A^*)$ with respect to θ^*. Put, for $\alpha \in K$,

$$\iota^*(\alpha) = {}^t\iota(\alpha^\rho). \tag{5}$$

Then we see that ι^* is an isomorphism of K into $\mathrm{End}_{\mathbf{Q}}(A^*)$; and $\iota^*(\alpha)$ is represented by $T(\alpha)$. It follows that (A^*, ι^*) is of type $(K; \{\psi_\nu\})$. Now let $E(z, w)$ be the Riemann form on $\mathbf{C}^m/D(\mathfrak{n})$ obtained from an element ζ of K, as in Theorem 4, and X a divisor on A corresponding to E. Since we have

$$E(z, w) = \langle T(\zeta)z, w\rangle,$$

the homomorphism φ_X of A into A^* corresponds to the mapping of $\mathbf{C}^m/D(\mathfrak{n})$ onto $\mathbf{C}^m/D(\mathfrak{n}^*)$ given by $T(\zeta)$. We see easily that

$$\varphi_X\iota(\alpha) = \iota^*(\alpha)\varphi_X \tag{6}$$

for every $\alpha \in K$.

7. Transformations and Multiplications

7.1. Definitions. Let $\mathfrak{R}$ be an algebra over $\mathbf{Q}$ with an identity element 1. We shall understand by a *lattice* in $\mathfrak{R}$ a free $\mathbf{Z}$-submodule of $\mathfrak{R}$ of rank $[\mathfrak{R} : \mathbf{Q}]$. We call a subring $\mathfrak{o}$ of $\mathfrak{R}$ an *order* in $\mathfrak{R}$ if it is a lattice in $\mathfrak{R}$ and contains the identity element of $\mathfrak{R}$. Let $\mathfrak{a}$ be a lattice in $\mathfrak{R}$; let $\mathfrak{o}_r$ (resp. $\mathfrak{o}_l$) be the set of all elements α of $\mathfrak{R}$ such that $\mathfrak{a}\alpha \subset \mathfrak{a}$ (resp. $\alpha\mathfrak{a} \subset \mathfrak{a}$). Then, $\mathfrak{o}_r$ and $\mathfrak{o}_l$ are orders in $\mathfrak{R}$. We call $\mathfrak{o}_r$ (resp. $\mathfrak{o}_l$) the *right* (resp. *left*) *order* of $\mathfrak{a}$.

Given $\mathfrak{R}$ as above, let (A, ι) be an abelian variety of type $(\mathfrak{R})$. Recall that ι is an isomorphism of $\mathfrak{R}$ into $\mathrm{End}_{\mathbf{Q}}(A)$ such that $\iota(1)$ is the identity 1_A of $\mathrm{End}_{\mathbf{Q}}(A)$. Put

$$\mathfrak{r} = \iota^{-1}[\mathrm{End}(A) \cap \iota(\mathfrak{R})].$$

It is easy to see that $\mathfrak{r}$ is an order in $\mathfrak{R}$. We call $\mathfrak{r}$ the *order* of (A, ι). We say that (A, ι) is defined over a field k if k is a field of definition for A and every element of $\iota(\mathfrak{r})$. Let (A', ι') be another abelian variety of type $(\mathfrak{R})$. A homomorphism (resp. an isomorphism) λ of A into A' is called a *homomorphism* (resp. an *isomorphism*) of (A, ι) into (A', ι'), or an $\mathfrak{R}$-*homomorphism* (resp. $\mathfrak{R}$-*isomorphism*) of A into A', if it satisfies

$$\lambda\iota(\alpha) = \iota'(\alpha)\lambda$$

for every $\alpha \in \mathfrak{R}$. An *endomorphism* or an *automorphism* of (A, ι) is similarly defined.

Let (A, ι) and (A', ι') be an abelian varieties of type $(\mathfrak{R})$; let $\mathfrak{r}$ be the order of (A, ι) and $\mathfrak{a}$ a lattice in $\mathfrak{R}$ contained in $\mathfrak{r}$. A homomorphism λ of (A, ι) onto (A', ι') is called an $\mathfrak{a}$-*multiplication of* (A, ι) *onto* (A', ι') if there exist a field k of definition for (A, ι), (A', ι') and λ, and a generic point x of A over k, such that $k(\lambda x)$ is the composite of all the fields $k(\iota(\alpha)x)$ for $\alpha \in \mathfrak{a}$. We note that if λ is an $\mathfrak{a}$-multiplication, then, for any field of definition k_1 for (A, ι), (A', ι') and λ, and for any generic point y of A over k_1, $k_1(\lambda y)$ is the composite of all the fields $k_1(\iota(\alpha)y)$ for $\alpha \in \mathfrak{a}$. It is easy to see that every $\mathfrak{a}$-multiplication is an $\mathfrak{r}\mathfrak{a}$-multiplication. (A', ι') is called an $\mathfrak{a}$-*transform* of (A, ι) if there exists an $\mathfrak{a}$-multiplication λ of (A, ι) onto (A', ι'). We call also the system $(A', \iota'; \lambda)$ an $\mathfrak{a}$-transform of (A, ι). By our definition, every $\mathfrak{a}$-transform of (A, ι) is of the same dimension as (A, ι) and every $\mathfrak{a}$-multiplication is an isogeny.

PROPOSITION 7. *Let $\mathfrak{R}$ be an algebra over $\mathbf{Q}$ with an identity element and (A, ι) an abelian variety of type $(\mathfrak{R})$; let $\mathfrak{r}$ be the order of (A, ι) and $\mathfrak{a}$ a lattice in $\mathfrak{R}$ contained in $\mathfrak{r}$. Then, there exists an $\mathfrak{a}$-transform $(A', \iota'; \lambda)$ of (A, ι); $(A', \iota'; \lambda)$ is uniquely determined by (A, ι) and $\mathfrak{a}$ up to an $\mathfrak{R}$-isomorphism; and the order of (A', ι') contains the right order of $\mathfrak{a}$. Moreover, if k is a field of definition for (A, ι), we can find (A', ι') and λ so that they are defined over k.*

PROOF. The uniqueness follows immediately from our definition. Let k be a field of definition for (A, ι) and x a generic point of A over k. Take a basis $\{\alpha_1, \dots, \alpha_d\}$ of $\mathfrak{a}$ over $\mathbf{Z}$. Let A' denote the locus of $\iota(\alpha_1)x \times \cdots \times \iota(\alpha_d)x$ over k in the product of d copies of A. Then, A' is an abelian variety with the origin $0 \times \cdots \times 0$, where 0 denotes the origin of A. Define a rational mapping λ of A onto A' by

$$\lambda x = \iota(\alpha_1)x \times \cdots \times \iota(\alpha_d)x$$

with respect to k; then λ is a homomorphism of A onto A'. We see easily that $k(\lambda x)$ is the composite of the fields $k(\iota(\alpha)x)$ for $\alpha \in \mathfrak{a}$, since $\{\alpha_1, \dots, \alpha_d\}$ is a basis of $\mathfrak{a}$ over $\mathbf{Z}$. Let $\mathfrak{r}'$ be the right order of $\mathfrak{a}$. Let β be an element of $\mathfrak{r}'$. Put $\beta_i = \alpha_i\beta$; then the β_i are contained in $\mathfrak{a}$. We can find a positive integer g such that $g\beta \in \mathfrak{r}$. Let y be a point of A such that $gy = x$; we have then

$$\iota(\beta_1)x \times \cdots \times \iota(\beta_d)x = \lambda(\iota(g\beta)y),$$

so that the point $\iota(\beta_1)x \times \cdots \times \iota(\beta_d)x$ is contained in A'. Since the elements β_i are contained in $\mathfrak{a}$, the field $k(\iota(\beta_1)x, \dots, \iota(\beta_d)x)$ is contained in $k(\lambda x)$. Hence we can define a rational mapping μ of A' into itself by $\mu(\lambda x) = \iota(\beta_1)x \times \cdots \times \iota(\beta_d)x$, with respect to k. Then μ is an endomorphism of A', since it maps the origin onto the origin. Denote this endomorphism μ by $\iota'(\beta)$. We have then

$$\iota'(\beta)\lambda x = \iota(\alpha_1\beta)x \times \cdots \times \iota(\alpha_d\beta)x.$$

It follows from this relation that $\beta \to \iota'(\beta)$ is an isomorphism of $\mathfrak{r}'$ into $\mathrm{End}(A')$; and $\iota'(1)$ is the identity element of $\mathrm{End}(A')$; so (A', ι') is an abelian variety of type $(\mathfrak{R})$ and the order of (A', ι') contains $\mathfrak{r}'$. If α is contained in $\mathfrak{r} \cap \mathfrak{r}'$, we have $\iota'(\alpha)\lambda x = \lambda\iota(\alpha)x$ by our construction. This completes the proof.

The notation being as in Proposition 7, we denote by $\mathfrak{g}(\mathfrak{a}, A)$, or simply by $\mathfrak{g}(\mathfrak{a})$, the set of points t on A such that $\iota(\alpha)t = 0$ for every $\alpha \in \mathfrak{a}$.

PROPOSITION 8. *The symbols $\mathfrak{R}$, (A, ι), $\mathfrak{r}$, $\mathfrak{a}$ being as in Proposition 7, let $(A', \iota'; \lambda)$ be an $\mathfrak{a}$-transform of (A, ι) and k a field of definition for (A, ι), (A', ι') and λ. Then, for every point t on A, the field $k(\lambda t)$ is the composite of all the fields $k(\iota(\alpha)t)$ for $\alpha \in \mathfrak{a}$; and the kernel of λ is $\mathfrak{g}(\mathfrak{a}, A)$.*

PROOF. By the uniqueness, it is sufficient to prove our conclusion for the $\mathfrak{a}$-transform $(A', \iota'; \lambda)$ constructed in the proof of Proposition 7; but this is easily seen by the relation $\lambda t = \iota(\alpha_1)t \times \cdots \times \iota(\alpha_d)t$.

Given (A, ι) and $\mathfrak{r}$ as above, let α be an invertible element of $\mathfrak{R}$ contained in $\mathfrak{r}$. Define an isomorphism ι' of $\mathfrak{R}$ into $\mathrm{End}_{\mathbf{Q}}(A)$ by $\iota'(\gamma) = \iota(\alpha\gamma\alpha^{-1})$. Then, we see easily that $\iota(\alpha)$ is an $\mathfrak{r}\alpha$-multiplication of (A, ι) onto (A, ι'). If $\mathfrak{R}$ is commutative, we have, of course, $(A, \iota) = (A, \iota')$.

PROPOSITION 9. *The symbols* $\mathfrak{R}$, (A, ι), $\mathfrak{r}$, $\mathfrak{a}$ *being as in Proposition 7, let* $\mathfrak{r}_1$ *be the right order of* $\mathfrak{a}$, *and* $\mathfrak{b}$ *a lattice in* $\mathfrak{R}$ *contained in* $\mathfrak{r}_1$. *Let* $(A_1, \iota_1; \lambda)$ *be an* $\mathfrak{a}$*-transform of* (A, ι) *and* $(A_2, \iota_2; \mu)$ *a* $\mathfrak{b}$*-transform of* (A_1, ι_1). *Then* $(A_2, \iota_2; \mu\lambda)$ *is an* $\mathfrak{ab}$*-transform of* (A, ι).

PROOF. Obviously $\mu\lambda$ is an $\mathfrak{R}$-homomorphism of A onto A_2. Let k be a field of definition for (A, ι), (A_1, ι_1), (A_2, ι_2), λ, μ; take a generic point x of A over k and put $y = \lambda x$. Then $k(\mu y)$ is the composite of the fields $k(\iota_1(\beta)y)$ for $\beta \in \mathfrak{b}$. We can find a positive integer m such that $m\mathfrak{b} \subset \mathfrak{r}$ and a point z of A such that $mz = x$. We have then $\iota_1(\beta)y = \lambda\iota(m\beta)z$, and, by Proposition 8, $k(\lambda\iota(m\beta)z)$ is the composite of the fields $k(\iota(\alpha)\iota(m\beta)z)$ for $\alpha \in \mathfrak{a}$. As we have $\mathfrak{ab} \subset \mathfrak{a}\mathfrak{r}_1 \subset \mathfrak{a}$, $\alpha\beta$ is contained in $\mathfrak{r}$, so that we have $\iota(\alpha)\iota(m\beta)z = \iota(\alpha\beta)x$ for every $\alpha \in \mathfrak{a}$, $\beta \in \mathfrak{b}$. Hence $k(\mu\lambda x)$ is the composite of the fields $k(\iota(\alpha\beta)x)$ for $\alpha \in \mathfrak{a}$ and $\beta \in \mathfrak{b}$. Since $\mathfrak{ab}$ is generated by the elements $\alpha\beta$, this proves that $\mu\lambda$ is an $\mathfrak{ab}$-multiplication.

7.2. From now on, we assume that $\mathfrak{R}$ *is a simple algebra over* $\mathbf{Q}$ and denote by $N(\xi)$ the reduced norm of $\xi \in \mathfrak{R}$ (cf. §5.1). (For our principal aim in Chapter IV, it is sufficient to consider the case where $\mathfrak{R}$ is an algebraic number field and $2(\dim A) = [\mathfrak{R} : \mathbf{Q}]$; so the reader who is interested only in this case may dispense with the trouble of considering the general case.) We first recall the ideal-theory in $\mathfrak{R}$; for details we refer to Deuring [7]. An order $\mathfrak{o}$ in $\mathfrak{R}$ is said to be *maximal* if there is no order containing $\mathfrak{o}$ other than itself. Let $\mathfrak{a}$ be a lattice in $\mathfrak{R}$ and $\mathfrak{o}$ an order in $\mathfrak{R}$. We call $\mathfrak{a}$ a *right* (resp. *left*) $\mathfrak{o}$*-ideal* if we have $\mathfrak{ao} \subset \mathfrak{a}$ (resp. $\mathfrak{oa} \subset \mathfrak{a}$). $\mathfrak{a}$ is said to be *normal* if both its right and left orders are maximal; it is known that if one of the left and right orders of $\mathfrak{a}$ is maximal, then $\mathfrak{a}$ is normal. Given normal lattices $\mathfrak{a}$ and $\mathfrak{b}$ in $\mathfrak{R}$, the product $\mathfrak{ab}$ is said to be *proper* if the right order of $\mathfrak{a}$ coincides with the left order of $\mathfrak{b}$. The set of all normal lattices in $\mathfrak{R}$ form a groupoid $\mathfrak{G}$ with respect to the operation of proper product; the maximal orders are the units of $\mathfrak{G}$ and the inverse of $\mathfrak{a}$ is given by

$$\mathfrak{a}^{-1} = \{\xi \mid \xi \in \mathfrak{R}, \mathfrak{a}\xi\mathfrak{a} \subset \mathfrak{a}\}.$$

$\mathfrak{o}_r$ and $\mathfrak{o}_l$ being the right and left orders of a lattice $\mathfrak{a}$ in $\mathfrak{R}$, if we have $\mathfrak{a} \subset \mathfrak{o}_r$, then $\mathfrak{a} \subset \mathfrak{o}_l$, and vice versa; we call such $\mathfrak{a}$ *integral*. If $\mathfrak{a}$ is integral and normal, the numbers of elements in the factor modules $\mathfrak{o}_r/\mathfrak{a}$ and $\mathfrak{o}_l/\mathfrak{a}$ are the same; we denote this number by $N_1(\mathfrak{a})$. Let $\mathfrak{Z}$ be the center of $\mathfrak{R}$; put

$$[\mathfrak{R} : \mathfrak{Z}] = f^2, \quad [\mathfrak{Z} : \mathbf{Q}] = d. \tag{1}$$

Then, we can prove that $N_1(\mathfrak{a})$ is the f-th power of a positive integer; we put $N(\mathfrak{a}) = N_1(\mathfrak{a})^{1/f}$. We can define $N(\mathfrak{a})$ in a natural manner for every normal lattice in $\mathfrak{R}$ which is not necessarily integral; and if the product $\mathfrak{a}\mathfrak{b}$ is proper, we have

$$N(\mathfrak{a}\mathfrak{b}) = N(\mathfrak{a})N(\mathfrak{b}).$$

If ξ is an invertible element in $\mathfrak{R}$, we have, for every maximal order $\mathfrak{o}$,

$$N(\mathfrak{o}\xi) = N(\xi\mathfrak{o}) = |N(\xi)|.$$

Now consider an abelian variety (A, ι) of type $(\mathfrak{R})$. Let n be the dimension of A and f, d be as in (1). Then, by Proposition 2, fd divides $2n$; putting $2n = mfd$, we call m the *index* of (A, ι). We call (A, ι) *principal* if the order of (A, ι) is maximal. By Proposition 7, if (A, ι) is principal, then, for every integral left $\mathfrak{o}$-ideal $\mathfrak{a}$, an $\mathfrak{a}$-transform of (A, ι) is also principal. In the following treatment, (A, ι), (A', ι') etc. will denote abelian varieties of type $(\mathfrak{R})$ which are assumed to be principal.

PROPOSITION 10. *Let $\mathfrak{o}$ be the order of (A, ι) and $\mathfrak{a}$ an integral left $\mathfrak{o}$-ideal; let $(A_1, \iota_1; \lambda_{\mathfrak{a}})$ be an $\mathfrak{a}$-transform of (A, ι). Then we have*

$$\nu(\lambda_{\mathfrak{a}}) = N(\mathfrak{a})^m,$$

where m is the index of (A, ι).

PROOF. Let $\mathfrak{o}'$ be the right order of $\mathfrak{a}$. Then we can find an integral left $\mathfrak{o}'$-ideal $\mathfrak{b}$ such that $\mathfrak{a}\mathfrak{b}$ is a principal ideal $\mathfrak{o}\gamma$ and $(N(\mathfrak{b}), \nu(\lambda_{\mathfrak{a}})) = 1$ (cf. [7] VI, Satz 27). Let $(A_2, \iota_2; \lambda_{\mathfrak{b}})$ be a $\mathfrak{b}$-transform of (A_1, ι_1); then $(A_2, \iota_2; \lambda_{\mathfrak{b}}\lambda_{\mathfrak{a}})$ is an $\mathfrak{o}\gamma$-transform of (A, ι) by virtue of Proposition 9. On the other hand, putting $\iota'(\alpha) = \iota(\gamma\alpha\gamma^{-1})$ for $\alpha \in \mathfrak{R}$, we see that $(A, \iota'; \iota(\gamma))$ is an $\mathfrak{o}\gamma$-transform of (A, ι). Hence there exists an isomorphism η of (A_2, ι_2) onto (A, ι') such that $\eta\lambda_{\mathfrak{b}}\lambda_{\mathfrak{a}} = \iota(\gamma)$. It follows that

$$\nu(\lambda_{\mathfrak{b}})\nu(\lambda_{\mathfrak{a}}) = \nu(\iota(\gamma)).$$

By Proposition 2, we have $\nu(\iota(\gamma)) = N(\gamma)^m = N(\mathfrak{a})^m N(\mathfrak{b})^m$, so that

$$\nu(\lambda_{\mathfrak{a}})\nu(\lambda_{\mathfrak{b}}) = N(\mathfrak{a})^m N(\mathfrak{b})^m. \tag{2}$$

Since $\nu(\lambda_{\mathfrak{a}})$ is prime to $N(\mathfrak{b})$, $\nu(\lambda_{\mathfrak{a}})$ must divide $N(\mathfrak{a})^m$. Applying this result to $\mathfrak{b}$, we see that $\nu(\lambda_{\mathfrak{b}})$ divides $N(\mathfrak{b})^m$. Therefore equality (2) shows $\nu(\lambda_{\mathfrak{a}}) = N(\mathfrak{a})^m$.

PROPOSITION 11. *Let $\mathfrak{o}$ be the order of (A, ι); let $\mathfrak{a}$ and $\mathfrak{b}$ be integral left $\mathfrak{o}$-ideals. Let $(A_1, \iota_1; \lambda_{\mathfrak{a}})$ and $(A_2, \iota_2; \lambda_{\mathfrak{b}})$ be respectively an $\mathfrak{a}$-transform and* *a* $\mathfrak{b}$-*transform of* (A, ι). *Then the following three conditions are equivalent to one another:*

(1) $\mathfrak{a} \supset \mathfrak{b}$.
(2) *There exist a field of definition* k *for* (A, ι), (A_1, ι_1), (A_2, ι_2), $\lambda_\mathfrak{a}$, $\lambda_\mathfrak{b}$ *and a generic point* x *of* A *over* k *such that* $k(\lambda_\mathfrak{a} x) \supset k(\lambda_\mathfrak{b} x)$.
(3) *There exists a homomorphism* μ *of* (A_1, ι_1) *onto* (A_2, ι_2) *such that* $\mu\lambda_\mathfrak{a} = \lambda_\mathfrak{b}$.

PROOF. It is easy to see that $(1) \Rightarrow (2) \Leftrightarrow (3)$. Suppose that $\mathfrak{a} \not\supset \mathfrak{b}$; and put $\mathfrak{c} = \mathfrak{a} + \mathfrak{b}$. We have then $\mathfrak{c} \underset{\neq}{\supseteq} \mathfrak{a}$, so that $N(\mathfrak{c}) < N(\mathfrak{a})$. Hence, if $\lambda_\mathfrak{c}$ is a $\mathfrak{c}$-multiplication of (A, ι), we have $k(\lambda_\mathfrak{c} x) \underset{\neq}{\supseteq} k(\lambda_\mathfrak{a} x)$ by virtue of Proposition 10. On the other hand, as we have $\mathfrak{c} = \mathfrak{a} + \mathfrak{b}$, the field $k(\lambda_\mathfrak{c} x)$ is the composite of $k(\lambda_\mathfrak{a} x)$ and $k(\lambda_\mathfrak{b} x)$; so we must have $k(\lambda_\mathfrak{a} x) \not\supset k(\lambda_\mathfrak{b} x)$. This proves $(2) \Rightarrow (1)$.

PROPOSITION 12. *The notation being as in Proposition 11, let* g *be a positive integer such that* $g\mathfrak{a}^{-1}\mathfrak{b}$ *is integral. Then, there exists a* $(g\mathfrak{a}^{-1}\mathfrak{b})$-*multiplication* μ *of* (A_1, ι_1) *onto* (A_2, ι_2) *such that* $\mu\lambda_\mathfrak{a} = g\lambda_\mathfrak{b}$.

PROOF. Put $\mathfrak{c} = g\mathfrak{a}^{-1}\mathfrak{b}$. Let $(A_3, \iota_3; \lambda_\mathfrak{c})$ be a $\mathfrak{c}$-transform of (A_1, ι_1). Since both (A_2, ι_2) and (A_3, ι_3) are $g\mathfrak{b}$-transforms of (A, ι), we obtain an isomorphism η of (A_3, ι_3) onto (A_2, ι_2) such that $\eta\lambda_\mathfrak{c}\lambda_\mathfrak{a} = g\lambda_\mathfrak{b}$. We see easily that $\eta\lambda_\mathfrak{c}$ is a $\mathfrak{c}$-multiplication; this proves our proposition.

Now we impose the following condition on our abelian varieties of type $(\mathfrak{R})$.

(C) *If* $\mathfrak{Z}$ *denotes the center of* $\mathfrak{R}$, *the commutor of* $\iota(\mathfrak{Z})$ *in* $\mathrm{End}_\mathbf{Q}(A)$ *is contained in* $\iota(\mathfrak{R})$.

This is trivially satisfied if $\iota(\mathfrak{R}) = \mathrm{End}_\mathbf{Q}(A)$. If the index of (A, ι) is 1, (A, ι) satisfies (C). In fact, the degrees f and d being as in (1), $\mathfrak{R}$ contains a subfield $\mathfrak{F}$ such that $[\mathfrak{F} : \mathfrak{Z}] = f$. If (A, ι) is of index 1, we have $2 \dim(A) = [\mathfrak{F} : \mathbf{Q}]$. By Proposition 3, $\mathrm{End}_\mathbf{Q}(A)$ must be simple; and if we denote by $\mathfrak{K}$ the center of $\mathrm{End}_\mathbf{Q}(A)$, $\iota(\mathfrak{F})$ contains $\mathfrak{K}$. It follows that $\iota(\mathfrak{Z})$ contains $\mathfrak{K}$. As we have $[\mathrm{End}_\mathbf{Q}(A) : \iota(\mathfrak{F})] = [\iota(\mathfrak{F}) : \mathfrak{K}]$ and $[\mathfrak{R} : \mathfrak{F}] = [\mathfrak{F} : \mathfrak{Z}]$, we get $[\mathrm{End}_\mathbf{Q}(A) : \iota(\mathfrak{R})] = [\iota(\mathfrak{Z}) : \mathfrak{K}]$. Let $\mathfrak{L}$ be the commutor of $\iota(\mathfrak{Z})$ in $\mathrm{End}_\mathbf{Q}(A)$; then $\mathfrak{L}$ contains $\iota(\mathfrak{R})$; and by a property of central simple algebra we have $[\mathrm{End}_\mathbf{Q}(A) : \mathfrak{L}] = [\iota(\mathfrak{Z}) : \mathfrak{K}]$. This shows $\mathfrak{L} = \iota(\mathfrak{R})$. Hence (A, ι) satisfies (C).

We note that if (A, ι) satisfies (C), every $\mathfrak{a}$-transform of (A, ι), for an integral lattice $\mathfrak{a}$, satisfies (C).

PROPOSITION 13. *Suppose that* (A, ι) *satisfies* (C). *Let* $\mathfrak{o}$ *be the order of* (A, ι) *and* $\mathfrak{a}$ *an integral left* $\mathfrak{o}$-*ideal; let* $(A_1, \iota_1; \lambda_\mathfrak{a})$ *be an* $\mathfrak{a}$-*transform of* (A, ι).

Denote by ι^* and ι_1^* the restrictions of ι and ι_1 to the center $\mathfrak{Z}$ of $\mathfrak{R}$. Then, every homomorphism of (A, ι^*) onto (A_1, ι_1^*) is a $\mathfrak{c}$-multiplication of (A, ι) for an integral left $\mathfrak{o}$-ideal $\mathfrak{c}$. Moreover, the set of all homomorphisms of (A, ι^*) into (A_1, ι_1^*) coincides with $\lambda_\mathfrak{a} \cdot \iota(\mathfrak{a}^{-1})$.

PROOF. Take a positive integer g such that $g\mathfrak{a}^{-1}$ is integral and put $\mathfrak{b} = g\mathfrak{a}^{-1}$. By Proposition 12, there exists a $\mathfrak{b}$-multiplication $\lambda_\mathfrak{b}$ of (A_1, ι_1) onto (A, ι) such that $\lambda_\mathfrak{b}\lambda_\mathfrak{a} = g1_A$; we fix $\mathfrak{b}$ and $\lambda_\mathfrak{b}$. Let μ be a homomorphism of (A, ι^*) into (A_1, ι_1^*). Then, $\mu\lambda_\mathfrak{b}$ is an endomorphism of (A_1, ι_1^*). By virtue of (C), $\mu\lambda_\mathfrak{b}$ must be of the form $\iota_1(\gamma)$ for an element $\gamma \in \mathfrak{o}_1$, where $\mathfrak{o}_1$ denotes the left order of $\mathfrak{b}$. Applying Proposition 11 to the ideals $\mathfrak{b}$ and $\mathfrak{o}_1\gamma + \mathfrak{b}$, we see that $\gamma \in \mathfrak{b}$. Now suppose that μ is an isogeny; then γ must be an invertible element of $\mathfrak{R}$. Put $\mathfrak{c} = \mathfrak{b}^{-1}\gamma$; let $(A_2, \iota_2; \lambda_\mathfrak{c})$ be a $\mathfrak{c}$-transform of (A, ι). By the same argument as in the proof of Proposition 10, we obtain an isomorphism η of A_2 onto A_1 such that $\eta\lambda_\mathfrak{c}\lambda_\mathfrak{b} = \iota_1(\gamma) = \mu\lambda_\mathfrak{b}$. As $\lambda_\mathfrak{b}$ is an isogeny, we have $\eta\lambda_\mathfrak{c} = \mu$. It follows that μ is a $\mathfrak{c}$-multiplication; this proves the first assertion. Now let $\mathfrak{H}$ denote the module of homomorphisms of (A, ι^*) into (A_1, ι_1^*). We have proved above $\mathfrak{H} \cdot \lambda_\mathfrak{b} \subset \iota_1(\mathfrak{b})$. Let β be an element of $\mathfrak{b}$. If k is a field of definition for (A, ι), (A_1, ι_1) and $\lambda_\mathfrak{b}$, and x is a generic point of A_1 over k, we have $k(\lambda_\mathfrak{b}x) \supset k(\iota_1(\beta)x)$. Hence there exists a homomorphism λ of A into A_1 such that $\lambda\lambda_\mathfrak{b} = \iota_1(\beta)$. We see easily that λ commutes with the operation of $\mathfrak{Z}$. This shows $\mathfrak{H} \cdot \lambda_\mathfrak{b} = \iota_1(\mathfrak{b})$. Multiplying this relation by $\lambda_\mathfrak{a}$, we get $\mathfrak{H} \cdot g1_A = \iota_1(\mathfrak{b})\lambda_\mathfrak{a} = g \cdot \lambda_\mathfrak{a} \cdot \iota(\mathfrak{a}^{-1})$, so that $\mathfrak{H} = \lambda_\mathfrak{a} \cdot \iota(\mathfrak{a}^{-1})$; this completes our proof.

PROPOSITION 14. *The notation and assumptions being as in Proposition 13, let $\mathfrak{b}$ be an integral left $\mathfrak{o}$-ideal and $(A_2, \iota_2; \lambda_\mathfrak{b})$ be a $\mathfrak{b}$-transform of (A, ι); denote by ι_2^* the restriction of ι_2 to $\mathfrak{Z}$. Then, (A_1, ι_1^*) is isomorphic to (A_2, ι_2^*) if and only if there exists an invertible element γ of $\mathfrak{R}$ such that $\mathfrak{a} = \mathfrak{b}\gamma$.*

PROOF. Let $\mathfrak{H}_i$ denote, for $i = 1, 2$, the set of all homomorphisms of (A, ι^*) into (A_i, ι_i^*). The $\mathfrak{H}_i$ are considered as right $\mathfrak{o}$-modules. If (A_1, ι_1^*) is isomorphic to (A_2, ι_2^*), $\mathfrak{H}_1$ must be $\mathfrak{o}$-isomorphic to $\mathfrak{H}_2$. By Proposition 13, this amounts to saying that $\mathfrak{a}^{-1}$ is isomorphic to $\mathfrak{b}^{-1}$ as right $\mathfrak{o}$-modules; this implies that $\mathfrak{a}$ and $\mathfrak{b}$ are isomorphic as left $\mathfrak{o}$-modules. This proves the "only if" part of our proposition. Conversely, suppose that there exists an invertible element γ in $\mathfrak{R}$ such that $\mathfrak{a}\gamma = \mathfrak{b}$. Take a positive integer g such that $g\gamma$ is contained in the right order of $\mathfrak{a}$. Then, by the same argument as in the proof of Proposition 10, we can find an isomorphism η of A_1 onto A_2 such that $\eta\iota_1(g\gamma)\lambda_\mathfrak{a} = g\lambda_\mathfrak{b}$. It is easy to see that η commutes with the operation of $\mathfrak{Z}$. This proves the "if" part.

7.3. Now we consider the case where $\mathfrak{R}$ is an algebraic number field F. In this case condition (C) is reduced to the following form.

(C′) *The commutor of* $\iota(F)$ *in* $\mathrm{End}_{\mathbf{Q}}(A)$ *is* $\iota(F)$ *itself.*

This is satisfied if $\iota(F) = \mathrm{End}_{\mathbf{Q}}(A)$ or if (A, ι) is of index 1.

Assuming condition (C′) to be satisfied, we give the following definition: Given an ideal-class c of F, we call (A', ι') a c-transform of (A, ι) if (A', ι') is an $\mathfrak{a}$-*transform* of (A, ι) for some integral ideal $\mathfrak{a}$ in c; if that is so, for *every* integral ideal $\mathfrak{c}$ in c, (A', ι') is a $\mathfrak{c}$-transform of (A, ι). Let c and d be ideal-classes of F; let (A_c, ι_c) and (A_d, ι_d) be respectively a c-transform and a d-transform of (A, ι). Then, by Proposition 12, (A_d, ι_d) is a $c^{-1}d$-transform of (A_c, ι_c). Furthermore, by Proposition 14, (A_c, ι_c) and (A_d, ι_d) are isomorphic if and only if $c = d$.

7.4. Let us consider the case of characteristic 0. $(F; \{\varphi_i\})$ being a CM-type, let (A, ι) be an abelian variety of type $(F; \{\varphi_i\})$; by our definition, (A, ι) is of index 1. By Theorem 2, (A, ι) is represented by a complex torus $\mathbf{C}^n/D(\mathfrak{a})$ for a free $\mathbf{Z}$-submodule $\mathfrak{a}$ of rank $2n$ in F, the notation being as in that theorem. We observe that (A, ι) is principal if and only if $\mathfrak{a}$ is an ideal (not necessarily integral) of F. It is easy to see that, for every ideal-class c of F, a c-transform of (A, ι) is also of type $(F; \{\varphi_i\})$.

PROPOSITION 15. *The notation being as in Theorem 2, let* $\mathfrak{a}$ *and* $\mathfrak{b}$ *be two ideals of* F*; let* (A_1, ι_1) *and* (A_2, ι_2) *be abelian varieties of type* $(F; \{\varphi_i\})$*, respectively represented by the complex tori* $\mathbf{C}^n/D(\mathfrak{a})$ *and* $\mathbf{C}^n/D(\mathfrak{b})$*. If* γ *is an element of* $\mathfrak{a}^{-1}\mathfrak{b}$ *other than 0, the diagonal matrix* $S(\gamma)$ *with the diagonal elements* $\gamma^{\varphi_1}, \dots, \gamma^{\varphi_n}$ *represents a* $(\gamma\mathfrak{b}^{-1}\mathfrak{a})$*-multiplication of* (A_1, ι_1) *into* (A_2, ι_2)*; conversely, every homomorphism of* (A_1, ι_1) *onto* (A_2, ι_2) *corresponds to some* $S(\gamma)$ *such that* $\gamma \in \mathfrak{a}^{-1}\mathfrak{b}$.

PROOF. If $\gamma \in \mathfrak{a}^{-1}\mathfrak{b}$, we have $S(\gamma)D(\mathfrak{a}) \subset D(\mathfrak{b})$, so that $S(\gamma)$ gives a homomorphism of $\mathbf{C}^n/D(\mathfrak{a})$ into $\mathbf{C}^n/D(\mathfrak{b})$. Hence $S(\gamma)$ represents a homomorphism λ of A_1 into A_2. Since λ commutes with the operation of F, λ is a homomorphism of (A_1, ι_1) into (A_2, ι_2). Suppose that $\gamma \neq 0$. We see easily that Ker(λ) corresponds to $D(\gamma^{-1}\mathfrak{b})/D(\mathfrak{a})$ and

$$D(\gamma^{-1}\mathfrak{b}) = \{u \mid u \in \mathbf{C}^n, S(\alpha)u \in D(\mathfrak{a}) \text{ for every } \alpha \in \gamma\mathfrak{a}\mathfrak{b}^{-1}\}.$$

This implies Ker(λ) $= \mathfrak{g}(\gamma\mathfrak{a}\mathfrak{b}^{-1}, A_1)$. By Proposition 8, it follows that λ is a $(\gamma\mathfrak{a}\mathfrak{b}^{-1})$-multiplication of (A_1, ι_1) onto (A_2, ι_2). The last assertion of our proposition follows from this and Proposition 13.

PROPOSITION 16. *Let (A, ι) and (A', ι') be two abelian varieties which are principal and of the same CM-type $(F; \{\varphi_i\})$. Then (A', ι') is a c-transform of (A, ι) for an ideal-class c of F.*

This is an easy consequence of Proposition 15. Furthermore, on account of Proposition 14, we obtain

PROPOSITION 17. *Let $(F; \{\varphi_i\})$ be a CM-type and h the number of ideal-classes of F. Then, there are exactly h abelian varieties of type $(F; \{\varphi_i\})$, which are principal and not isomorphic to each other.*

7.5. Ideal-section points. Now coming back to the case of arbitrary characteristic, denote by $\mathfrak{o}$ the ring of integers in the number field F; let (A, ι) be an abelian variety of type (F) that is principal. Let $\mathfrak{a}$ be an ideal of $\mathfrak{o}$ and $(A', \iota'; \lambda_\mathfrak{a})$ an $\mathfrak{a}$-transform of (A, ι). We observe that $\nu_i(\lambda_\mathfrak{a})$ and $\nu_s(\lambda_\mathfrak{a})$ depends only upon (A, ι) and $\mathfrak{a}$, and not on the choice of $(A', \iota'; \lambda_\mathfrak{a})$; so we denote them by $N_i(\mathfrak{a}, A)$ and $N_s(\mathfrak{a}, A)$. Let c be an ideal-class of F and (A_c, ι_c) a c-transform of (A, ι). Then we have

$$N_i(\mathfrak{a}, A) = N_i(\mathfrak{a}, A_c), \quad N_s(\mathfrak{a}, A) = N_s(\mathfrak{a}, A_c).$$

In fact, let $(A'_c, \iota'_c; \lambda'_\mathfrak{a})$ be an $\mathfrak{a}$-transform of (A_c, ι_c). Take an integral ideal $\mathfrak{b}$ in the class c, prime to the characteristic p; let $\lambda_\mathfrak{b}$ and $\lambda'_\mathfrak{b}$ be respectively $\mathfrak{b}$-multiplications of (A, ι) onto (A_c, ι_c) and of (A', ι') onto (A'_c, ι'_c). Then both $\lambda'_\mathfrak{b}\lambda_\mathfrak{a}$ and $\lambda'_\mathfrak{a}\lambda_\mathfrak{b}$ are $\mathfrak{ab}$-multiplications of (A, ι) onto (A'_c, ι'_c); so there exists an automorphism η of (A'_c, ι'_c) such that $\lambda'_\mathfrak{b}\lambda_\mathfrak{a} = \eta\lambda'_\mathfrak{a}\lambda_\mathfrak{b}$. We have then $\nu_i(\lambda'_\mathfrak{b})\nu_i(\lambda_\mathfrak{a}) = \nu_i(\lambda'_\mathfrak{a})\nu_i(\lambda_\mathfrak{b})$. By our assumption that $\mathfrak{b}$ is prime to p, the degree $\nu(\lambda_\mathfrak{b}) = \nu(\lambda'_\mathfrak{b}) = N(\mathfrak{b})^m$ is prime to p, where m denotes the index of (A, ι); hence we have $\nu_i(\lambda_\mathfrak{b}) = \nu_i(\lambda'_\mathfrak{b}) = 1$, so that $\nu_i(\lambda_\mathfrak{a}) = \nu_i(\lambda'_\mathfrak{a})$. This proves the above relations.

Now let $\mathcal{S} = \{(A, \iota)\}$ be a system of abelian varieties of type (F), whose members are transforms of each other by ideals of $\mathfrak{o}$. Then, $N_i(\mathfrak{a}, A)$ and $N_s(\mathfrak{a}, A)$ for $(A, \iota) \in \mathcal{S}$ does not depend on the choice of (A, ι); so we denote them by $N_i(\mathfrak{a}, \mathcal{S})$ and $N_s(\mathfrak{a}, \mathcal{S})$ or simply by $N_i(\mathfrak{a})$ and $N_s(\mathfrak{a})$ when we fix our attention to a given system $\mathcal{S}$. We can easily verify

$$N_i(\mathfrak{a})N_s(\mathfrak{a}) = N(\mathfrak{a})^m,$$

m denoting the index of the members of $\mathcal{S}$, and

$$N_i(\mathfrak{ab}) = N_i(\mathfrak{a})N_i(\mathfrak{b}), \quad N_s(\mathfrak{ab}) = N_s(\mathfrak{a})N_s(\mathfrak{b}).$$

If $\lambda_\mathfrak{a}$ is an $\mathfrak{a}$-multiplication of (A, ι), $\nu_s(\lambda_\mathfrak{a})$ is the order of the kernel $\mathrm{Ker}(\lambda_\mathfrak{a})$ of $\lambda_\mathfrak{a}$. As we have $\mathrm{Ker}(\lambda_\mathfrak{a}) = \mathfrak{g}(\mathfrak{a}, A)$, this shows that $N_s(\mathfrak{a})$ is the order of

$\mathfrak{g}(\mathfrak{a}, A)$; in particular, if $\mathfrak{a}$ is prime to the characteristic of the fields of definition for A, we get $N_i(\mathfrak{a}) = 1$, and hence $\mathfrak{g}(\mathfrak{a}, A)$ is of order $N(\mathfrak{a})^m$.

PROPOSITION 18. *Let* $\mathfrak{a}$ *and* $\mathfrak{b}$ *be ideals of* $\mathfrak{o}$. *Then we have*

$$\mathfrak{g}(\mathfrak{a} + \mathfrak{b}, A) = \mathfrak{g}(\mathfrak{a}, A) \cap \mathfrak{g}(\mathfrak{b}, A),$$

$$\mathfrak{g}(\mathfrak{a} \cap \mathfrak{b}, A) = \mathfrak{g}(\mathfrak{a}, A) + \mathfrak{g}(\mathfrak{b}, A).$$

Moreover, if $\mathfrak{a}$ *is prime to* $\mathfrak{b}$, $\mathfrak{g}(\mathfrak{a}\mathfrak{b}, A)$ *is the direct sum of* $\mathfrak{g}(\mathfrak{a}, A)$ *and* $\mathfrak{g}(\mathfrak{b}, A)$.

PROOF. The first equality is obvious. We see easily that

$$\mathfrak{g}(\mathfrak{a} \cap \mathfrak{b}, A) \supset \mathfrak{g}(\mathfrak{a}, A) + \mathfrak{g}(\mathfrak{b}, A).$$

By an elementary theorem of group-theory, the order of $\mathfrak{g}(\mathfrak{a}, A) + \mathfrak{g}(\mathfrak{b}, A)$ is equal to

$$[\mathfrak{g}(\mathfrak{a}, A) : \{0\}][\mathfrak{g}(\mathfrak{b}, A) : \{0\}]/[\mathfrak{g}(\mathfrak{a}, A) \cap \mathfrak{g}(\mathfrak{b}, A) : \{0\}].$$

As we have $\mathfrak{g}(\mathfrak{a} + \mathfrak{b}, A) = \mathfrak{g}(\mathfrak{a}, A) \cap \mathfrak{g}(\mathfrak{b}, A)$, this number is equal to $N_s(\mathfrak{a})N_s(\mathfrak{b})N_s(\mathfrak{a} + \mathfrak{b})^{-1}$. On the other hand, by means of the relation $\mathfrak{a}\mathfrak{b} = (\mathfrak{a} \cap \mathfrak{b})(\mathfrak{a} + \mathfrak{b})$, we have $N_s(\mathfrak{a})N_s(\mathfrak{b})N_s(\mathfrak{a} + \mathfrak{b})^{-1} = N_s(\mathfrak{a} \cap \mathfrak{b})$. Hence both sides of the above inclusion-relation have the same order; this proves the second equality. If $\mathfrak{a}$ is prime to $\mathfrak{b}$, we have $\mathfrak{a} \cap \mathfrak{b} = \mathfrak{a}\mathfrak{b}$ and $\mathfrak{a} + \mathfrak{b} = \mathfrak{o}$, so that

$$\mathfrak{g}(\mathfrak{a}, A) \cap \mathfrak{g}(\mathfrak{b}, A) = \mathfrak{g}(\mathfrak{o}, A) = \{0\}.$$

This implies the last assertion.

PROPOSITION 19. *Let* $\mathfrak{p}$ *be a prime ideal of* $\mathfrak{o}$. *Then,* $N_i(\mathfrak{p})$ *and* $N_s(\mathfrak{p})$ *are powers of* $N(\mathfrak{p})$. *In particular, if* (A, ι) *is of index 1, we have* $N_i(\mathfrak{p}) = 1$ *or* $N_s(\mathfrak{p}) = 1$.

PROOF. We can easily verify that $\mathfrak{g}(\mathfrak{p}, A)$ is invariant under the operation of $\mathfrak{o}$; moreover, $\mathfrak{g}(\mathfrak{p}, A)$ is considered as an $(\mathfrak{o}/\mathfrak{p})$-module. Since $\mathfrak{p}$ is a prime ideal, $\mathfrak{o}/\mathfrak{p}$ is a finite field with $N(\mathfrak{p})$ elements; and $\mathfrak{g}(\mathfrak{p}, A)$ is a vector space over the field $\mathfrak{o}/\mathfrak{p}$. This proves the first assertion. If (A, ι) is of index 1, we must have $N_i(\mathfrak{p})N_s(\mathfrak{p}) = N(\mathfrak{p})$; this proves the second assertion.

As an example, we shall determine $N_i(\mathfrak{a})$ in the case of dimension 1. Suppose that $[F : \mathbf{Q}] = 2$ and (A, ι) is of dimension 1; then the index of (A, ι) is 1. Let k be a field of definition for (A, ι). If the characteristic of k is 0, we have $N_i(\mathfrak{a}) = 1$ for every $\mathfrak{a}$; so there is no problem. Suppose that k is of characteristic $p \neq 0$. If $\mathfrak{a}$ is prime to p, we have $N_i(\mathfrak{a}) = 1$; so we have only to consider

$N_i(\mathfrak{p})$ for the prime ideals $\mathfrak{p}$ dividing p. Since F is of degree 2, there can occur three cases:

(i) $(p) = \mathfrak{p}_1\mathfrak{p}_2$, $\mathfrak{p}_1 \neq \mathfrak{p}_2$,
(ii) $(p) = \mathfrak{p}$,
(iii) $(p) = \mathfrak{p}^2$,

where $\mathfrak{p}_1$, $\mathfrak{p}_2$ and $\mathfrak{p}$ denote prime ideals of F. By Proposition 7 of §2, we have $N_i((p)) = \nu_i(p1_A) = p$ or p^2. Hence, in cases (ii) and (iii), we must have $N_i(\mathfrak{p}) > 1$; so by Proposition 19, we have $N_i(\mathfrak{p}) = N(\mathfrak{p})$, $\nu_i(p1_A) = p^2$. For the same reason, in case (i), we cannot have $N_i(\mathfrak{p}_1) = N_i(\mathfrak{p}_2) = 1$. Assume that $N_i(\mathfrak{p}_1) = N_i(\mathfrak{p}_2) = p$. Let λ_1 and λ_2 be respectively a $\mathfrak{p}_1$-multiplication and a $\mathfrak{p}_2$-multiplication of (A, ι). Then, as A is of dimension 1, if x is a generic point of A over k, the fields $k(\lambda_1 x)$ and $k(\lambda_2 x)$ must be contained in $k(x^p)$. On the other hand, as we have $\mathfrak{o} = \mathfrak{p}_1 + \mathfrak{p}_2$, the field $k(x)$ is the composite of $k(\lambda_1 x)$ and $k(\lambda_2 x)$; we have thus arrived at a contradiction. Hence we must have $N_i(\mathfrak{p}_\alpha) = 1$ for one of $\mathfrak{p}_1$ and $\mathfrak{p}_2$, say $\mathfrak{p}_1$. Then, we have $N_i(\mathfrak{p}_2) = p$ and hence $\nu_i(p1_A) = p$. This completes the analysis of case (i).

Returning to the general case, we call a point t of $\mathfrak{g}(\mathfrak{a}, A)$ a proper $\mathfrak{a}$-section point on A if $\iota(\alpha)t = 0$ implies $\alpha \in \mathfrak{a}$.

PROPOSITION 20. *Given a principal (A, ι) of index 1, let $\mathfrak{m}$ be an ideal of $\mathfrak{o}$ and t a proper $\mathfrak{m}$-section point of A. Then, for every point t' of $\mathfrak{g}(\mathfrak{m})$, there exists an element α of $\mathfrak{o}$ such that $t' = \iota(\alpha)t$; and the mapping $\alpha \to \iota(\alpha)t$ gives an $\mathfrak{o}$-isomorphism of $\mathfrak{o}/\mathfrak{m}$ onto $\mathfrak{g}(\mathfrak{m})$. The point $\iota(\alpha)t$ is a proper $\mathfrak{m}$-section point on A if and only if α is prime to $\mathfrak{m}$.*

PROOF. It is easy to see that the mapping $\alpha \to \iota(\alpha)t$ is an $\mathfrak{o}$-homomorphism of $\mathfrak{o}$ into $\mathfrak{g}(\mathfrak{m})$; as t is a proper $\mathfrak{m}$-section point, the kernel of this homomorphism is $\mathfrak{m}$. Hence the module $\mathfrak{o}/\mathfrak{m}$ is isomorphic to a submodule of $\mathfrak{g}(\mathfrak{m})$. On the other hand, $\mathfrak{o}/\mathfrak{m}$ is of order $N(\mathfrak{m})$ and the order of $\mathfrak{g}(\mathfrak{m})$ is $N_s(\mathfrak{m})$, which is not greater than $N(\mathfrak{m})$. It follows that the mapping $\alpha \to \iota(\alpha)t$ is an isomorphism of $\mathfrak{o}/\mathfrak{m}$ onto $\mathfrak{g}(\mathfrak{m})$; all the assertions easily follow from this fact.

PROPOSITION 21. *(A, ι) and $\mathfrak{m}$ being as in Proposition 20, there exists a proper $\mathfrak{m}$-section point on A if and only if $N_i(\mathfrak{m}, A) = 1$. In particular, if $\mathfrak{m}$ is prime to the characteristic of the fields of definition for A, there exists a proper $\mathfrak{m}$-section point.*

PROOF. By Proposition 20, if there exists a proper $\mathfrak{m}$-section point, then the order of $\mathfrak{g}(\mathfrak{m})$ is equal to $N(\mathfrak{m})$, so that $N_i(\mathfrak{m}) = 1$. Conversely, suppose that $N_i(\mathfrak{m}) = 1$. Let $\mathfrak{m} = \mathfrak{p}_1^{e_1} \cdots \mathfrak{p}_r^{e_r}$ be the factorization of $\mathfrak{m}$ into prime ideals $\mathfrak{p}_u$. By Proposition 18, $\mathfrak{g}(\mathfrak{m})$ is the direct sum of the $\mathfrak{g}(\mathfrak{p}_u^{e_u})$. We have clearly $N_i(\mathfrak{p}_u) = 1$, and hence $\mathfrak{g}(\mathfrak{p}_u^f)$ is of order $N(\mathfrak{p}_u^f)$. Therefore, we can find a point

t_u in $\mathfrak{g}(\mathfrak{p}_u^{e_u})$ which is not contained in $\mathfrak{g}(\mathfrak{p}_u^{e_u - 1})$. Let α be an element of $\mathfrak{o}$ such that $\iota(\alpha)t_u = 0$. Put $\alpha\mathfrak{o} + \mathfrak{p}_u^{e_u} = \mathfrak{p}_u^f$; then we see that t_u is contained in $\mathfrak{g}(\mathfrak{p}_u^f)$, so that $e_u = f$; this implies that α is contained in $\mathfrak{p}_u^{e_u}$. Hence t_u is a proper $\mathfrak{p}_u^{e_u}$-section point. Put $t = t_1 + \cdots + t_r$. Then we can easily verify that t is a proper $\mathfrak{m}$-section point. This proves our proposition.

PROPOSITION 22. *Let (A, ι) be principal and of index 1, and let $\mathfrak{h}$ be a finite subgroup of A such that $\iota(\mathfrak{o})\mathfrak{h} \subset \mathfrak{h}$. Then, there exists an ideal $\mathfrak{a}$ of $\mathfrak{o}$ such that $\mathfrak{h} = \mathfrak{g}(\mathfrak{a}, A)$.*

PROOF. Let $\{t_1, \ldots, t_h\}$ be a system of generators of $\mathfrak{h}$ over $\iota(\mathfrak{o})$. Let $\mathfrak{a}_u$, for each u, denote the set of elements $\alpha \in \mathfrak{o}$ such that $\iota(\alpha)t_u = 0$. Then, $\mathfrak{a}_u$ is an ideal of $\mathfrak{o}$ and t_u is a proper $\mathfrak{a}_u$-section point. Hence, by Proposition 20, we have $\mathfrak{g}(\mathfrak{a}_u) = \iota(\mathfrak{o})t_u$. Putting $\mathfrak{a} = \mathfrak{a}_1 \cap \cdots \cap \mathfrak{a}_h$, we obtain, by Proposition 18,

$$\mathfrak{g}(\mathfrak{a}) = \mathfrak{g}(\mathfrak{a}_1) + \cdots + \mathfrak{g}(\mathfrak{a}_h) = \iota(\mathfrak{o})t_1 + \cdots + \iota(\mathfrak{o})t_h = \mathfrak{h}.$$

This proves our proposition.

PROPOSITION 23. *Let (A, ι) and (A', ι') be abelian varieties of type (F), which are principal and of index 1. Let λ be a homomorphism of (A, ι) onto (A', ι') such that $\nu_i(\lambda) = 1$. Then, λ is an $\mathfrak{a}$-multiplication of (A, ι) onto (A', ι') for an ideal $\mathfrak{a}$ of $\mathfrak{o}$.*

PROOF. Put $\mathfrak{h} = \mathrm{Ker}(\lambda)$ and apply the argument of the proof of Proposition 22 to this case. We obtain then $\mathrm{Ker}(\lambda) = \mathfrak{g}(\mathfrak{a}, A)$; moreover by Proposition 21, we have $N_i(\mathfrak{a}_u) = 1$ for each u, so that $N_i(\mathfrak{a}) = 1$. Hence, if $(A_1, \iota_1; \lambda_\mathfrak{a})$ is an $\mathfrak{a}$-transform of (A, ι), we have $\nu_i(\lambda_\mathfrak{a}) = 1$. The equality $\mathrm{Ker}(\lambda) = \mathfrak{g}(\mathfrak{a}, A)$ shows that λ and $\lambda_\mathfrak{a}$ have the same kernel. It follows from this and the relation $\nu_i(\lambda) = \nu_i(\lambda_\mathfrak{a}) = 1$ that there exists an isomorphism η of A' onto A_1 such that $\eta\lambda = \lambda_\mathfrak{a}$. Since both λ and $\lambda_\mathfrak{a}$ are $\mathfrak{o}$-homomorphisms, η is an $\mathfrak{o}$-isomorphism. This proves that $(A', \iota'; \lambda)$ is an $\mathfrak{a}$-transform of (A, ι).

PROPOSITION 24. *Let $\mathfrak{m}$ be an ideal of $\mathfrak{o}$ and c an ideal-class of F. Assuming (A, ι) to be principal and of index 1, let (A', ι') be a c-transform of (A, ι) and t a proper $\mathfrak{m}$-section point on A. Then, for every t' in $\mathfrak{g}(\mathfrak{m}, A')$, there exist an ideal $\mathfrak{a}$ in c and an $\mathfrak{a}$-multiplication $\lambda_\mathfrak{a}$ of (A, ι) onto (A', ι') such that $t' = \lambda_\mathfrak{a} t$; and $\lambda_\mathfrak{a} t$ is a proper $(\mathfrak{a}, \mathfrak{m})^{-1}\mathfrak{m}$-section point of A'. In particular, $\lambda_\mathfrak{a} t$ is a proper $\mathfrak{m}$-section point of A' if and only if $\mathfrak{a}$ is prime to $\mathfrak{m}$; and $\lambda_\mathfrak{a} t = 0$ if and only if $\mathfrak{a} \subset \mathfrak{m}$.*

PROOF. Let $\mathfrak{a}$ be an ideal in c and $\lambda_\mathfrak{a}$ an $\mathfrak{a}$-multiplication of (A, ι) onto (A', ι'). We first prove that $\lambda_\mathfrak{a} t$ is a proper $(\mathfrak{a}, \mathfrak{m})^{-1}\mathfrak{m}$-section point on A'. Take an integral ideal $\mathfrak{b}$ in the class c^{-1}, prime to $\mathfrak{m}$, and a $\mathfrak{b}$-multiplication $\lambda_{\mathfrak{b}}$ of (A', ι') onto (A, ι). Then there exists an element γ of $\mathfrak{o}$ such that $\lambda_{\mathfrak{b}}\lambda_{\mathfrak{a}} = \iota(\gamma)$ and $\mathfrak{a}\mathfrak{b} = \gamma\mathfrak{o}$. If we have $\iota'(\mu)\lambda_{\mathfrak{a}}t = 0$ for an element $\mu \in \mathfrak{o}$, we have $\iota(\mu\gamma)t = 0$, so that $\mu\gamma \in \mathfrak{m}$, namely $\mu\mathfrak{a}\mathfrak{b} \subset \mathfrak{m}$. As $\mathfrak{b}$ is prime to $\mathfrak{m}$, we get $\mu\mathfrak{a} \subset \mathfrak{m}$; this shows that $\mu \in (\mathfrak{a}, \mathfrak{m})^{-1}\mathfrak{m}$. Conversely, if we have $\mu \in (\mathfrak{a}, \mathfrak{m})^{-1}\mathfrak{m}$, then $\mu\mathfrak{a} \subset \mathfrak{m}$ and hence $\mu\gamma \in \mu\mathfrak{a}\mathfrak{b} \subset \mathfrak{m}$, so that $\lambda_{\mathfrak{b}}\iota'(\mu)\lambda_{\mathfrak{a}}t = 0$; this shows that $\iota'(\mu)\lambda_{\mathfrak{a}}t \in \mathfrak{g}(\mathfrak{b}, A')$. On the other hand, as t is contained in $\mathfrak{g}(\mathfrak{m}, A)$, we have clearly $\iota'(\mu)\lambda_{\mathfrak{a}}t \in \mathfrak{g}(\mathfrak{m}, A')$. Since $\mathfrak{b}$ is prime to $\mathfrak{m}$, the intersection of $\mathfrak{g}(\mathfrak{b}, A')$ and $\mathfrak{g}(\mathfrak{m}, A')$ must be $\{0\}$; it follows that $\iota'(\mu)\lambda_{\mathfrak{a}}t = 0$. We have thus proved that $\iota'(\mu)\lambda_{\mathfrak{a}}t = 0$ if and only if μ is contained in $(\mathfrak{a}, \mathfrak{m})^{-1}\mathfrak{m}$, namely, $\lambda_{\mathfrak{a}}t$ is a proper $(\mathfrak{a}, \mathfrak{m})^{-1}\mathfrak{m}$-section point. In particular, $\lambda_{\mathfrak{a}}t$ is a proper $\mathfrak{m}$-section point if and only if $\mathfrak{a}$ is prime to $\mathfrak{m}$; and $\lambda_{\mathfrak{a}}t = 0$ if and only if $\mathfrak{a} \subset \mathfrak{m}$. Now fix an integral ideal $\mathfrak{c}$ in c, prime to $\mathfrak{m}$, and a $\mathfrak{c}$-multiplication $\lambda_{\mathfrak{c}}$ of (A, ι) onto (A', ι'). Then $\lambda_{\mathfrak{c}}t$ is a proper $\mathfrak{m}$-section point; hence, for every t' in $\mathfrak{g}(\mathfrak{m}, A')$, there exists, by Proposition 20, an element α of $\mathfrak{o}$ such that $t' = \iota'(\alpha)\lambda_{\mathfrak{c}}t$. This proves the first assertion of our proposition, since $\iota'(\alpha)\lambda_{\mathfrak{c}}$ is an $\alpha\mathfrak{c}$-multiplication of (A, ι) onto (A', ι'). The rest of the proposition is already proved.

7.6. Let (A, ι) be an abelian variety of type (F), defined over a field k, and σ an isomorphism of k onto a field k^σ. Let $\mathfrak{r}$ be the order of (A, ι). Put, for every $\alpha \in \mathfrak{r}$, $\iota^\sigma(\alpha) = \iota(\alpha)^\sigma$ (cf. §1.5); then ι^σ is uniquely extended to an isomorphism of F into $\mathrm{End}_{\mathbf{Q}}(A^\sigma)$ which we denote also by ι^σ. We obtain thus an abelian variety (A^σ, ι^σ) of type (F), defined over k^σ; $\mathfrak{r}$ is the order of (A^σ, ι^σ). Let $\mathfrak{a}$ be an ideal of $\mathfrak{r}$ and $(A_1, \iota_1; \lambda)$ an $\mathfrak{a}$-transform of (A, ι), defined over k. Then, we see easily that $(A_1^\sigma, \iota_1^\sigma; \lambda^\sigma)$ is an $\mathfrak{a}$-transform of (A^σ, ι^σ). If (A, ι) is principal, so is (A^σ, ι^σ); and if t is a proper $\mathfrak{a}$-section point on A, rational over k, then t^σ is a proper $\mathfrak{a}$-section point on A^σ. Now suppose that k is of characteristic $p \neq 0$. Then, for every power $q = p^f$ with $f > 0$, we obtain an abelian variety (A^q, ι^q) of type (F); let π be the q-th power homomorphism of A onto A^q, defined in §1.5. We see easily that $\pi\iota(\alpha) = \iota^q(\alpha)\pi$ for every $\alpha \in F$, so that π is a homomorphism of (A, ι) onto (A^q, ι^q). In particular, if (A, ι) is defined over a finite field with q elements, the q-th power endomorphism π of A is an endomorphism of (A, ι); hence, if further (A, ι) satisfies condition (C′) of §7.3, there exists an element γ of $\mathfrak{r}$ such that $\pi = \iota(\gamma)$.

8. The Reflex of a CM-Type

The purpose of this section is to investigate the algebraic structure of a CM-type. For the sake of simplicity, we assume that the fields appearing in this section are all contained in the field $\mathbf{C}$ of complex numbers.

8.1. Group-theoretic characterization of CM-types. We begin with

LEMMA 3. *Let L be a Galois extension of $\mathbf{Q}$, G the Galois group of L over $\mathbf{Q}$ and ρ the element of G such that ξ^{ρ} is the complex conjugate of ξ for every $\xi \in L$. Let K and K_0 be two subfields of L such that $[K : K_0] = 2$, and H, H_0 be respectively the subgroups of G corresponding to K, K_0. Then, the following two conditions are equivalent:*

(i) *K_0 is totally real and K is totally imaginary.*
(ii) *$H_0 = H \cup H\sigma\rho\sigma^{-1}$ for every $\sigma \in G$.*

If these conditions are satisfied, we have $\rho H\tau = H\tau\rho = \sigma\rho\sigma^{-1}H\tau = H\tau\sigma\rho\sigma^{-1}$ for every $\sigma \in G$, $\tau \in G$.

PROOF. If σ is an element of G, the subfields K_0^{σ}, K^{σ} of L correspond to the subgroups $\sigma^{-1}H_0\sigma$, $\sigma^{-1}H\sigma$ of G. If K_0 is totally real and K is totally imaginary, then K_0^{σ} is a real field and K^{σ} is an imaginary field, so that ρ fixes every element of K_0^{σ} and does not fix some element of K^{σ}. Hence we have $\rho \in \sigma^{-1}H_0\sigma$ and $\rho \notin \sigma^{-1}H\sigma$, so that $\sigma\rho\sigma^{-1} \in H_0$ and $\sigma\rho\sigma^{-1} \notin H$. By the assumption $[K : K_0] = 2$, we have $[H_0 : H] = 2$; it follows that $H_0 = H \cup H\sigma\rho\sigma^{-1}$. Conversely, if the relation $H_0 = H \cup H\sigma\rho\sigma^{-1}$ holds for every $\sigma \in G$, we can easily see, following up the above argument in the opposite direction, that K_0 is totally real and K is totally imaginary. Now suppose that conditions (i) and (ii) are satisfied. Then, we see that both $H\sigma\rho\sigma^{-1}$ and $\sigma\rho\sigma^{-1}H$ are equal to the set $H_0 - H$ for every $\sigma \in G$. We have hence

$$\rho H = H\rho = \sigma\rho\sigma^{-1}H = H\sigma\rho\sigma^{-1}.$$

It follows that $\rho H\sigma = H\sigma\rho$. Let μ be an element of G. Then, K^{μ} is totally imaginary and K_0^{μ} is totally real; and the subgroup $\mu^{-1}H\mu$ of G corresponds to K^{μ}. Therefore, applying the formula $\rho H\sigma = H\sigma\rho$ to $\mu^{-1}H\mu$, we have $\rho(\mu^{-1}H\mu)\sigma = (\mu^{-1}H\mu)\sigma\rho$. Transform this by the inner automorphism $\gamma \to \mu\gamma\mu^{-1}$ and put $\tau = \mu\sigma\mu^{-1}$. We have then $\mu\rho\mu^{-1}H\tau = H\tau\mu\rho\mu^{-1}$. This completes the proof.

PROPOSITION 25. *Let F be an extension of $\mathbf{Q}$ of degree $2n$, $\{\varphi_1, \ldots, \varphi_n\}$ a set of n distinct isomorphisms of F into $\mathbf{C}$. Let L be a Galois extension of $\mathbf{Q}$ containing F, and G the Galois group of L over $\mathbf{Q}$. Denote by ρ the element of G such that ξ^{ρ} is the complex conjugate of ξ for every $\xi \in L$, and by S the set of all the elements of G inducing some φ_i on F. Then, $(F; \{\varphi_i\})$ is a CM-type if and only if we have*

$$G = S \cup S\sigma\rho\sigma^{-1}, \; S\sigma\rho\sigma^{-1} = \sigma\rho\sigma^{-1}S \tag{1}$$

for every $\sigma \in G$.

PROOF. If $(F; \{\varphi_i\})$ is a CM-type, then, by Theorem 1 of §5, F has two subfields K and K_0 satisfying conditions (CM1, 2) in that theorem. The symbols L, G, ρ, S being as in our proposition, let H be the subgroup of G corresponding to K. Let $\{\psi_1, \ldots, \psi_m\}$ be the set of all distinct isomorphisms of K into $\mathbf{C}$ obtained by restricting φ_i to K. Then there are no two members of $\{\psi_j\}$ which are complex conjugate of each other; and it is easy to see that $\{\psi_j\}$ gives the set of all distinct isomorphisms of K_0 into $\mathbf{C}$. It follows that $[K_0 : \mathbf{Q}] = m$. We observe also that $\{\varphi_i\}$ is the set of all the isomorphisms of F into $\mathbf{C}$ inducing some ψ_j on K. Hence S is the set of all the elements of G inducing some ψ_j on K. Take, for every j, an element τ_j of G inducing ψ_j on K. Then we have $S = \cup H\tau_j$. Every element of G coincides with τ_j or $\tau_j\rho$ on K. Hence we have

$$G = \bigcup_j (H\tau_j \cup H\tau_j\rho) = S \cup S\rho.$$

By Lemma 3, we have $\rho H\tau_j = H\tau_j\rho = \sigma\rho\sigma^{-1}H\tau_j = H\tau_j\sigma\rho\sigma^{-1}$; it follows that $\rho S = S\rho = \sigma\rho\sigma^{-1}S = S\sigma\rho\sigma^{-1}$. Thus we have proved the "only if" part of our proposition. Conversely, suppose that relation (1) holds for every $\sigma \in G$. Put $\rho' = \sigma\rho\sigma^{-1}$. As the number of the elements in S is the half of the order of G, the sets S and $\rho'S = S\rho'$ have no common element and $\rho'S = G - S$. Put

$$H' = \{\gamma \mid \gamma \in G, \gamma S = S\}.$$

We have then, $\rho'H'\rho S = \rho'H'\rho'S = \rho'H'S\rho' = \rho'S\rho' = \rho'\rho'S = S$, so that we have $\rho'H'\rho \subset H'$, $\rho'H'\rho' \subset H'$. It follows from this that $\rho'H' = H'\rho' = H'\rho$. Since S does not coincide with ρS, the element ρ is not contained in H'. Put now

$$H_0' = H' \cup H'\rho.$$

We can easily verify that H_0' is a subgroup of G and $[H_0' : H'] = 2$; and we have $H_0' = H' \cup H'\sigma\rho\sigma^{-1}$ for every $\sigma \in G$. Hence, if we denote by K' and K_0' the subfields of L corresponding to H' and H_0', respectively, K' is totally imaginary and K_0' is totally real by virtue of Lemma 3. If γ is an element of G leaving invariant every element of F, then every element of γS coincides with one of the φ_i on F, so that we have $\gamma S \subset S$ and hence $\gamma \in H'$; this implies $F \supset K'$. As we have $H'S = S$, S is expressed as a union of cosets: $S = \cup H'\mu_\alpha$. We see that $\{\mu_\alpha\}$ gives the half of the isomorphisms of K' into $\mathbf{C}$; and $H'\mu_\alpha\rho$ does not coincide with any $H'\mu_\beta$, since S and $S\rho$ have no common element; in other words, for any α and β, μ_α does not coincide with the complex conjugate of μ_β on K'; namely, the system $\{K', K_0', \{\varphi_i\}\}$ satisfies conditions (CM1, 2) of Theorem 1. Hence $(F; \{\varphi_i\})$ is a CM-type. This completes the proof.

8.2. Primitive CM-types. We call a CM-type *primitive* if the abelian varieties of that type are simple; recall that any two abelian varieties of the same

CM-type are isogenous to each other (Corollary of Theorem 2). The following proposition is a criterion for the primitiveness of a CM-type.

PROPOSITION 26. *Given a CM-type* $(F; \{\varphi_i\})$, *let* L, G, ρ, S, *be as in Proposition 25 and* H_1 *the subgroup of* G *corresponding to* F. *Put*

$$H' = \{\gamma \in G \mid \gamma S = S\}.$$

Then $(F; \{\varphi_i\})$ *is primitive if and only if* $H_1 = H'$.

PROOF. Using the same notation as in the proof of Proposition 25, we see that F contains the field K' corresponding to H'. If $(F; \{\varphi_i\})$ is primitive, we must have $F = K'$ by virtue of Theorem 3 of §6, so that $H_1 = H'$. If $(F; \{\varphi_i\})$ is not primitive, F contains two subfields K and K_0, satisfying conditions (CM1, 2) of Theorem 1, such that $F \neq K$; this follows from the discussion in §5.2. Denoting by H the subgroup of G corresponding to K, there exist elements τ_j of G such that $S = \cup H\tau_j$, as is seen in the proof of Proposition 25. We have then $HS = S$, and hence $H \subset H'$. By the relation $F \supsetneqq K$, we have $H \supsetneqq H_1$, so that $H_1 \neq H'$; this completes the proof.

We shall now give another criterion with no use of Galois group. If $(K; \{\varphi_i\})$ is a primitive CM-type, Theorem 3 of §6 shows that K is a totally imaginary quadratic extension of a totally real field K_0. We have seen in the proof of that theorem that there exists an element ζ of K such that $K = K_0(\zeta)$, $-\zeta^2$ is totally positive and $\mathrm{Im}(\zeta^{\varphi_i}) > 0$ for every i. Conversely, suppose a totally real field K_0 and a totally positive element η of K_0 are given; let ζ be a number such that $-\zeta^2 = \eta$; and put $K = K_0(\zeta)$. Let $\{\varphi_i\}$ be the set of all the isomorphisms φ of K into $\mathbf{C}$ such that $\mathrm{Im}(\zeta^{\varphi}) > 0$. Then it can be easily seen that $(K; \varphi_i\})$ is a CM-type. We shall denote this CM-type by $K_0((\zeta))$. We have $K_0((\zeta)) = K_0((\zeta'))$ if and only if ζ/ζ' is a totally positive element of K_0.

PROPOSITION 27. $K_0((\zeta))$ *is primitive if and only if the following two conditions are satisfied:*

(i) $K_0(\zeta) = \mathbf{Q}(\zeta)$.
(ii) *For any conjugate* ζ' *of* ζ *other than* ζ *itself, over* $\mathbf{Q}$, ζ'/ζ *is not totally positive.*

PROOF. First we note that ζ'/ζ'' is totally real for any two conjugates ζ' and ζ'' of ζ over $\mathbf{Q}$, since ζ' and ζ'' are purely imaginary numbers. Put $F = K = K_0(\zeta)$; let $\{\varphi_i\}$ be the set of all the isomorphisms φ of F into $\mathbf{C}$ such that $\mathrm{Im}(\zeta^{\varphi}) > 0$. Define L, G, ρ, S for $(F; \{\varphi_i\})$ as in Proposition 25 and denote by H the subgroup of G corresponding to K; put $H' = \{\gamma \mid \gamma \in G, \gamma S = S\}$.

Then we see easily that $S = \{\sigma \mid \sigma \in G, \mathrm{Im}(\zeta^\sigma) > 0\}$. If $\gamma \in H'$, we have $\gamma\sigma \in S$ for every $\sigma \in S$, so that $\mathrm{Im}(\zeta^{\gamma\sigma}) > 0$ for every $\sigma \in S$. As $\zeta^{\gamma\sigma}/\zeta^\sigma$ is real, we see that $\zeta^{\gamma\sigma}/\zeta^\sigma > 0$. If $\sigma \notin S$, then $\sigma \in S\rho$, $\gamma\sigma \in \gamma S\rho = S\rho$, and hence $\gamma\sigma \notin S$, so that $\mathrm{Im}(\zeta^{\gamma\sigma}) < 0$, $\mathrm{Im}(\zeta^\sigma) < 0$. Thus we have $\zeta^{\gamma\sigma}/\zeta^\sigma > 0$ for every $\sigma \in G$, namely ζ^γ/ζ is totally positive. Conversely, if ζ^γ/ζ is totally positive for an element γ of G, $\mathrm{Im}(\zeta^{\gamma\sigma})$ has the same sign as $\mathrm{Im}(\zeta^\sigma)$ for every $\sigma \in G$; it follows that $\gamma S = S$. Therefore, H' is the set of all the elements $\gamma \in G$ such that ζ^γ/ζ is totally positive. By Proposition 26, $K_0((\zeta)) = (K; \{\varphi_i\})$ is primitive if and only if $H' = H$. Our proposition is an immediate consequence of these facts.

8.3. Now we define the reflex of a CM-type.

PROPOSITION 28. *The symbols* $(F; \{\varphi_i\})$, L, G, ρ, S, *being the same as in Proposition 26, put*

$$S^* = \{\sigma^{-1} \mid \sigma \in S\}, \qquad H^* = \{\gamma \mid \gamma \in G, \gamma S^* = S^*\}.$$

Let K^* *be the subfield of* L *corresponding to* H^* *and* $\{\psi_j\}$ *the set of all the isomorphisms of* K^* *into* $\mathbf{C}$ *obtained from the elements of* S^*. *Then,* $(K^*; \{\psi_j\})$ *is a primitive CM-type and we have*

$$K^* = \mathbf{Q}\left(\sum_i \xi^{\varphi_i} \mid \xi \in F\right).$$

$(K^*; \{\psi_j\})$ *is determined only by* $(F; \{\varphi_i\})$ *and independent of the choice of* L.

PROOF. Let σ be an element of G. Then, by relation (1) of Proposition 25, we have $G = S^* \cup S^*\sigma\rho\sigma^{-1}$, $S^*\sigma\rho\sigma^{-1} = \sigma\rho\sigma^{-1}S^*$. We see easily that S^* is the set of all the elements of G inducing on K^* some ψ_j and $\{\psi_j\}$ is the half of all the isomorphisms of K^* into $\mathbf{C}$. Hence, by Proposition 25, $(K^*; \{\psi_j\})$ is a CM-type, and is primitive by virtue of Proposition 26. Let γ be an element of H^*. We have then $\gamma^{-1}S^* = S^*$, so that $S\gamma = S$; hence $\{\varphi_1\gamma, \dots, \varphi_n\gamma\}$ coincides with $\{\varphi_1, \dots, \varphi_n\}$ as a whole. We have therefore $(\sum_i \xi^{\varphi_i})^\gamma = \sum_i \xi^{\varphi_i}$ for every $\xi \in F$. Conversely, suppose that an element γ of G fixes $\sum_i \xi^{\varphi_i}$ for every $\xi \in F$. Then, we have, for every integer a,

$$\sum_i (\xi^{\varphi_i\gamma})^a = \sum_i (\xi^a)^{\varphi_i\gamma} = \sum_i (\xi^a)^{\varphi_i} = \sum_i (\xi^{\varphi_i})^a.$$

By an elementary theorem of algebra, we see that $\{\xi^{\varphi_1\gamma}, \dots, \xi^{\varphi_n\gamma}\}$ coincides with $\{\xi^{\varphi_1}, \dots, \xi^{\varphi_n}\}$ as a whole; this shows that $\{\varphi_1, \dots, \varphi_n\}$ coincides with $\{\varphi_1\gamma, \dots, \varphi_n\gamma\}$ on F as a whole; in other words we have $S\gamma = S$, so that

$\gamma \in H^*$. Thus we have proved that H^* is the set of all the elements of G leaving invariant every element of $\mathbf{Q}(\sum_i \xi^{\varphi_i} \mid \xi \in F)$; this implies $K^* = \mathbf{Q}(\sum_i \xi^{\varphi_i} \mid \xi \in F)$. The last assertion of our proposition follows easily from the definition.

We call the CM-type $(K^*; \{\psi_j\})$ of the above proposition the *reflex* of $(F; \{\varphi_i\})$. By the proposition, the reflex of every CM-type is primitive. If a CM-type $(F; \{\varphi_i\})$ is primitive, then $(F; \{\varphi_i\})$ coincides with the reflex of its reflex; this follows immediately from Propositions 26 and 28. For every type $(F; \{\varphi_i\})$, we can find two subfields K and K_0 of F satisfying conditions (CM1, 2) of Theorem 1. Let $\{\chi_j\}$ be the set of distinct isomorphisms of K into $\mathbf{C}$ induced by the φ_i. Then it is easy to see that $(K; \{\chi_j\})$ is a CM-type; and we observe that $(F; \{\varphi_i\})$ and $(K; \{\chi_j\})$ have the same reflex; this is also an immediate consequence of the definition.

PROPOSITION 29. *Let $(F, \{\varphi_i\})$ be a CM-type and $(K^*; \{\psi_j\})$ its reflex; let ρ denote complex conjugation. If $\beta = \prod_j \alpha^{\psi_j}$ with $\alpha \in K^*$, then $\beta \in F$ and $\beta\beta^\rho = N_{K^*/\mathbf{Q}}(\alpha)$. Further let $\mathfrak{a}$ be an ideal in K^*. Then there exists an ideal $\mathfrak{b}$ in F such that $\mathfrak{b}\mathfrak{o}_L = \prod_j \mathfrak{a}^{\psi_j}\mathfrak{o}_L$ and $\mathfrak{b}\mathfrak{b}^\rho = N_{K^*/\mathbf{Q}}(\mathfrak{a})\mathfrak{o}_F$, where $\mathfrak{o}_L$ resp. $\mathfrak{o}_F$ denotes the maximal order of L resp. F.*

PROOF. Define L, G, S as before. Let H be the subgroup of G corresponding to F. Then, by the definition of S, we have $HS = S$, so that $S^*H = S^*$, if we define S^* as in Proposition 28. It follows that, for every $\sigma \in H$, $\{\psi_j\sigma\}$ coincides with $\{\psi_j\}$ as a whole. Hence we have $\beta^\sigma = \prod \alpha^{\psi_j\sigma} = \prod \alpha^{\psi_j} = \beta$ for every $\sigma \in H$; so β is contained in F. As we have $G = S^* \cup S^*\rho$, $\{\psi_j, \psi_j\rho\}$ gives the set of all distinct isomorphisms of K^* into $\mathbf{C}$; this proves the relation $\beta\beta^\rho = N_{K^*/\mathbf{Q}}(\alpha)$. Now, as for the assertions concerning ideals, it is sufficient to prove it in the case where $\mathfrak{a}$ is an integral ideal. Take a non-zero element α of K^*, divisible by $\mathfrak{a}$. We can find an element γ of K^* divisible by $\mathfrak{a}$ such that $\gamma\mathfrak{a}^{-1}$ is prime to $N_{K^*/\mathbf{Q}}(\alpha)$. Put $\beta = \prod_j \alpha^{\psi_j}$ and $\delta = \prod_j \gamma^{\psi_j}$. Then, by what we have already proved, β and δ are contained in F. Put $\mathfrak{c} = \prod_j \mathfrak{a}^{\psi_j}\mathfrak{o}_L$. It is clear that both β and δ are divisible by $\mathfrak{c}$. As $\gamma\mathfrak{a}^{-1}$ is prime to $N_{K^*/\mathbf{Q}}(\alpha)$, we see that $\delta\mathfrak{c}^{-1}$ is prime to β. It follows that $\mathfrak{c}$ is the greatest common divisor of β and δ. This proves that $\mathfrak{c} = \mathfrak{o}_L\mathfrak{b}$ with an ideal $\mathfrak{b}$ in F. The relation $\mathfrak{b}\mathfrak{b}^\rho = N_{K^*/\mathbf{Q}}(\mathfrak{a})$ can be proved by the same argument as for $\beta\beta^\rho = N_{K^*/\mathbf{Q}}(\alpha)$.

8.4. Examples. We shall now give some examples of CM-types.

(1) First we consider a CM-type $(F; \{\varphi_i\})$ such that F is a Galois extension of $\mathbf{Q}$. We can take F itself as L of Proposition 25; the subgroup of the Galois group G corresponding to F is then the subgroup consisting only of the identity; and we have $S = \{\varphi_1, \ldots, \varphi_n\}$, $S^* = \{\varphi_1^{-1}, \ldots, \varphi_n^{-1}\}$, where we consider the

φ_i as elements of G. Hence, if F is abelian over $\mathbf{Q}$, the subgroup H^* defined in Proposition 28 coincides with the subgroup $H' = \{\gamma \in G \mid \gamma S = S\}$. If moreover $(F; \{\varphi_i\})$ is primitive, H' must be the identity-subgroup, so that the field corresponding to H^* is F. Thus we get the following result: if F is abelian over $\mathbf{Q}$ and if $(F; \{\varphi_i\})$ is primitive, the reflex of $(F; \{\varphi_i\})$ is $(F; \{\varphi_i^{-1}\})$. In the classical case, F is an imaginary quadratic field; so the reflex is the same as itself.

Now we give an example of a primitive CM-type $(F; \{\varphi_i\})$ where F is a cyclotomic field. Let p be an odd prime and $\zeta = e^{2\pi i/p}$. The automorphisms of $\mathbf{Q}(\zeta)$ are given by $\zeta \to \zeta^a$ for the integers a such that $1 \leq a \leq p-1$. Put $n = (p-1)/2$ and denote by φ_i the automorphism $\zeta \to \zeta^i$ for $i = 1, \ldots, n$. Then we see that there are no two automorphisms among φ_i which are complex conjugate of each other; and $\mathbf{Q}(\zeta)$ is a totally imaginary quadratic extension of the totally real field $\mathbf{Q}(\zeta + \zeta^{-1})$. Hence $(\mathbf{Q}(\zeta); \{\varphi_i\})$ is a CM-type. Let γ be an automorphism of $\mathbf{Q}(\zeta)$ such that $\{\gamma\varphi_1, \ldots, \gamma\varphi_n\} = \{\varphi_1, \ldots, \varphi_n\}$; let a be an integer such that $\zeta^\gamma = \zeta^a$. Then, $\{1 \cdot a, \ldots, n \cdot a \bmod (p)\}$ coincides with $\{1, \ldots, n \bmod (p)\}$ as a whole, so that we have

$$1 \cdot a + \cdots + n \cdot a \equiv 1 + \cdots + n \bmod (p).$$

As $1 + \cdots + n = (p^2-1)/8$ is relatively prime to p, we have $a \equiv 1 \bmod (p)$, so that γ is the identity. Therefore, our CM-type $(\mathbf{Q}(\zeta); \{\varphi_i\})$ is primitive.

There can exist many CM-types with the same field F. Consider for example $F = \mathbf{Q}(\zeta)$ with $p = 13$. We normalize $S = \{\varphi_1, \ldots, \varphi_n\}$ taking the identity as φ_1. Then, we obtain 32 CM-types $(F; \{\varphi_i\})$. By an easy calculation, we see that two of them are non-primitive and the remaining primitive 30 CM-types are divided into 5 families in the following way: each family consists of 6 types; and any two types belong to the same family if and only if they are transformed onto each other by an automorphism of F. We can similarly treat the case where F is cyclic over $\mathbf{Q}$.

(2) Let K_0 be a real quadratic field and ξ a number such that $-\xi^2$ is a totally positive number of K_0. Then we obtain a CM-type $K_0((\xi))$. Put $K = K_0(\xi)$. For the sake of simplicity, suppose $\mathrm{Im}(\xi) > 0$; this amounts to assuming that S contains the identity. If K is a Galois extension of $\mathbf{Q}$, the Galois group G is an abelian group of degree 4, so that G is cyclic or the product of two cyclic groups of order 2.

(A) The case where G is the product of two cyclic groups of order 2. Denoting by ρ the element of G such that α^ρ is the complex conjugate of α for every $\alpha \in K$, the subgroup of G corresponding to K_0 is $\{1, \rho\}$. Put $S = \{1, \sigma\}$. Then we have $G = \{1, \sigma, \rho, \sigma\rho\}$; and the elements γ such that $\gamma S = S$ form the subgroup $\{1, \sigma\}$. Hence $K_0((\xi))$ is not primitive. Let K^* be the subfield of K corresponding to the subgroup $\{1, \sigma\}$. Then we see that K^* is an imaginary quadratic field, K is the composite of K_0 and K^*, and the reflex of $K_0((\xi))$ is $(K^*, \{1\})$.

(B) The case where G is cyclic. ρ and $S = \{1, \sigma\}$ being as above, we have $G = \{1, \sigma, \sigma^2 = \rho, \sigma^3\}$. We see that there is no element γ other than the identity such that $\gamma S = S$, so that the CM-type $K_0((\xi)) = (K; \{1, \sigma\})$ is primitive. The reflex is $(K; \{1, \sigma^{-1}\})$ by the result of (1). We can easily verify that $K_0((\xi^\sigma)) = (K; \{1, \sigma^{-1}\})$.

(C) The case where K is not normal over $\mathbf{Q}$ (The case of Hecke [20]). Put $K_0((\xi)) = (K; \{1, \varphi\})$. As K_0 is a real quadratic field, there exists a positive integer d such that $K_0 = \mathbf{Q}(\sqrt{d})$; and $-\xi^2$ is expressed in the form $-\xi^2 = x + y\sqrt{d}$, where x and y are rational numbers; as $-\xi^2$ is totally positive, we have $x + y\sqrt{d} > 0$, $x - y\sqrt{d} > 0$, and $\xi = i\sqrt{x+y\sqrt{d}}$, $\xi^\varphi = i\sqrt{x-y\sqrt{d}}$. We see that four elements $\pm\xi$, $\pm\xi^\varphi$ are the conjugates of ξ over $\mathbf{Q}$. As K is not normal over $\mathbf{Q}$, ξ^φ is not contained in K, so that $K_0(\xi^\varphi) \neq K_0(\xi)$. Put $d' = x^2 - y^2 d$; then, we have $d' > 0$. On the other hand, we have $d' = (\xi\xi^\varphi)^2$; hence $\sqrt{d'}$ is not contained in K_0, since we have $K_0(\xi^\varphi) \neq K_0(\xi)$. Therefore, $\mathbf{Q}(\sqrt{d'})$ is a real quadratic field different from $\mathbf{Q}(\sqrt{d}) = K_0$. Put now $L = \mathbf{Q}(\xi, \xi^\varphi)$; then we see that L is normal over $\mathbf{Q}$, $L = K(\sqrt{d'})$ and hence $[L : \mathbf{Q}] = 8$. The Galois group of L over $\mathbf{Q}$ consists of eight automorphisms which map the couple (ξ, ξ^φ) onto $(\pm\xi, \pm\xi^\varphi)$, $(\pm\xi, \mp\xi^\varphi)$, $(\pm\xi^\varphi, \pm\xi)$, $(\pm\xi^\varphi, \mp\xi)$. Denote by σ and τ respectively the elements of G which send (ξ, ξ^φ) onto $(\xi^\varphi, -\xi)$ and (ξ^φ, ξ). Then we see that $\sigma^2 = \rho$, $\sigma^4 = \tau^2 = 1$, $\tau\sigma = \sigma^3\tau$ and G is generated by σ and τ. We can easily verify that the subgroup of G corresponding to $K = \mathbf{Q}(\xi)$ is $\{1, \sigma\tau\}$ and $S = \{1, \sigma, \tau, \sigma\tau\}$; moreover, we have $\{1, \sigma\tau\} = \{\gamma \mid \gamma \in G, \gamma S = S\}$, which proves that $(K; \{1, \varphi\})$ is primitive. If we define H^* as in Proposition 28, then we have $H^* = \{1, \tau\}$; and the subfield of L corresponding to H^* is $\mathbf{Q}(\xi + \xi^\varphi)$. As we have $S^* = H^* \cup H^*\sigma\tau$, the reflex of $(\mathbf{Q}(\xi), \{1, \varphi\})$ is $(\mathbf{Q}(\xi + \xi^\varphi), \{1, \sigma\tau\})$; the latter is also written as $\mathbf{Q}(\sqrt{d'})$ $((\xi + \xi^\varphi))$.

Some more examples of CM-types can be found in [S70, §1], [S77b], and [Do].

8.5. Fields of definition for (A, ι).

PROPOSITION 30. *Let $(F; \{\varphi_i\})$ be a CM-type and $(K^*; \{\psi_j\})$ its reflex; let (A, ι) be an abelian variety of type $(F; \{\varphi_i\})$ and k a field of definition for A. Then, if every element of $\iota(F) \cap \mathrm{End}(A)$ is defined over k, we have $k \supset K^*$. Conversely, if $k \supset K^*$ and if $(F; \{\varphi_i\})$ is primitive, every element of* $\mathrm{End}(A)$ *is defined over k.*

PROOF. Let S be a representation of $\mathrm{End}_{\mathbf{Q}}(A)$ by invariant differential forms on A. Then, for every $\xi \in F$, $\xi^{\varphi_1}, \dots, \xi^{\varphi_n}$ are the characteristic roots of $S(\iota(\xi))$. Suppose that every element of $\iota(F) \cap \mathrm{End}(A)$ is defined over k. Then we can find a basis of invariant differential forms on A with respect to which $S(\iota(\xi))$

has coefficients in k for every $\xi \in F$ (cf. §2.8); hence the trace of $S(\iota(\xi))$ is contained in k, namely, we have $\sum_i \xi^{\varphi_i} \in k$ for every $\xi \in F$. This proves $k \supset K^*$. Conversely, suppose that $k \supset K^*$ and $(F; \{\varphi_i\})$ is primitive, namely A is simple. Then, by Proposition 6 of §5.1, $\mathrm{End}_{\mathbf{Q}}(A)$ coincides with $\iota(F)$. Let σ be an isomorphism of $\mathbf{C}$ into itself which is the identity on k; we have then $A^\sigma = A$. Our proposition is proved if we show that $\lambda^\sigma = \lambda$ for every $\lambda \in \mathrm{End}(A)$. First we see that $\lambda \to \lambda^\sigma$ gives an automorphism of $\mathrm{End}_{\mathbf{Q}}(A)$. As we have $\mathrm{End}_{\mathbf{Q}}(A) = \iota(F)$, there exists an automorphism τ of F such that $\iota(\xi^\tau) = \iota(\xi)^\sigma$ for every $\xi \in F$. As is remarked in §5.2, we can find n invariant differential forms $\omega_1, \ldots, \omega_n$ on A such that for every $\xi \in F$,

$$\delta\iota(\xi)\omega_i = \xi^{\varphi_i}\omega_i \quad (1 \le i \le n). \tag{2}$$

We have then $\delta\iota(\xi)^\sigma\omega_i^\sigma = \xi^{\varphi_i\sigma}\omega_i^\sigma$ $(1 \le i \le n)$; this shows that $S(\iota(\xi)^\sigma)$ has the characteristic roots $\xi^{\varphi_1\sigma}, \ldots, \xi^{\varphi_n\sigma}$. On the other hand, $S(\iota(\xi^\tau))$ has the characteristic roots $\xi^{\tau\varphi_1}, \ldots, \xi^{\tau\varphi_n}$; so $\{\xi^{\varphi_1\sigma}, \ldots, \xi^{\varphi_n\sigma}\}$ coincides with $\{\xi^{\tau\varphi_1}, \ldots, \xi^{\tau\varphi_n}\}$ as a whole. Since we have assumed $k \supset K^*$, σ fixes every element of K^*; hence, for every $\xi \in F$, we have $\sum \xi^{\varphi_i\sigma} = (\sum \xi^{\varphi_i})^\sigma = \sum \xi^{\varphi_i}$; so by the same argument as in the proof of Proposition 28, we see that $\{\xi^{\varphi_1\sigma}, \ldots, \xi^{\varphi_n\sigma}\}$ coincides with $\{\xi^{\varphi_1}, \ldots, \xi^{\varphi_n}\}$ as a whole. Consequently, $\{\xi^{\varphi_1}, \ldots, \xi^{\varphi_n}\}$ coincides with $\{\xi^{\tau\varphi_1}, \ldots, \xi^{\tau\varphi_n}\}$ as a whole; this implies that $\{\varphi_1, \ldots, \varphi_n\}$ coincides with $\{\tau\varphi_1, \ldots, \tau\varphi_n\}$ as a whole. Now define L, G, S as in Proposition 25; and let H_1 denote the subgroup of G corresponding to F. Take an element of G inducing τ on F, and denote it again by τ. Then as τ gives an automorphism of F, we have $\tau H_1 = H_1\tau$. Take for every i an element γ_i of G inducing φ_i on F. We have then $S = \cup_i H_1\gamma_i$. The above result shows that $\{H_1\tau\gamma_1, \ldots, H_1\tau\gamma_n\}$ coincides with $\{H_1\gamma_1, \ldots, H_1\gamma_n\}$ as a whole. We have therefore,

$$S = \bigcup_i H_1\gamma_i = \bigcup_i H_1\tau\gamma_i = \bigcup_i \tau H_1\gamma_i = \tau S.$$

By our assumption that $(F; \{\varphi_i\})$ is primitive, τ must be contained in H_1 by virtue of Proposition 26. This proves that τ is the identity on F; so we have $\iota(\xi)^\sigma = \iota(\xi)$ for every $\xi \in F$; this completes the proof.

PROPOSITION 31. *Let $(F; \{\varphi_i\})$ be a CM-type and $(K^*; \{\psi_j\})$ its reflex; let (A, ι) be an abelian variety of type $(F; \{\varphi_i\})$ and k a field of definition for (A, ι). Let σ be an isomorphism of k into* $\mathbf{C}$. *Then, σ fixes every element of K^* if and only if there exists a homomorphism of (A, ι) onto (A^σ, ι^σ).*

PROOF. We note that k must contain K^* by virtue of Proposition 30. Extend σ to an isomorphism of $\mathbf{C}$ into itself and denote it again by σ. Take n invariant differential forms ω_i on A for which relation (2) holds for every $\xi \in F$. Then, the ω_i^σ form a basis of invariant differential forms on A^σ; and we have

$\delta\iota^\sigma(\xi)\omega_i^\sigma = \xi^{\varphi_i\sigma}\omega_i^\sigma$. This shows that (A^σ, ι^σ) is of type $(F; \{\varphi_i\sigma\})$. Suppose that there exists a homomorphism λ of (A, ι) onto (A^σ, ι^σ). By Theorem 1 of §2, $\delta\lambda$ is an isomorphism of $\mathfrak{D}_0(A^\sigma)$ onto $\mathfrak{D}_0(A)$; putting $\omega_i' = (\delta\lambda)^{-1}\omega_i$, we see easily $\delta\iota^\sigma(\xi)\omega_i' = \xi^{\varphi_i}\omega_i'$ for every $\xi \in F$. Hence (A^σ, ι^σ) is of type $(F; \{\varphi_i\})$. It follows that $\{\varphi_i\sigma\}$ coincides with $\{\varphi_i\}$ as a whole; so we have, for every $\xi \in F$, $(\sum \xi^{\varphi_i})^\sigma = \sum \xi^{\varphi_i\sigma} = \sum \xi^{\varphi_i}$; namely, σ leaves invariant every element of K^*. This proves the "if" part of the proposition. Conversely, suppose that σ leaves invariant every element of K^*; then, by the same argument as in the proof of Proposition 28, we see that $\{\varphi_i\sigma\}$ coincides with $\{\varphi_i\}$ as a whole; so (A^σ, ι^σ) is of the same type $(F; \{\varphi_i\})$ as (A, ι). By Corollary of Theorem 2 of §6 and Remark below it, we can obtain a homomorphism of (A, ι) onto (A^σ, ι^σ). This completes our proof.

CHAPTER III

Reduction of Constant Fields

The aim of §§9–12 is to complement the theory of reduction modulo $\mathfrak{p}$ of algebraic varieties, given in [33], with a particular interest in abelian varieties, toward the later use. We will first recall definitions and results from the general theory with a slight modification, and then proceed to the main subject. For omitted proofs in §9, we refer to [33].

9. Reduction of Varieties and Cycles

9.1. Places. Let K be a field. We denote by K_∞ the union $K \cup \{\infty\}$ of the set K and one additional element ∞; and we define, besides the operation in K, the operation in K_∞ as follows:

$$a \pm \infty = \infty, \quad a/\infty = 0 \quad \text{for } a \in K,$$

$$a \cdot \infty = a/0 = \infty \quad \text{for } a \in K_\infty, a \neq 0.$$

Let k and k' be two fields; we call a mapping φ of k_∞ onto k'_∞ a *place* of k if it satisfies

$$\varphi(a+b) = \varphi(a) + \varphi(b), \quad \varphi(ab) = \varphi(a)\varphi(b);$$

k' is called the *residue field* of φ. Put

$$\mathfrak{o} = \{x \mid x \in k, \varphi(x) \neq \infty\},$$

$$\mathfrak{p} = \{x \mid x \in k, \varphi(x) = 0\}.$$

Then $\mathfrak{o}$ is a valuation ring of k, $\mathfrak{p}$ is the maximal ideal of $\mathfrak{o}$ and $\mathfrak{o}/\mathfrak{p}$ is isomorphic to k'; so we denote the place also by $\mathfrak{p}$ and $\varphi(x)$ by $\mathfrak{p}(x)$. If there is no fear of confusion, we denote $\mathfrak{p}(x)$ by $\tilde{x}$; similarly, the residue field is denoted by $\mathfrak{p}(k)$ or $\tilde{k}$. We call $\mathfrak{o}$ the *ring of* $\mathfrak{p}$*-integers*. If $F(X)$ is a polynomial with coefficients in $\mathfrak{o}$, $\tilde{F}(X)$ or $F_\mathfrak{p}(X)$ denotes the polynomial whose coefficients are the images of the corresponding coefficients of F by $\mathfrak{p}$. The following lemma is fundamental and well-known. A proof is given in Weil [49].

LEMMA 1. *Let S be a subring of a field k, containing the identity of k. Then, every homomorphism of S into a field k′, which maps the identity of k onto the identity of k′, can be extended to a place of k whose residue field is contained in the algebraic closure of k′.*

We call a place $\mathfrak{p}$ of a field k *trivial* if $\mathfrak{p}$ is the identity mapping of k onto itself. We say that a place $\mathfrak{p}_1$ of a field k_1 is an *extension* of a place $\mathfrak{p}$ of a field k if $k_1 \supset k$ and $\mathfrak{p}_1(a) = \mathfrak{p}(a)$ for every $a \in k$.

9.2. Specializations over a place. For the sake of simplicity, we fix two universal domains $\mathbf{K}$ and $\tilde{\mathbf{K}}$ with the same or different characteristics and deal only with the places defined on a subfield of $\mathbf{K}$ taking values in $\tilde{\mathbf{K}}_\infty$; this restriction is kept until the end of §11. We denote by P^n, $\tilde{P}^n$, S^n, $\tilde{S}^n$ respectively the projective spaces and affine spaces, of dimension n, defined with respect to $\mathbf{K}$ and $\tilde{\mathbf{K}}$. Let $(x) = (x_1, \dots, x_n)$ be a set of n elements in $\mathbf{K}_\infty$ and $(\xi) = (\xi, \dots, \xi_n)$ a set of n elements in $\tilde{\mathbf{K}}_\infty$. We say that (ξ) is a *specialization* of (x) over a place $\mathfrak{p}$ of a field k, and write

$$(x) \to (\xi) \text{ ref. } \mathfrak{p},$$

if there exists an extension $\mathfrak{p}'$ of $\mathfrak{p}$ such that $\mathfrak{p}'(x_i) = \xi_i$ for every i. When $\mathfrak{p}$ is trivial, we write $(x) \to (\xi)$ ref. k. If the x_i are contained in k_∞, then (x) has a unique specialization over $\mathfrak{p}$, which we denote by $\mathfrak{p}(x)$ or $(\tilde{x})$. We can easily verify

$$(x) \to (x') \text{ ref. } k,\ (x') \to (\xi) \text{ ref. } \mathfrak{p} \quad \Rightarrow \quad (x) \to (\xi) \text{ ref. } \mathfrak{p},$$

$$(x) \to (\xi) \text{ ref. } \mathfrak{p},\ (\xi) \to (\xi') \text{ ref. } \tilde{k} \quad \Rightarrow \quad (x) \to (\xi') \text{ ref. } \mathfrak{p}.$$

If we have $(x) \to (\xi)$ ref. $\mathfrak{p}$ and (y) is a set of elements in $\mathbf{K}_\infty$, then there exists a set of elements (η) in $\tilde{\mathbf{K}}_\infty$ such that $(x, y) \to (\xi, \eta)$ ref. $\mathfrak{p}$. This is an immediate consequence of Lemma 1.

We call a set of elements in $\mathbf{K}_\infty$ or $\tilde{\mathbf{K}}_\infty$ *finite* if it does not contain ∞; so, if (x) is finite, (x) is considered as a point of S^n, and the same for $\tilde{S}^n$. Let (ξ) be a specialization of (x) over a place $\mathfrak{p}$ of a field k. Then, if (ξ) is finite, so is (x). Assuming (ξ) to be finite, we denote by $[(x) \to (\xi); \mathfrak{p}]$ (or $[(x) \to (\xi); k]$, if $\mathfrak{p}$ is trivial), the set of elements of the form $a(x)/b(x)$ in $k(x)$, where a and b are polynomials in $\mathfrak{o}[X]$ such that $\tilde{b}(\xi) \neq 0$, $\mathfrak{o}$ being the ring of $\mathfrak{p}$-integers. It is easy to see that $[(x) \to (\xi); \mathfrak{p}]$ is a local ring having $k(x)$ as its quotient field.

9.3. Reduction of affine varieties. We call a place $\mathfrak{p}$ *discrete* if the ring of $\mathfrak{p}$-integers is a discrete valuation ring of rank 1. The theory in [33] concerns reduction of varieties or cycles with respect to a discrete place. It is not difficult to extend the theory to non-discrete places, as is indicated in the appendix of

[36], by means of the following lemma, which is a restatement of Proposition 26 of [33].

LEMMA 2. *Let* $\mathfrak{p}$ *be a discrete place of a field* k *and* (ξ) *a specialization of* (x) *over* $\mathfrak{p}$. *Then there exists a discrete place* $\mathfrak{p}_1$ *of* $k(x)$, *which is an extension of* $\mathfrak{p}$, *such that* $\mathfrak{p}_1(x) = (\xi)$.

COROLLARY. *Let* $\mathfrak{p}$ *be a place (not necessarily discrete) and* (ξ) *a specialization of* (x) *over* $\mathfrak{p}$. *Then there exist a field* k_1 *containing* (x) *and a discrete place* $\mathfrak{p}_1$ *of* k_1 *such that* $\mathfrak{p}_1(x) = (\xi)$.

PROOF. Let k_0 be the prime field contained in k and $\mathfrak{p}_0$ the restriction of $\mathfrak{p}$ to k_0; then we see that $\mathfrak{p}_0$ is discrete and (ξ) is a specialization of (x) over $\mathfrak{p}_0$. Hence by Lemma 2, there exists a discrete place $\mathfrak{p}_1$ of $k_0(x)$ such that $\mathfrak{p}_1(x) = (\xi)$.

In the present treatment, however, we restrict ourselves to discrete places, in order to avoid the complication arising from the generalization; therefore, from now on, *a place or an extension of a place means a discrete one*, except when the contrary is specifically stated.

Let k be a field, $\mathfrak{p}$ a place of k, and $\mathfrak{o}$ the ring of $\mathfrak{p}$-integers. Until the end of §11, we use these symbols always in this sense. Now let V be an algebraic set in the affine space S^n, defined over k. Denote by $\mathfrak{A}$ the set of polynomials $F(X)$ in $\mathfrak{o}[X]$ such that $F(x) = 0$ for every $(x) \in V$; then $\mathfrak{A}$ is an ideal of $\mathfrak{o}[X]$. We denote by $\mathfrak{p}(V)$ or $\tilde{V}$ the set of points (ξ) in $\tilde{S}^n$ such that $\tilde{F}(\xi) = 0$ for all $F \in \mathfrak{A}$. $\mathfrak{p}(V)$ is clearly an algebraic set defined over $\tilde{k}$. We can prove

$$\mathfrak{p}(V) = \{(\alpha) \mid (\alpha) \in \tilde{S}^n, (a) \to (\alpha) \text{ ref. } \mathfrak{p} \quad \text{for some } (a) \in V\}.$$

We call $\mathfrak{p}(V)$ *the reduction of* V *modulo* $\mathfrak{p}$. In general, $\mathfrak{X}$ being an algebro-geometric object defined with respect to k, we call the algebro-geometric object obtained by considering $\mathfrak{X}$ modulo $\mathfrak{p}$ (for which a precise definition will be given in each case) the *reduction of* $\mathfrak{X}$ *modulo* $\mathfrak{p}$, and denote it by $\mathfrak{p}(\mathfrak{X})$ or $\tilde{\mathfrak{X}}$. Let $\mathfrak{p}'$ be an extension of $\mathfrak{p}$; then we can easily verify $\mathfrak{p}(V) = \mathfrak{p}'(V)$. We have also $\mathfrak{p}(\mathfrak{X}) = \mathfrak{p}'(\mathfrak{X})$ for every algebro-geometric object $\mathfrak{X}$, whose reduction modulo $\mathfrak{p}$ is to be defined later. The following relations can be easily verified:

$$\mathfrak{p}(V \cup W) = \mathfrak{p}(V) \cup \mathfrak{p}(W), \qquad \mathfrak{p}(V \cap W) \subset \mathfrak{p}(V) \cap \mathfrak{p}(W),$$

$$\mathfrak{p}(V \times W) = \mathfrak{p}(V) \times \mathfrak{p}(W).$$

It may happen that $\tilde{V}$ is an empty set even if V is not empty. If the components of V are all of the same dimension r, and if $\tilde{V}$ is not empty, then the components of $\tilde{V}$ are all of dimension r.

We shall now define the reduction of cycles in S^n modulo $\mathfrak{p}$. We begin with the cycles of dimension 0. Let Y be a cycle of dimension 0 in S^n, rational over k; let $Y = \sum_{i=1}^{s} m_i a_i$ be its reduced expression, where the m_i are rational integers and the a_i are points of S^n. Consider a specialization

$$(a_1, \ldots, a_s) \to (\alpha_1, \ldots, \alpha_s) \text{ ref. } \mathfrak{p}$$

and remove those α_i having ∞ as one of their coordinates. Then the sum $\sum' m_i\alpha_i$ of the remaining α_i gives a cycle in $\tilde{S}^n$; and it can be proved that this cycle is uniquely determined by Y and $\mathfrak{p}$, and does not depend upon the choice of the above specialization. We call the cycle the *reduction of Y modulo* $\mathfrak{p}$ and denote it by $\mathfrak{p}(Y)$ or $\tilde{Y}$.

Now consider a cycle Z in S^n, of dimension r, rational over k. Let t_{ij}, for $1 \leq i \leq r, 0 \leq j \leq n$, be $r(n+1)$ independent variables over $\bar{k}$, and $\tilde{t}_{ij}$, for $1 \leq i \leq r, 0 \leq j \leq n$, be $r(n+1)$ independent variables over $\tilde{k}$. Let L and $\tilde{L}$ denote the linear varieties defined respectively by

$$\sum_{j=1}^{n} t_{ij} X_j - t_{i0} = 0 \quad (1 \leq i \leq r),$$

$$\sum_{j=1}^{n} \tilde{t}_{ij} X_j - \tilde{t}_{i0} = 0 \quad (1 \leq i \leq r).$$

Then the intersection-product $Z \cdot L$ is defined and is a cycle of dimension 0 in S^n, rational over $k(t_{ij})$. We see that the specialization $(t_{ij}) \to (\tilde{t}_{ij})$ ref. $\mathfrak{p}$ gives a discrete extension $\mathfrak{p}'$ of $\mathfrak{p}$ in $k(t_{ij})$; so we can consider $\mathfrak{p}'(Z \cdot L)$. Now, it can be proved that there exists a cycle $\tilde{Z}$ of dimension r in $\tilde{S}^n$, rational over $\tilde{k}$, such that $\tilde{Z} \cdot \tilde{L} = \mathfrak{p}'(Z \cdot L)$; and such a cycle $\tilde{Z}$ is uniquely determined by Z and $\mathfrak{p}$. We call $\tilde{Z}$ the *reduction of Z modulo* $\mathfrak{p}$ and denote it by $\mathfrak{p}(Z)$. If Z is positive, the components of $\tilde{Z}$ are the components of the reduction of the support of Z.

Let V be an algebraic variety of dimension r in S^n, defined over k and α a point of $\tilde{V}$. We say that α is *simple* on V if there exist $n-r$ polynomials $F_1(X), \ldots, F_{n-r}(X)$ in $\mathfrak{o}[X]$, vanishing on V, such that

$$\operatorname{rank}(\partial \tilde{F}_i / \partial X_j(\alpha)) = n - r.$$

Let $\mathfrak{U}$ be a subvariety of $\tilde{V}$; we say that $\mathfrak{U}$ is *simple* on V if $\mathfrak{U}$ has a point which is simple on V. If x is a generic point of V over k and if a point ξ of $\tilde{V}$ is simple on V, then $[x \to \xi; \mathfrak{p}]$ is a regular local ring by virtue of the theorem in [24, p. 206], so that it is integrally closed. Let $\mathfrak{V}$ be a component of $\tilde{V}$. Then, $\mathfrak{V}$ is simple on V if and only if $\mathfrak{V}$ is of multiplicity 1 in the cycle $\mathfrak{p}(V)$. Suppose that a component $\mathfrak{V}$ of $\tilde{V}$ is simple on V; then, a point ξ of $\mathfrak{V}$ is simple on V if and only if ξ is simple on $\mathfrak{V}$; if x is generic on V over k and ξ is generic on

$\mathfrak{V}$ over the algebraic closure of $\tilde{k}$, then $[x \to \xi; \mathfrak{p}]$ is a discrete valuation ring of rank 1 and every prime element of $\mathfrak{o}$ is a prime element of $[x \to \xi; \mathfrak{p}]$.

Let V and W be two affine varieties, defined over k, and T a birational correspondence between V and W, defined over k. Let $x \times y$ be a generic point of T over k and ξ a point of $\tilde{V}$. We say that T is *regular* at ξ if the coordinates of y are all contained in $[x \to \xi; \mathfrak{p}]$; if that is so, there exists one and only one point η on $\tilde{W}$ such that $\xi \times \eta \in \tilde{T}$. Let $\mathfrak{p}'$ be an extension of $\mathfrak{p}$; then T is regular at a point ξ of $\tilde{V}$ with respect to $\mathfrak{p}$ if and only if T is so with respect to $\mathfrak{p}'$. Given a point $\xi \times \eta$ in $\tilde{T}$, we say that ξ and η are *regularly corresponding points* by T if T is regular at ξ and at η.

9.4. Reduction of abstract varieties. Let $[V_\alpha; F_\alpha; T_{\beta\alpha}]$ be an abstract variety, defined over k; let, for each α, $\mathfrak{F}_\alpha$ be an algebraic subset in $\tilde{V}_\alpha$, other than $\tilde{V}_\alpha$, defined over $\tilde{k}$, containing $\tilde{F}_\alpha$. We call the system

$$[V_\alpha; F_\alpha; \mathfrak{F}_\alpha; T_{\beta\alpha}]$$

a $\mathfrak{p}$-*variety* if the following condition is satisfied:

(V) *If ξ is a point in $\tilde{V}_\alpha - \mathfrak{F}_\alpha$ and η is a point in $\tilde{V}_\beta - \mathfrak{F}_\beta$ such that $\xi \times \eta$ is in $\tilde{T}_{\beta\alpha}$, then ξ and η are regularly corresponding points by $T_{\beta\alpha}$.*

Given a $\mathfrak{p}$-variety $V = [V_\alpha; F_\alpha; \mathfrak{F}_\alpha; T_{\beta\alpha}]$, let $[U_{\alpha(\lambda)}; F_{\alpha(\lambda)} \cap U_{\alpha(\lambda)}; R_{\mu\lambda}]$ be a subvariety of $[V_\alpha; F_\alpha; T_{\beta\alpha}]$, defined over k. Then the system

$$[U_{\alpha(\lambda)}; F_{\alpha(\lambda)} \cap U_{\alpha(\lambda)}; \mathfrak{F}_{\alpha(\lambda)} \cap \tilde{U}_{\alpha(\lambda)}; R_{\mu\lambda}]$$

defines a $\mathfrak{p}$-variety, which we call a subvariety of the $\mathfrak{p}$-variety V. Let $W = [W_\gamma; G_\gamma; \mathfrak{G}_\gamma; S_{\delta\gamma}]$ be another $\mathfrak{p}$-variety and $[V_\alpha \times W_\gamma; H_{\alpha\gamma}; U_{\beta\delta,\alpha\gamma}]$ the product-variety of the abstract varieties $[V_\alpha; F_\alpha; T_{\beta\alpha}]$ and $[W_\gamma; G_\gamma; S_{\delta\gamma}]$. Put $\mathfrak{H}_{\alpha\gamma} = (\mathfrak{F}_\alpha \times \tilde{W}_\gamma) \cup (\tilde{V}_\alpha \times \mathfrak{G}_\gamma)$. Then the system $[V_\alpha \times W_\gamma; H_{\alpha\gamma}; \mathfrak{H}_{\alpha\gamma}; U_{\beta\delta,\alpha\gamma}]$ defines a $\mathfrak{p}$-variety which will be called the product variety of the $\mathfrak{p}$-varieties V and W. Every projective variety defined over k defines a uniquely determined $\mathfrak{p}$-variety with empty $\mathfrak{F}_\alpha$.

Let $V = [V_\alpha; F_\alpha; \mathfrak{F}_\alpha; T_{\beta\alpha}]$ be a $\mathfrak{p}$-variety and the x_α corresponding generic points of the V_α over k by the $T_{\beta\alpha}$; let $(\xi_1, \ldots, \xi_h)$ be a specialization of $(x_1, \ldots, x_h)$ over $\mathfrak{p}$. By a full set of representatives attached to $\tilde{V}$, we understand the set $(\xi_{\alpha_1}, \ldots, \xi_{\alpha_s})$ of all the ξ_α which are finite and not in $\mathfrak{F}_\alpha$. We say that V is $\mathfrak{p}$-*complete* if no full set of representatives attached to $\tilde{V}$ is empty. The following facts can easily be proved: (i) the underlying abstract variety of a $\mathfrak{p}$-complete $\mathfrak{p}$-variety is complete; (ii) every subvariety of a $\mathfrak{p}$-complete $\mathfrak{p}$-variety is $\mathfrak{p}$-complete; (iii) the product of $\mathfrak{p}$-complete $\mathfrak{p}$-varieties is $\mathfrak{p}$-complete; (iv) every projective variety defined over k defines a $\mathfrak{p}$-complete $\mathfrak{p}$-variety.

Given V as above, let $\mathfrak{W} = [\mathfrak{W}_\lambda; \mathfrak{G}_\lambda; \mathfrak{S}_{\mu\lambda}]$ be an abstract variety defined over an extension $\tilde{k}_1$ of $\tilde{k}$. We say that $\mathfrak{W}$ is a *variety in* $\tilde{V}$ if the following conditions are satisfied: (i) there exists a full set of representatives $(\xi_{\alpha(1)}, \ldots, \xi_{\alpha(s)})$ attached to $\tilde{V}$ such that, for every λ, $\mathfrak{W}_\lambda$ is the locus of $\xi_{\alpha(\lambda)}$ over $\tilde{k}_1$; (ii) $\mathfrak{G}_\lambda = \mathfrak{W}_\lambda \cap \mathfrak{F}_{\alpha(\lambda)}$; (iii) $\mathfrak{S}_{\mu\lambda}$ is the variety in $\tilde{T}_{\alpha(\mu)\alpha(\lambda)}$ with the projection $\mathfrak{W}_\lambda$ on $\tilde{V}_{\alpha(\lambda)}$ and $\mathfrak{W}_\mu$ on $\tilde{V}_{\alpha(\mu)}$. We call $\mathfrak{W}_\lambda$ the *representatives* of $\mathfrak{W}$ in $V_{\alpha(\lambda)}$. For every variety $\mathfrak{W}_\alpha$ in $\tilde{V}_\alpha$ which is not contained in $\mathfrak{F}_\alpha$, there exists one and only one variety $\mathfrak{W}$ in $\tilde{V}$ such that $\mathfrak{W}_\alpha$ is a representative of $\mathfrak{W}$ in V_α. By a *point* in $\tilde{V}$ we understand a 0-dimensional variety in $\tilde{V}$. It is obvious that all the representatives of a point in $\tilde{V}$ form a full set of representatives attached to $\tilde{V}$ and vice versa. Let x be a point in V and ξ a point in $\tilde{V}$. We say that ξ is a *specialization of x over* $\mathfrak{p}$ if x and ξ have representatives x_α and ξ_α in some V_α such that ξ_α is a specialization of x_α over $\mathfrak{p}$; since $[x_\alpha \to \xi_\alpha; \mathfrak{p}]$ is independent of α, we denote it by $[x \to \xi; \mathfrak{p}]$. A variety $\mathfrak{W}$ in $\tilde{V}$ is called *simple* on V if a representative $\mathfrak{W}_\alpha$ of $\mathfrak{W}$ is simple on V_α.

Let $V = [V_\alpha; F_\alpha; \mathfrak{F}_\alpha; T_{\beta\alpha}]$ be a $\mathfrak{p}$-variety of dimension n. V is called $\mathfrak{p}$-*simple* if there exists one and only one variety $\mathfrak{V}$ of dimension n in $\tilde{V}$ and if every representative $\mathfrak{V}_\alpha$ of $\mathfrak{V}$ in V_α is of multiplicity 1 in the cycle $\mathfrak{p}(V_\alpha)$. If that is so, $\mathfrak{V}$ is an abstract variety defined over $\tilde{k}$. We call the abstract variety $\mathfrak{V}$ the *reduction of V modulo* $\mathfrak{p}$ and denote it by $\mathfrak{p}(V)$.

V and $\mathfrak{V}$ being as above, we see that every subvariety of $\mathfrak{V}$ is a variety in $\tilde{V}$ and vice versa; and $\mathfrak{W}$ is simple on $\mathfrak{V}$ if and only if $\mathfrak{W}$ is simple on V. Let W be an algebraic subset of V, rational over k. We denote by $\mathfrak{p}(W)$ or $\tilde{W}$ the set of points of $\mathfrak{V}$ which are specializations of points in W. Then $\mathfrak{p}(W)$ is an algebraic subset of $\mathfrak{V}$, rational over $\tilde{k}$. Let $X = \sum_i m_i A_i$ be a cycle on V, rational over k, where the A_i are subvarieties of V. Let k' be a field of definition for the A_i containing k and $\mathfrak{p}'$ an extension of $\mathfrak{p}$ in k'. Take, for each i, a representative $A_{i\alpha}$ of A_i in V_α and put $\mathfrak{p}'(A_{i\alpha}) = \sum_j l_j \mathfrak{B}_{j\alpha}$, where the $\mathfrak{B}_{j\alpha}$ are varieties in $\tilde{V}_\alpha$. Put $\mathfrak{A}_i = \sum' l_j \mathfrak{B}_j$, where $\mathfrak{B}_j$ is the variety in $\mathfrak{V}$ having $\mathfrak{B}_{j\alpha}$ as the representative in V_α and the sum is taken over all $\mathfrak{B}_j$ which is simple on $\mathfrak{V}$. Then the cycle $\sum m_i \mathfrak{A}_i$ on $\mathfrak{V}$ is rational over $\tilde{k}$ and determined only by X and $\mathfrak{p}$. We denote it by $\mathfrak{p}(X)$ or $\tilde{X}$. The reduction of cycles preserves the operation on cycles:

PROPOSITION 1. (1) *Let V be a $\mathfrak{p}$-simple $\mathfrak{p}$-variety; let X and Y be positive cycles on V, rational over k such that the intersection-products $X \cdot Y$ and $\mathfrak{p}(X) \cdot \mathfrak{p}(Y)$ are defined. Then we have*

$$\mathfrak{p}(X \cdot Y) = \mathfrak{p}(X) \cdot \mathfrak{p}(Y).$$

(2) *Let V and W be $\mathfrak{p}$-simple $\mathfrak{p}$-varieties, X a cycle on V and Y a cycle on W, both rational over k. Then $V \times W$ is $\mathfrak{p}$-simple and we have*

$$\mathfrak{p}(X \times Y) = \mathfrak{p}(X) \times \mathfrak{p}(Y).$$

(3) *Let V and W be $\mathfrak{p}$-simple $\mathfrak{p}$-varieties; denote by $\mathfrak{V}$ and $\mathfrak{W}$ the reduction of V and W modulo $\mathfrak{p}$, respectively. Suppose that W is $\mathfrak{p}$-complete and $\mathfrak{W}$ has no multiple point. If X is a cycle on $V \times W$, rational over k, then we have*

$$\mathfrak{p}(\mathrm{pr}_V(X)) = \mathrm{pr}_{\mathfrak{V}}(\mathfrak{p}(X)).$$

9.5. Remark. Here, we straighten out a few incomplete points in the paper [33]. First of all, the proof of Proposition 17 of this paper was not given, because it was viewed as an easy translation of [44, Chapter V, Proposition 19]. The first part, which asserts $\partial\bar{F}/\partial Z(\eta, \zeta) \neq 0$, can indeed be proved in the same fashion. However, it is hardly possible to prove the remaining part by the same argument as in [44]. A complete proof is given in Theorem and Lemma 1 on p. 206 of [24].

Let us now list some more corrections to [33]:

P. 150, the bottom line: "Obviously, (η) is" should read "Obviously, (ξ) is."

P. 151, line 1: (τ_{ij}, τ_i) should read $(\delta_{ij}, \varepsilon_i)$.

P. 155: Corollary to Theorem 10 should read as follows:

COROLLARY. *Let V be a variety defined over k and $\mathfrak{V}$ a component of $\bar{V}$. If $\mathfrak{V}$ is simple on V, then $\mu(V, \mathfrak{V}) = 1$ and $[\mathfrak{V} : \kappa]_i = 1$.*

We also note here a correction to [24]:

P. 191, Proposition 2: Insert the sentence "Suppose that f is defined along $\tilde{V}$" after "$f(x)f(y)^{-1}$."

10. Reduction of Rational Mappings and Differential Forms

10.1. Reduction of rational mappings. Let V and W be two $\mathfrak{p}$-varieties and f a rational mapping of V into W defined over k. Let ξ be a point in $\tilde{V}$ and x a generic point of V over k; put $f(x) = y$. We say that f is *defined* at ξ if there exists a point η on $\tilde{W}$ such that

$$[y \to \eta; \mathfrak{p}] \subset [x \to \xi; \mathfrak{p}];$$

this definition is independent of the choice of x and any extension of $\mathfrak{p}$. It is easy to see that η is uniquely determined by f and ξ. We shall write $f(\xi) = \eta$.

V, W and f being as above, suppose that V and W are $\mathfrak{p}$-simple. Let T be the graph of f, $x \times y$ a generic point of T over k and ξ a generic point of $\tilde{V}$ over $\tilde{k}$. Since $[x \to \xi; \mathfrak{p}]$ is a valuation ring, f is defined at ξ whenever W is $\mathfrak{p}$-complete. Suppose that f is defined at some point in $\tilde{V}$; then f is defined at ξ; put $f(\xi) = \eta$. Define a rational mapping $\tilde{f}$ of $\tilde{V}$ into $\tilde{W}$ by $\tilde{f}(\xi) = \eta$ with respect to $\tilde{k}$. We call $\tilde{f}$ the *reduction of f modulo* $\mathfrak{p}$. We see easily that if f is defined at $\alpha \in \tilde{V}$, $\tilde{f}$ is also defined at α and $f(\alpha) = \tilde{f}(\alpha)$. Denote by $\mathfrak{T}$

the graph of $\tilde{f}$. Then clearly $\tilde{T} \supset \mathfrak{T}$, and $\mathfrak{T}$ is the only component of $\tilde{T}$ whose projection on $\tilde{V}$ is $\tilde{V}$.

PROPOSITION 2. *Let V and W be two $\mathfrak{p}$-simple $\mathfrak{p}$-varieties, and f a rational mapping of V into W, defined over k. Suppose that f is everywhere defined on $\tilde{V}$. Let $\tilde{f}$ be the reduction of f modulo $\mathfrak{p}$; denote by T and $\mathfrak{T}$ the graphs of f and $\tilde{f}$, respectively. If $\mathfrak{T}$ is simple on $V \times W$, we have $\mathfrak{p}(T) = \mathfrak{T}$, where T and $\mathfrak{T}$ are considered as cycles.*

PROOF. The letters x, y, ξ, η being as above, let $\alpha \times \beta$ be a point of $\tilde{T}$. As f is defined at α, we have $\beta = f(\alpha) = \tilde{f}(\alpha)$, so that $\alpha \times \beta$ is a point of $\mathfrak{T}$. Hence $\mathfrak{T}$ is the only component of $\tilde{T}$. Put $\mathfrak{p}(T) = m\mathfrak{T}$, where m is a positive integer. If W is $\mathfrak{p}$-complete and $\mathfrak{p}(W)$ has no multiple point, we have $m = 1$ by (3) of Proposition 1. We can prove $m = 1$ without such assumptions on W, applying Theorem 12 of [33] to a representative of T.

Let U, V and W be $\mathfrak{p}$-simple $\mathfrak{p}$-varieties; let f and g be rational mappings of U into V and of V into W, respectively. Suppose that f is defined at a point ξ on $\tilde{U}$ and g is defined at $f(\xi)$. Then it is easy to see that the rational mapping $g \circ f$ is defined at ξ and $(g \circ f)(\xi) = g(f(\xi))$. We see easily $g(f(\xi)) = g(\tilde{f}(\xi)) = \tilde{g}(\tilde{f}(\xi))$. This shows

$$\widetilde{g \circ f} = \tilde{g} \circ \tilde{f}.$$

10.2. Reduction of functions Let V be a $\mathfrak{p}$-simple variety and f a generalized function on V, defined over k, namely, a rational mapping of V into the projective space P^1 of dimension 1. Then $\tilde{f}$ gives a rational mapping of $\tilde{V}$ into P^1. We say that f is $\mathfrak{p}$-*finite* if $\tilde{f}$ is not the constant ∞.

PROPOSITION 3. *Let V be a $\mathfrak{p}$-simple $\mathfrak{p}$-variety, and f a function on V, defined over k. If f is $\mathfrak{p}$-finite and $\tilde{f}$ is not the constant 0, we have*

$$\mathfrak{p}(\mathrm{div}(f)) = \mathrm{div}(\tilde{f}).$$

This is a restatement of Theorem 20 of [33].

Now denote by k^* the completion of k with respect to $\mathfrak{p}$ and by $\mathfrak{o}^*$, $\mathfrak{p}^*$ the closure of $\mathfrak{o}$, $\mathfrak{p}$ in k^*; let μ denote the normalized discrete valuation of k^*; namely, μ is the mapping of k^* into $\mathbf{Z} \cup \{\infty\}$ such that

$$\mu(xy) = \mu(x) + \mu(y), \qquad \mu(x + y) \geq \mathrm{Min}\{\mu(x), \mu(y)\},$$

$$\mathfrak{o}^* = \{a \mid a \in k^*, \mu(a) \geq 0\},$$

and $\mu(a) = 1$ for every prime element a of $\mathfrak{o}^*$.

LEMMA 3. *Let $\mathcal{M}$ be a vector space over k^* of dimension n. Let λ be a mapping of $\mathcal{M}$ into $\mathbf{R} \cup \{\infty\}$ such that, for every $f, g \in \mathcal{M}$ and every $a \in k^*$:*

(i) $\lambda(f) = \infty \Leftrightarrow f = 0$.
(ii) $\lambda(f + g) \geq \operatorname{Min}\{\lambda(f), \lambda(g)\}$.
(iii) $\lambda(af) = \mu(a) + \lambda(f)$.

Put $\mathcal{M}_0 = \{f \mid f \in \mathcal{M}, \lambda(f) \geq 0\}$, $\mathcal{N} = \{f \mid f \in \mathcal{M}, \lambda(f) \geq 1\}$. Then $\mathcal{M}_0$ is a free $\mathfrak{o}^*$-module of rank n, and $\mathcal{M}_0/\mathcal{N}$ is a vector space of dimension n over the residue field $\mathfrak{p}(k)$.

PROOF. It is easy to see that $\mathcal{M}_0$ forms an $\mathfrak{o}^*$-module, $\mathcal{M} = k^*\mathcal{M}_0$ and $\mathcal{M}_0/\mathcal{N}$ can be viewed as a vector space over $\tilde{k}$. Let $f_1, \ldots, f_m$ be elements of $\mathcal{M}_0$ which define linearly independent elements over $\tilde{k}$ in $\mathcal{M}_0/\mathcal{N}$. Suppose $\sum_i c_i f_i = 0$ with $c_i \in k^*$. If some c_i is nonzero, division by c_j with the least value of $\mu(c_j)$ allows us to assume that $c_i \in \mathfrak{o}^*$ for every i and $c_j = 1$ for some j. But that produces a nontrivial linear relation for the f_i modulo $\mathcal{N}$. Thus they must be linearly independent over k^*, and hence $m \leq n$. Take now the f_i so that they define a basis of $\mathcal{M}_0/\mathcal{N}$ over $\tilde{k}$. We are going to show that $\mathcal{M}_0 = \mathfrak{o}^* f_1 + \cdots + \mathfrak{o}^* f_m$. Take and fix a prime element π of $\mathfrak{o}^*$. Let g be an element of $\mathcal{M}_0$. Then there exist m elements a_{oi} of $\mathfrak{o}^*$ such that $g - \sum_i a_{oi} f_i \in \mathcal{N}$. Put $g = \sum_i a_{oi} f_i + \pi g_1$; then g_1 is contained in $\mathcal{M}_0$. Applying the same argument to g_1, we find m elements a_{1i} of $\mathfrak{o}^*$ and an element g_2 of $\mathcal{M}_0$ such that $g_1 = \sum_i a_{1i} f_i + \pi g_2$. Repeating this procedure, we obtain m sequences $\{a_{\nu i}\}$ of elements of $\mathfrak{o}^*$ and a sequence $\{g_\nu\}$ of elements of $\mathcal{M}_0$ such that $g_\nu = \sum_i a_{\nu i} f_i + \pi g_{\nu+1}$. Since k^* is complete, the series $\sum_{\nu=0}^{\infty} a_{\nu i} \pi^\nu$, for each i, has a meaning and defines an element b_i of $\mathfrak{o}^*$. Then, it can be easily verified that $g = \sum b_i f_i$. This proves that $\mathcal{M}_0 = \mathfrak{o}^* f_1 + \cdots + \mathfrak{o}^* f_m$. As we have $\mathcal{M} = k^*\mathcal{M}_0$, we must have $m = n$; this proves our lemma.

Let V be a $\mathfrak{p}$-simple $\mathfrak{p}$-variety and X a divisor of V, rational over k. Consider the set $L(X; k)$ of the functions f on V, defined over k, such that $\operatorname{div}(f) \succ -X$. Denote by $L_0(X; k)$ the set of $\mathfrak{p}$-finite elements in $L(X; k)$. Suppose that $L(X; k)$ is of a finite dimension n over k. We shall now prove that $L_0(X; k)$ is a free $\mathfrak{o}$-module of rank n. The symbols k^*, $\mathfrak{o}^*$ and $\mathfrak{p}^*$ being as above, we can choose our universal domain $\mathbf{K}$ so that it contains k^* as a subfield. Let $k(V)$ and $k^*(V)$ denote the fields of functions on V defined over k and over k^*, respectively. Then, $k(V)$ and k^* are linearly disjoint over k; so we can define a tensor product $\mathcal{M}$ of $L(X; k)$ and k^* over k as a submodule of $k^*(V)$. Let x be a generic point of V over k^* and ξ a generic point of $\tilde{V}$ over $\mathfrak{p}^*(k^*)$. Then $[x \to \xi; \mathfrak{p}^*]$ is a discrete valuation ring and every prime element of $\mathfrak{o}$ gives a prime element of that valuation ring. Therefore, considering the isomorphism between $k^*(x)$ and $k^*(V)$, we obtain a discrete valuation λ of $k^*(V)$, which satisfies the conditions of Lemma 3. Hence, by that lemma, if we denote by

$\mathcal{M}_0$ the set of $\mathfrak{p}^*$-finite elements in $\mathcal{M}$, $\mathcal{M}_0$ has a basis $\{g_1, \dots, g_n\}$ over $\mathfrak{o}^*$. Let $\{h_1, \dots, h_n\}$ be a basis of $L(X; k)$ over k; then the g_i are expressed in the form $g_i = \sum c_{ij} h_j$ with c_{ij} in k^*. As k is dense in k^*, there exist elements d_{ij} in k such that the matrix $(d_{ij})(c_{ij})^{-1}$ is congruent with the unit matrix modulo $\mathfrak{p}^*$. Put $f_i = \sum d_{ij} h_j$. Then we see that $\{f_1, \dots, f_n\}$ gives a basis of $\mathcal{M}_0$ over $\mathfrak{o}^*$, and hence a basis of $L_0(X; k)$ over $\mathfrak{o}$. Thus we have proved that $L_0(X; k)$ is a free $\mathfrak{o}$-module of rank n. By Proposition 3, we have $\operatorname{div}(\tilde{f}_i) \succ -\mathfrak{p}(X)$. As the $\tilde{f}_i$ are linearly independent over $\tilde{k}$, we obtain the inequality

$$l(X) \leq l(\mathfrak{p}(X)). \tag{1}$$

10.3. Local parameters at a point of $\tilde{V}$. Let V be a $\mathfrak{p}$-simple $\mathfrak{p}$-variety of dimension r; let ξ be a simple point on $\tilde{V}$; then ξ is simple on V. We call a set of r functions $\varphi_1, \dots, \varphi_r$ in $k(V)$ a *system of local parameters* for V at ξ, defined over k, if the following conditions are satisfied:

(L_0 1) *$k(V)$ is separably algebraic over $k(\varphi_1, \dots, \varphi_r)$.*
(L_0 2) *The φ_i are all defined and finite at ξ.*
(L_0 3) *For every f in $k(V)$, defined and finite at ξ, $\partial f/\partial \varphi_i$ is defined and finite at ξ for every i.* (For the notation $\partial/\partial \varphi_i$, see §2.2.)

PROPOSITION 4. *Given V and ξ as above, let x be a generic point of V over k and V_α, x_α, ξ_α be representatives of V, x, ξ; and let S^n be the ambient space for V_α. Then, r elements $\varphi_1, \dots, \varphi_r$ in $k(V)$ form a system of local parameters at ξ if and only if the following conditions are satisfied:*

(L'_0 1) *The φ_i are defined and finite at ξ.*
(L'_0 2) *There exists a set of n polynomials $F_i(X_1, \dots, X_n, T_1, \dots, T_r)$ in $\mathfrak{o}[X_1, \dots, X_n, T_1, \dots, T_r]$ such that $F_i(x_\alpha, t) = 0$ for $1 \leq i \leq n$ and*

$$\det\left(\frac{\partial \tilde{F}_i}{\partial X_j}(\xi_\alpha, \tau)\right) \neq 0,$$

where $t_i = \varphi_i(x)$, $\tau_i = \varphi_i(\xi)$ for $1 \leq i \leq r$.

PROOF. We first prove the "if" part. Let $(x_{\alpha 1}, \dots, x_{\alpha n})$ be the coordinates of x_α and f_i the function on V defined by $f_i(x) = x_{\alpha i}$, with respect to k. By Corollary of Theorem 1 of Weil [44, Chapter I], ($\mathrm{L}'_0$2) implies that $k(x)$ is separably algebraic over $k(t)$; so $k(V) = k(f_1, \dots, f_n)$ is separably algebraic over $k(\varphi_1, \dots, \varphi_r)$. Differentiating the equations $F_h(x_\alpha, t) = 0$, we have

$$\sum_{i=1}^{n} \frac{\partial F_h}{\partial X_i}(x_\alpha, t)\frac{\partial f_i}{\partial \varphi_j}(x) + \frac{\partial F_h}{\partial T_j}(x_\alpha, t) = 0 \quad (1 \leq h \leq n, 1 \leq j \leq r). \tag{2}$$

As we have $\det(\partial\tilde{F}_h/\partial X_j(\xi_\alpha, \tau)) \neq 0$, we observe that the functions $\partial f_i/\partial\varphi_j$ are all defined and finite at ξ. If g is an element of $k(V)$ defined and finite at ξ, we get a representation $g(x) = G(x_\alpha)/H(x_\alpha)$, where $G(X)$ and $H(X)$ are polynomials in $\mathfrak{o}[X]$ such that $\tilde{H}(\xi_\alpha) \neq 0$. We have then

$$(3)\qquad \frac{\partial g}{\partial\varphi_j}(x) = H(x_\alpha)^{-2} \cdot\left\{H(x_\alpha)\sum_{i=1}^{n}\frac{\partial G}{\partial X_i}(x_\alpha)\frac{\partial f_i}{\partial\varphi_j}(x) - G(x_\alpha)\sum_{i=1}^{n}\frac{\partial H}{\partial X_i}(x_\alpha)\frac{\partial f_i}{\partial\varphi_j}(x)\right\}$$

for $1 \leq j \leq n$. This shows that the functions $\partial g/\partial\varphi_j$ are all defined and finite at ξ. The "if" part is thereby proved. Conversely, suppose that the φ_i satisfy conditions ($L_0$1-3). As ξ_α is simple on V_α, there exist $n-r$ polynomials $A_\nu(X)$ in $\mathfrak{o}[X]$ such that $A_\nu(x_\alpha) = 0$ and

$$(4)\qquad \operatorname{rank}\left(\frac{\partial\tilde{A}_\nu}{\partial X_i}(\xi_\alpha)\right) = n - r.$$

From the relation $A_\nu(x_\alpha) = 0$, it follows that

$$(5)\qquad \sum_{i=1}^{n}\frac{\partial\tilde{A}_\nu}{\partial X_i}(\xi_\alpha)\frac{\partial f_i}{\partial\varphi_j}(\xi) = 0 \qquad (1 \leq \nu \leq n-r,\ 1 \leq j \leq r).$$

By (L_0 2), each φ_i has an expression $\varphi_i(x) = B_i(x_\alpha)/C_i(x_\alpha)$, where B_i and C_i are polynomials in $\mathfrak{o}[X]$ such that $\tilde{C}_i(\xi_\alpha) \neq 0$; then, differentiating $B_h(x_\alpha) - \varphi_h(x)C_h(x_\alpha) = 0$ and substituting ξ for x, we have

$$(6)\qquad \sum_{i=1}^{n}\left[\frac{\partial\tilde{B}_h}{\partial X_i}(\xi_\alpha) - \varphi_h(\xi)\frac{\partial\tilde{C}_h}{\partial X_i}(\xi_\alpha)\right]\frac{\partial f_i}{\partial\varphi_j}(\xi) = \delta_{hj}\tilde{C}_h(\xi_\alpha),$$

where $\delta_{hj} = 0$ or 1 according as $h \neq j$ or $h = j$. Put

$$\begin{aligned} F_h(X,T) &= B_h(X) - T_h C_h(X) && (1 \leq h \leq r),\\ F_{r+\nu}(X,T) &= A_\nu(X) && (1 \leq \nu \leq n-r). \end{aligned}$$

We have then $F_i(x_\alpha, t) = 0$ for $1 \leq i \leq n$, and, by the relations (4), (5), (6),

$$\det\left(\frac{\partial\tilde{F}_i}{\partial X_j}(\xi_\alpha, \tau)\right) \neq 0.$$

This completes proof.

We have to show the existence of a system of local parameters. Let the notation be the same as in Proposition 4. As ξ_α is simple on V_α, there exist $n-r$ polynomials $G_i(X)$ in $\mathfrak{o}[X]$ such that $G_i(x_\alpha)=0$ and

$$\operatorname{rank}\left(\frac{\partial \tilde{G}_i}{\partial X_j}(\xi_\alpha)\right)=n-r.$$

We can find r linear forms $H_i(X)=\sum_{j=1}^{n} c_{ij}X_j$ $(1\leq i\leq r)$ with c_{ij} in $\mathfrak{o}$, such that

$$\det\begin{bmatrix}\partial \tilde{G}_i/\partial X_j(\xi_\alpha)\\ \tilde{c}_{ij}\end{bmatrix}\neq 0.$$

Let φ_i, for each i, be the function on V defined by $\varphi_i(x)=\sum_{j=1}^{n} c_{ij}x_{\alpha j}$ with respect to k, where $(x_{\alpha 1},\dots,x_{\alpha n})$ denotes the coordinates of x_α. Put

$$\begin{aligned} F_h(X,T)&=T_h-\sum c_{hj}X_j && (1\leq h\leq r),\\ F_{r+i}(X,T)&=G_i(X) && (1\leq i\leq n-r).\end{aligned}$$

Then it can be easily verified that the set $\{\varphi_1,\dots,\varphi_r\}$ satisfies conditions (L_0'1-2) of the above proposition for these F_i; so the φ_i are local parameters at ξ. This means in particular that suitably chosen r coordinate functions $x_{\alpha j}$ form a system of local parameters at ξ.

PROPOSITION 5. *Let V be a $\mathfrak{p}$-simple $\mathfrak{p}$-variety, ξ a simple point on $\tilde{V}$, and $\{\varphi_1,\dots,\varphi_r\}$ a system of local parameters for V at ξ, defined over k. Then $\{\tilde{\varphi}_1,\dots,\tilde{\varphi}_r\}$ is a system of local parameters for $\tilde{V}$ at ξ. Moreover, if f is an element of $k(V)$, defined and finite at ξ, then f and the $\partial f/\partial\varphi_i$ are all $\mathfrak{p}$-finite and we have, for every i,*

$$\widetilde{\frac{\partial f}{\partial \varphi_i}}=\frac{\partial \tilde{f}}{\partial \tilde{\varphi}_i}. \tag{7}$$

PROOF. The first assertion follows from Proposition 4. Let f be an element of $k(V)$, defined and finite at ξ. Then by (L_0 3), the $\partial f/\partial\varphi_j$ are defined and finite at ξ. It follows that f and the $\partial f/\partial\varphi_i$ are all defined and finite at a generic point η of $\tilde{V}$ over $\tilde{k}$; this implies that f and the $\partial f/\partial\varphi_i$ are all $\mathfrak{p}$-finite. Considering equations (2) and (3) in the proof of Proposition 4 modulo the maximal ideal of the valuation ring $[x\to\eta;\,\mathfrak{p}]$, we obtain equality (7).

10.4. Reduction of differential forms. Given V as in §10.3, let ω be a differential form on V, defined over k. We say that ω is $\mathfrak{p}$-*finite* if it has an expression

$$\omega=\sum_{(i)} f_{(i)}dg_{i_1}\wedge\cdots\wedge dg_{i_s},$$

where the $f_{(i)}$ and the g_i are $\mathfrak{p}$-finite elements of $k(V)$. We say that ω is *finite* at a simple point ξ of $\tilde{V}$ if the $f_{(i)}$ and the g_i in the above expression can be taken in such a way that they are all defined and finite at ξ. Clearly ω is $\mathfrak{p}$-finite if and only if it is finite at a generic point on $\tilde{V}$ over $\tilde{k}$. Let $\{\varphi_1, \ldots, \varphi_r\}$ be a system of local parameters at ξ. We have then

$$\omega = \sum_{j_1 < \cdots < j_s} h_{j_1 \cdots j_s} d\varphi_{j_1} \wedge \cdots \wedge d\varphi_{j_s},$$

where

$$h_{j_1 \cdots j_s} = \sum_{(i)} \sum_{(l)} \varepsilon \begin{pmatrix} l_1 \cdots l_s \\ j_1 \cdots j_s \end{pmatrix} f_{(i)} \frac{\partial g_{i_1}}{\partial \varphi_{l_1}} \cdots \frac{\partial g_{i_s}}{\partial \varphi_{l_s}},$$

$\varepsilon\binom{l_1 \cdots l_s}{j_1 \cdots j_s}$ denoting the sign of the permutation $\binom{l_1 \cdots l_s}{j_1 \cdots j_s}$. By ($L_0$ 3), ω is finite at ξ if and only if the $h_{(j)}$ are defined and finite at ξ. Supposing that the $f_{(i)}$ and the g_i are $\mathfrak{p}$-finite, we have, by Proposition 5,

$$\tilde{h}_{j_1 \cdots j_s} = \sum_{(i)} \sum_{(l)} \varepsilon \begin{pmatrix} l_1 \cdots l_s \\ j_1 \cdots j_s \end{pmatrix} \tilde{f}_{(i)} \frac{\partial \tilde{g}_{i_1}}{\partial \tilde{\varphi}_{l_1}} \cdots \frac{\partial \tilde{g}_{i_s}}{\partial \tilde{\varphi}_{l_s}},$$

so that

(8) $$\tilde{f}_{(i)} d\tilde{g}_{i_1} \wedge \cdots \wedge d\tilde{g}_{i_s} = \sum_{j_1 < \cdots < j_s} \tilde{h}_{(j)} d\tilde{\varphi}_{j_1} \wedge \cdots \wedge d\tilde{\varphi}_{j_s}.$$

As the $h_{(j)}$ are determined only by ω and $\varphi_1, \ldots, \varphi_r$, relation (8) shows that the differential form $\sum_{(i)} \tilde{f}_{(i)} d\tilde{g}_{i_1} \wedge \cdots \wedge d\tilde{g}_{i_s}$ on $\tilde{V}$ is determined only by ω and is independent of the choice of the $f_{(i)}$ and the g_i. We denote this differential form by $\mathfrak{p}(\omega)$ or $\tilde{\omega}$ and call it the *reduction of ω modulo* $\mathfrak{p}$. We can easily verify the following facts.

(i) *If ω and ω' are $\mathfrak{p}$-finite, then $\omega + \omega'$ and $\omega \wedge \omega'$ are $\mathfrak{p}$-finite and we have* $\mathfrak{p}(\omega + \omega') = \mathfrak{p}(\omega) + \mathfrak{p}(\omega')$, $\mathfrak{p}(\omega \wedge \omega') = \mathfrak{p}(\omega) \wedge \mathfrak{p}(\omega')$.

(ii) *If ω is $\mathfrak{p}$-finite, then $d\omega$ is $\mathfrak{p}$-finite and we have*

$$d\mathfrak{p}(\omega) = \mathfrak{p}(d\omega).$$

(iii) *If ω is a differential form other than 0, there exists an element a of k such that $a\omega$ is $\mathfrak{p}$-finite and $\mathfrak{p}(a\omega) \neq 0$.*

The last assertion is a special case of the following proposition.

PROPOSITION 6. *Let V be a $\mathfrak{p}$-simple $\mathfrak{p}$-variety; let $\mathcal{M}$ be a vector space over k of differential forms on V of degree s, defined over k, and $\mathcal{M}_0$ the set of $\mathfrak{p}$-finite elements in $\mathcal{M}$. If $\mathcal{M}$ is of a finite dimension m over k, then $\mathcal{M}_0$ is a free $\mathfrak{o}$-module of rank m, and the set $\{\mathfrak{p}(\omega) \mid \omega \in \mathcal{M}_0\}$ is a vector space of dimension m over $\tilde{k}$.*

PROOF. Let $\{\varphi_1, \ldots, \varphi_r\}$ be a system of local parameters for V at a generic point of $\tilde{V}$ over $\tilde{k}$. Denote by λ the discrete valuation of the field $k(V)$ introduced in §10.2. For every differential form

$$\omega = \sum_{i_1 < \cdots < i_s} h_{(i)} d\varphi_{i_1} \wedge \cdots \wedge d\varphi_{i_s},$$

put

$$\lambda(\omega) = \operatorname*{Min}_{(i)}\{\lambda(h_{(i)})\}.$$

Then λ satisfies conditions (i)–(iii) of Lemma 3, and ω is $\mathfrak{p}$-finite if and only if $\lambda(\omega) \geq 0$. So our proposition is proved by the same argument as in the last part of §10.2

PROPOSITION 7. *Let V and W be $\mathfrak{p}$-simple $\mathfrak{p}$-varieties, T a rational mapping of V into W and ω a differential form on W, defined over k. Suppose that there exists a point ξ on $\tilde{V}$ such that T is defined at ξ and ω is finite at $T(\xi)$. Then $\omega \circ T$ is defined and finite at ξ, and*

$$\mathfrak{p}(\omega \circ T) = \mathfrak{p}(\omega) \circ \tilde{T}.$$

This follows easily from our definition.

PROPOSITION 8. *Let V be a $\mathfrak{p}$-simple $\mathfrak{p}$-variety of dimension r and ω a differential form on V defined over k. Suppose ω to be $\mathfrak{p}$-finite and $\mathfrak{p}(\omega) \neq 0$. Then we have*

$$\mathfrak{p}(\operatorname{div}(\omega)) \prec \operatorname{div}(\mathfrak{p}(\omega)),$$

where $\operatorname{div}(\eta)$ denotes the divisor of a differential form η. If ω is of degree r, then we have

$$\mathfrak{p}(\operatorname{div}(\omega)) = \operatorname{div}(\mathfrak{p}(\omega)).$$

(For the definition of the divisor of a differential form, we refer to Nakai [29].)

PROOF. Let $\mathfrak{A}$ be a simple subvariety of $\tilde{V}$ of dimension $r-1$; let $\{\varphi_1, \ldots, \varphi_r\}$ be a system of local parameters on V at some point of $\mathfrak{A}$. Then, ω has an expression

$$\omega = \sum_{i_1 < \cdots < i_s} f_{(i)} d\varphi_{i_1} \wedge \cdots \wedge d\varphi_{i_s},$$

where the $f_{(i)}$ are $\mathfrak{p}$-finite elements of $k(V)$; we have then

$$\mathfrak{p}(\omega) = \sum_{(i)} \tilde{f}_{(i)} d\tilde{\varphi}_{i_1} \wedge \cdots \wedge d\tilde{\varphi}_{i_s}.$$

Take an extension k' of k and an extension $\mathfrak{p}'$ of $\mathfrak{p}$ in k' such that the components of the divisors $(f_{(i)})$ are defined over k' and $\mathfrak{A}$ is defined over $\mathfrak{p}'(k')$. Let $A_1, \ldots, A_t$ be the components of $(f_{(i)})$ such that $\mathfrak{A}$ is a component of $\mathfrak{p}'(A_\nu)$. By Proposition 4, we see that the φ_i are local parameters for V along A_ν for every ν. Denote by $v_\nu(\alpha)$ the multiplicity of A_ν in the divisor $\operatorname{div}(\alpha)$ of a function or a differential form α, and similarly by $\tilde{v}$ the multiplicity of $\mathfrak{A}$; denote further by $\mu(A, \mathfrak{A})$ the multiplicity of $\mathfrak{A}$ in the cycle $\mathfrak{p}'(A)$. Then, by Proposition 3, we have, for every (i),

$$\tilde{v}(\tilde{f}_{(i)}) = \sum_{\nu=1}^{t} \mu(A_\nu, \mathfrak{A}) v_\nu(f_{(i)}) \geq \sum_{\nu=1}^{t} \mu(A_\nu, \mathfrak{A}) v_\nu(\omega).$$

This proves $\operatorname{div}(\mathfrak{p}(\omega)) \succ \mathfrak{p}(\operatorname{div}(\omega))$. If ω is of degree r, ω is written in the form $\omega = f d\varphi_1 \wedge \cdots \wedge d\varphi_r$, so that $\mathfrak{p}(\omega) = \tilde{f} d\tilde{\varphi}_1 \wedge \cdots \wedge d\tilde{\varphi}_r$. We have then

$$\tilde{v}(\tilde{f}) = \sum_{\nu} \mu(A_\nu, \mathfrak{A}) v_\nu(f) = \sum_{\nu} \mu(A_\nu, \mathfrak{A}) v_\nu(\omega);$$

this implies $\operatorname{div}(\mathfrak{p}(\omega)) = \mathfrak{p}(\operatorname{div}(\omega))$.

PROPOSITION 9. *Let V be a $\mathfrak{p}$-simple $\mathfrak{p}$-variety; suppose that V is $\mathfrak{p}$-complete and $\tilde{V}$ has no multiple point. Let ω be a differential form on V of the first kind, defined over k. If ω is $\mathfrak{p}$-finite, then $\mathfrak{p}(\omega)$ is of the first kind.*

PROOF. We first note that V has no multiple point. Now by Proposition 5 of Koizumi [23], a differential form η on a complete non-singular variety is of the first kind if and only if $\operatorname{div}(\eta) \succ 0$. Our proposition follows from this and from Proposition 8.

U being a complete non-singular variety, denote by $h_s(U)$ the number of linearly independent differential forms on U of degree s, of the first kind. Then, by Propositions 6 and 9, we get:

PROPOSITION 10. *Let V be a $\mathfrak{p}$-simple $\mathfrak{p}$-variety. Suppose that V is $\mathfrak{p}$-complete and $\tilde{V}$ has no multiple point. Then we have, for every s,*

$$h_s(V) \leq h_s(\tilde{V}).$$

We conclude this section by a simple application to curves:

PROPOSITION 11. *Let C be a $\mathfrak{p}$-simple curve. Suppose that C is $\mathfrak{p}$-complete and $\tilde{C}$ has no multiple point. Then C and $\tilde{C}$ have the same genus.*

PROOF. Take a differential form ω on C such that ω is $\mathfrak{p}$-finite and $\mathfrak{p}(\omega) \neq 0$. This is possible by Proposition 6. Let g and $\tilde{g}$ be the genera of C and $\tilde{C}$, respectively. Then, by Proposition 8, we have

$$2g - 2 = \deg(\omega) = \deg(\mathfrak{p}(\omega)) = 2\tilde{g} - 2,$$

and hence $g = \tilde{g}$.

11. Reduction of Abelian Varieties

11.1. Let A be an abelian variety defined over k. Denote by f the rational mapping of $A \times A$ into A defined by $f(x, y) = x + y$, and by g the rational mapping of A into itself defined by $g(x) = -x$. Suppose that a structure of $\mathfrak{p}$-variety is defined on A. We say that A has *no defect for* $\mathfrak{p}$ (with respect to this structure) if the following conditions (A1–3) are satisfied:

(A1) *A is $\mathfrak{p}$-simple and $\mathfrak{p}$-complete.*
(A2) *f is everywhere defined on $\tilde{A} \times \tilde{A}$.*
(A3) *g is everywhere defined on $\tilde{A}$.*

Under these conditions, $\tilde{A}$ becomes, in a natural way, an abelian variety defined over $\tilde{k}$. As is remarked in §10.1, $\tilde{f}$ and $\tilde{g}$ are everywhere defined on $\tilde{A} \times \tilde{A}$ and on $\tilde{A}$, and we have $\tilde{f}(\xi, \eta) = f(\xi, \eta)$, $\tilde{g}(\xi) = g(\xi)$ for $\xi \in \tilde{A}$, $\eta \in \tilde{A}$. Put $\xi + \eta = \tilde{f}(\xi, \eta)$. Then it can be easily verified that $\tilde{A}$ is a group variety with respect to this law of composition, defined over $\tilde{k}$, and $\tilde{g}(\xi)$ gives $-\xi$. As A is $\mathfrak{p}$-complete, $\tilde{A}$ is a complete variety, so that $\tilde{A}$ is an abelian variety; if 0 denotes the origin of A, then $\mathfrak{p}(0)$ is the origin of $\tilde{A}$. We call the abelian variety $\tilde{A}$ the *reduction of* the abelian variety A *modulo* $\mathfrak{p}$.

PROPOSITION 12. *Let A and B be two abelian varieties having no defect for $\mathfrak{p}$; denote by* $\mathrm{Hom}(A, B; k)$ *the set of all homomorphisms of A into B, defined over k, and by* $\mathrm{Hom}(\tilde{A}, \tilde{B}; \tilde{k})$ *the set of all homomorphisms of $\tilde{A}$ into $\tilde{B}$, defined over $\tilde{k}$. Then, for every $\lambda \in \mathrm{Hom}(A, B; k)$, the reduction $\tilde{\lambda}$ of the rational mapping λ modulo $\mathfrak{p}$ is an element of* $\mathrm{Hom}(\tilde{A}, \tilde{B}; \tilde{k})$*; and the graph of $\tilde{\lambda}$ is the reduction of the graph of λ modulo $\mathfrak{p}$, in the sense of reduction of cycles. The correspondence $\lambda \rightarrow \tilde{\lambda}$ defines an injection of the additive group* $\mathrm{Hom}(A, B; k)$ *into* $\mathrm{Hom}(\tilde{A}, \tilde{B}; \tilde{k})$. *If $A = B$, this is a ring-injection. If A and B have the same dimension, we have $\nu(\lambda) = \nu(\tilde{\lambda})$ for every $\lambda \in$* $\mathrm{Hom}(A, B; k)$.

PROOF. We shall first show that every $\lambda \in \mathrm{Hom}(A, B; k)$ is everywhere defined on $\tilde{A}$. Let ξ be a point of $\tilde{A}$ and η a generic point of $\tilde{A}$ over $\tilde{k}(\xi)$. Then, $\xi + \eta$ is generic on $\tilde{A}$ over $\tilde{k}$, so that λ is defined at η and $\xi + \eta$. Take two independent generic points x, y of A over k and define a rational mapping h of $A \times A$ into B by

$$h(x, y) = \lambda(x + y) - \lambda(y);$$

then h is defined at $\xi \times \eta$ since λ is defined at η and at $\xi + \eta$. Put $\zeta = h(\xi, \eta)$. We have clearly $h(x, y) = \lambda(x)$. Hence we have

$$[\lambda(x) \to \zeta; \mathfrak{p}] \subset [x \times y \to \xi \times \eta; \mathfrak{p}].$$

Since the ring on the left-hand side is contained in $k(x)$, we have

$$[\lambda(x) \to \zeta; \mathfrak{p}] \subset [x \to \xi; \mathfrak{p}]$$

by virtue of Proposition 7 of [33]; this shows that λ is defined at ξ. Thus λ is everywhere defined on $\tilde{A}$. Then, by Proposition 2, the graph $\Gamma_{\tilde{\lambda}}$ of $\tilde{\lambda}$ is the reduction of the graph Γ_λ of λ modulo $\mathfrak{p}$, both considered as cycles. It is easy to see that $\tilde{\lambda}$ is a homomorphism of $\tilde{A}$ into $\tilde{B}$ and $\lambda \to \tilde{\lambda}$ gives an additive mapping of $\mathrm{Hom}(A, B; k)$ into $\mathrm{Hom}(\tilde{A}, \tilde{B}; \tilde{k})$. If λ is not 0, we have $\dim_k(\lambda x) \geq 1$ for a generic point x of A over k. Let D be the locus of $\lambda(x)$ over k. As B is $\mathfrak{p}$-complete, $\mathfrak{p}(D)$ is not empty, so that there exists a specialization η of $\lambda(x)$ over $\mathfrak{p}$ such that $\dim_{\tilde{k}}(\eta) \geq 1$. As A is $\mathfrak{p}$-complete, there exists a point ξ such that $x \times \lambda(x) \to \xi \times \eta$ ref. $\mathfrak{p}$; we have then $\tilde{\lambda}(\xi) = \eta$; this shows that $\tilde{\lambda}$ is not 0. It follows that $\lambda \to \tilde{\lambda}$ is injective. Now assume that A and B have the same dimension. By the definition of $\nu(\lambda)$, we have $\mathrm{pr}_B(\Gamma_\lambda) = \nu(\lambda)B$, $\mathrm{pr}_{\tilde{B}}(\Gamma_{\tilde{\lambda}}) = \nu(\tilde{\lambda})\tilde{B}$. As we have $\mathfrak{p}(B) = \tilde{B}$, $\mathfrak{p}(\Gamma_\lambda) = \Gamma_{\tilde{\lambda}}$, we get, by (3) of Proposition 1, $\nu(\tilde{\lambda}) = \nu(\lambda)$. It is clear that, when $A = B$, $\lambda \to \tilde{\lambda}$ gives a ring-injection.

PROPOSITION 13. *The notation being as in Proposition 12, let λ be an element of* $\mathrm{Hom}(A, B; k)$. *Suppose that λ is an isogeny of A onto B and every element of* $\mathrm{Ker}(\lambda)$ *is rational over k. Then reduction of points modulo $\mathfrak{p}$ defines a homomorphism of* $\mathrm{Ker}(\lambda)$ *onto* $\mathrm{Ker}(\tilde{\lambda})$. *If $\nu_i(\tilde{\lambda}) = 1$, this homomorphism is an isomorphism.*

PROOF. Denote by Γ_λ and $\Gamma_{\tilde{\lambda}}$ the graphs of λ and $\tilde{\lambda}$, respectively. We have then

$$\mathrm{pr}_A[\Gamma_\lambda \cdot (A \times 0)] = \sum \alpha_t t, \qquad \mathrm{pr}_{\tilde{A}}[\Gamma_{\tilde{\lambda}} \cdot (\tilde{A} \times \tilde{0})] = \sum \alpha_\tau \tau,$$

where the sums are taken over all elements $t \in \mathrm{Ker}(\lambda)$, $\tau \in \mathrm{Ker}(\tilde{\lambda})$ with certain multiplicities α_t, α_τ, respectively. By the relation $\mathfrak{p}(\Gamma_\lambda) = \Gamma_{\tilde{\lambda}}$ and by Proposition 1, we obtain

$$\mathfrak{p}\{\mathrm{pr}_A[\Gamma_\lambda \cdot (A \times 0)]\} = \mathrm{pr}_{\tilde{A}}[\Gamma_{\tilde{\lambda}} \cdot (\tilde{A} \times \tilde{0})].$$

This shows that reduction modulo $\mathfrak{p}$ gives a surjective mapping of Ker(λ) onto Ker($\tilde{\lambda}$). It is clear that this mapping is a homomorphism. Hence the order of Ker(λ) is not less than the order of Ker($\tilde{\lambda}$), so that $\nu_s(\lambda) \geq \nu_s(\tilde{\lambda})$. If $\nu_i(\tilde{\lambda}) = 1$, we get, by the relation $\nu(\lambda) = \nu(\tilde{\lambda})$, the equality $\nu_s(\lambda) = \nu_s(\tilde{\lambda})$. This proves the last assertion.

PROPOSITION 14. *The notation being as in Proposition 12, let l be a prime other than the characteristic of $\tilde{k}$. Then, we can choose l-adic coordinate-systems of* $\mathfrak{g}_l(A)$, $\mathfrak{g}_l(B)$, $\mathfrak{g}_l(\tilde{A})$, $\mathfrak{g}_l(\tilde{B})$ *in such a way that:*

(i) *For every* $\lambda \in \text{Hom}(A, B; k)$, *we have* $M_l(\lambda) = M_l(\tilde{\lambda})$.
(ii) *For every divisor X on A, rational over k, we have* $E_l(X) = E_l(\tilde{X})$.

PROOF. Let k' be an extension of k over which every point of $\mathfrak{g}_l(A)$ and $\mathfrak{g}_l(B)$ are rational. Take an extension $\mathfrak{p}'$ of $\mathfrak{p}$ in k'; $\mathfrak{p}'$ may not be discrete. We will now consider the reduction of the points in $\mathfrak{g}_l(A)$ and $\mathfrak{g}_l(B)$ modulo $\mathfrak{p}'$. Since every point of $\mathfrak{g}_l(A)$ and $\mathfrak{g}_l(B)$ is rational over a finite extension of k, reduction modulo $\mathfrak{p}'$ of a point of $\mathfrak{g}_l(A)$ and $\mathfrak{g}_l(B)$ is in substance the same as reduction modulo a discrete place. Then, we see by Proposition 13 that reduction modulo $\mathfrak{p}'$ gives an isomorphism of $\mathfrak{g}_l(A)$ onto $\mathfrak{g}_l(\tilde{A})$ and an isomorphism of $\mathfrak{g}_l(B)$ onto $\mathfrak{g}_l(\tilde{B})$. Hence we can choose l-adic coordinate-systems of $\mathfrak{g}_l(A)$, $\mathfrak{g}_l(\tilde{A})$, $\mathfrak{g}_l(B)$, $\mathfrak{g}_l(\tilde{B})$ in such a way that a point x in $\mathfrak{g}_l(A)$, or $\mathfrak{g}_l(B)$, has the same l-adic coordinates as $\mathfrak{p}'(x)$. Then, we get obviously, with respect to these systems, $M_l(\lambda) = M_l(\tilde{\lambda})$ for every $\lambda \in \text{Hom}(A, B; k)$. Let k'' be the algebraic closure of k' and U_l the set of roots of unity in k'' whose orders are powers of l. Take an extension $\mathfrak{p}''$ of $\mathfrak{p}'$ in k''; then $\mathfrak{p}''(U_l)$ is the set of roots of unity in $\mathfrak{p}''(k'')$ whose orders are powers of l. Choose isomorphisms of U_l and $\mathfrak{p}''(U_l)$ onto $\mathbf{Q}_l/\mathbf{Z}_l$ in such a way that for every $\zeta \in U_l$, ζ and $\mathfrak{p}''(\zeta)$ have the same image in $\mathbf{Q}_l/\mathbf{Z}_l$. Then we can easily verify, following step by step the definition of the matrix $E_l(X)$, the relation $E_l(X) = E_l(\tilde{X})$ for every divisor X on A, rational over k.

11.2. Given k and $\mathfrak{p}$ as before, let F be an algebraic number field; and let (A, ι) be an abelian variety of type (F), defined over k. Put $\mathfrak{r} = \iota^{-1}[\text{End}(A) \cap \iota(F)]$. If A has no defect for $\mathfrak{p}$, then, for every $\lambda \in \iota(\mathfrak{r})$, we obtain, by reduction modulo $\mathfrak{p}$, an element $\tilde{\lambda}$ of End($\tilde{A}$). Put $\tilde{\iota}(\mu) = \widetilde{\iota(\mu)}$ for every $\mu \in \mathfrak{r}$. Then $\tilde{\iota}$ is an isomorphism of $\mathfrak{r}$ into End($\tilde{A}$) such that $\tilde{\iota}(1) = 1_{\tilde{A}}$; we can extend this isomorphism to an isomorphism of F into $\text{End}_{\mathbf{Q}}(\tilde{A})$, which we denote again by $\tilde{\iota}$. Thus we obtain an abelian variety $(\tilde{A}, \tilde{\iota})$ of type (F), defined over $\tilde{k}$. We call $(\tilde{A}, \tilde{\iota})$ the reduction of (A, ι) modulo $\mathfrak{p}$. $(\tilde{A}, \tilde{\iota})$ is clearly of the same index as (A, ι). If (A, ι) is principal, so is $(\tilde{A}, \tilde{\iota})$.

PROPOSITION 15. *Let (A, ι) be an abelian variety of type (F), defined over k, which is principal. Let $\mathfrak{a}$ be an integral ideal of F and $(A_1, \iota_1; \lambda)$ an $\mathfrak{a}$-transform of (A, ι), defined over k. Suppose that A and A_1 have no defect for $\mathfrak{p}$. Then, $(\tilde{A}_1, \tilde{\iota}_1; \tilde{\lambda})$ is an $\mathfrak{a}$-transform of $(\tilde{A}, \tilde{\iota})$.*

PROOF. It is clear that $\tilde{\lambda}$ commutes with the operation of F. Let x be a generic point of A over k and ξ a generic point of $\tilde{A}$ over $\tilde{k}$. For every $\alpha \in \mathfrak{a}$, we have $k(\iota(\alpha)x) \subset k(\lambda x)$, so that we obtain a rational mapping μ of A_1 into A, defined over k, such that $\mu(\lambda x) = \iota(\alpha)x$; μ is clearly a homomorphism. We have then $\tilde{\mu}\tilde{\lambda}\xi = \tilde{\iota}(\alpha)\xi$, and hence $\tilde{k}(\tilde{\iota}(\alpha)\xi) \subset \tilde{k}(\tilde{\lambda}\xi)$. Therefore, if $\tilde{\lambda}_1$ is an $\mathfrak{a}$-multiplication of $\tilde{A}$, defined over $\tilde{k}$, we have $\tilde{k}(\tilde{\lambda}\xi) \supset \tilde{k}(\tilde{\lambda}_1\xi)$. On the other hand, by Proposition 10 of §7.2 and Proposition 12 of §11.1, we have

$$[\tilde{k}(\xi) : \tilde{k}(\tilde{\lambda}\xi)] = \nu(\tilde{\lambda}) = \nu(\lambda) = N(\mathfrak{a})^m = \nu(\tilde{\lambda}_1) = [\tilde{k}(\xi) : \tilde{k}(\tilde{\lambda}_1\xi)],$$

where m denotes the index of (A, ι). Hence $\tilde{k}(\tilde{\lambda}\xi) = \tilde{k}(\tilde{\lambda}_1\xi)$. This shows that $\tilde{\lambda}$ is an $\mathfrak{a}$-multiplication.

PROPOSITION 16. *(A, ι) being as in Proposition 15, let $\mathfrak{a}$ be an integral ideal of F. Suppose that A has no defect for $\mathfrak{p}$ and every point of $\mathfrak{g}(\mathfrak{a}, A)$ is rational over k. Then, reduction modulo $\mathfrak{p}$ defines a homomorphism of $\mathfrak{g}(\mathfrak{a}, A)$ onto $\mathfrak{g}(\mathfrak{a}, \tilde{A})$. Moreover, if $\mathfrak{a}$ is prime to the characteristic of $\tilde{k}$, this homomorphism is an isomorphism.*

PROOF. We can find an integral ideal $\mathfrak{b}$, prime to $\mathfrak{a}$ and the characteristic p of $\tilde{k}$, such that $\mathfrak{ab}$ is a principal ideal (γ). By Proposition 18 of §7.5, we have

$$\mathfrak{g}((\gamma), A) = \mathfrak{g}(\mathfrak{a}, A) + \mathfrak{g}(\mathfrak{b}, A), \qquad \mathfrak{g}((\gamma), \tilde{A}) = \mathfrak{g}(\mathfrak{a}, \tilde{A}) + \mathfrak{g}(\mathfrak{b}, \tilde{A}).$$

Take an extension k' of k such that every point of $\mathfrak{g}((\gamma), A)$ is rational over k', and an extension $\mathfrak{p}'$ of $\mathfrak{p}$ in k'. Then, reduction modulo $\mathfrak{p}'$ gives homomorphisms of $\mathfrak{g}((\gamma), A), \mathfrak{g}(\mathfrak{a}, A), \mathfrak{g}(\mathfrak{b}, A)$ respectively into $\mathfrak{g}((\gamma), \tilde{A}), \mathfrak{g}(\mathfrak{a}, \tilde{A}), \mathfrak{g}(\mathfrak{b}, \tilde{A})$. By Proposition 13, $\mathfrak{g}((\gamma), A)$ is mapped *onto* $\mathfrak{g}((\gamma, \tilde{A})$; so $\mathfrak{g}(\mathfrak{a}, A)$ must be mapped onto $\mathfrak{g}(\mathfrak{a}, \tilde{A})$. If $\mathfrak{a}$ is prime to p, $\mathfrak{g}(\mathfrak{a}, A)$ and $\mathfrak{g}(\mathfrak{a}, \tilde{A})$ are of the same order $N(\mathfrak{a})^m$, where m is the index of (A, ι); hence reduction modulo $\mathfrak{p}$ gives an isomorphism of $\mathfrak{g}(\mathfrak{a}, A)$ onto $\mathfrak{g}(\mathfrak{a}, \tilde{A})$.

11.3. Consider now the case where both k and $\tilde{k}$ are of characteristic 0. Let k be a subfield of $\mathbf{C}$ and $\mathfrak{p}$ a place of k taking values in $\mathbf{C}$. Let $(F; \{\varphi_i\})$ be a CM-type and (A, ι) an abelian variety of type $(F; \{\varphi_i\})$, defined over k. Suppose that A has no defect for $\mathfrak{p}$, k contains $\cup F^{\varphi_i}$, and $\mathfrak{p}(\xi) = \xi$ for every $\xi \in \cup F^{\varphi_i}$. Under these assumptions, we shall prove that the reduction $(\tilde{A}, \tilde{\iota})$ of (A, ι) modulo $\mathfrak{p}$ is of type $(F; \{\varphi_i\})$. By the definition of CM-type, there exist invariant differential forms $\omega_1, \ldots, \omega_n$ on A such that

$$\delta(\iota(\alpha))\omega_i = \alpha^{\varphi_i}\omega_i \qquad (1 \leq i \leq n)$$

for every $\alpha \in \mathfrak{r}$, where n is the dimension of A and $\mathfrak{r}$ is the order of (A, ι). In view of the results of §2, we may assume that the ω_i are defined over a finite algebraic extension k' of k. Take an extension $\mathfrak{p}'$ of $\mathfrak{p}$ in k' and consider reduction modulo $\mathfrak{p}'$. By assertion (iii) of §10.4, which is a special case of Proposition 6, we may assume that the ω_i are $\mathfrak{p}'$-finite and $\mathfrak{p}'(\omega_i) \neq 0$ for every i. Then we get, by Proposition 7,

$$(1) \qquad \delta(\tilde{\iota}(\alpha))\mathfrak{p}'(\omega_i) = \mathfrak{p}'(\alpha^{\varphi_i})\mathfrak{p}'(\omega_i) \qquad (1 \leq i \leq n).$$

By our assumption, we have $\mathfrak{p}'(\alpha^{\varphi_i}) = \alpha^{\varphi_i}$; so relation (1) shows that $(\tilde{A}, \tilde{\iota})$ is of type $(F; \{\varphi_i\})$.

12. The Theory "For Almost All $\mathfrak{p}$"

12.1. Preliminary lemmas. The symbols k, $\mathfrak{o}$, $\mathfrak{p}$ being as before, let V be a variety of dimension r in the affine space S^n, defined over k. Let (x) be a generic point of V over k and the t_{ij}, for $0 \leq i \leq r$, $1 \leq j \leq n$, be $(r+1)n$ independent variables over $k(x)$. Put $y_i = \sum_{j=1}^n t_{ij}x_j$ for $0 \leq i \leq r$. Then we have $k(t, x) = k(t, y)$. As $(y_0, \ldots, y_r)$ is of dimension r over $k(t)$, there exists an irreducible polynomial $F(T, Y)$ in $k[T, Y]$ such that $F(t, y) = 0$. Substituting $\sum_{j=1}^n T_{ij}X_j$ for Y_i, we obtain from $F(T, Y)$ a polynomial $G(T, X)$ in (T_{ij}) and (X_j). We write $G(T, X)$ as a polynomial in (T_{ij}) with coefficients in $k[X]$ and denote by the $H_\alpha(X)$ for $1 \leq \alpha \leq s$ those coefficients. We call the set $\{H_\alpha(X)\}$ a *k-basic system* for V. We can take F in such a way that all its coefficients are contained in $\mathfrak{o}$ and at least one coefficient is a $\mathfrak{p}$-unit. When F is taken in this manner, we call $\{H_\alpha(X)\}$ a $\mathfrak{p}$-*basic system* for V.

LEMMA 4. *Let V be a variety of dimension r in S^n, defined over k, and $\{H_\alpha(X) \mid 1 \leq \alpha \leq s\}$ a $\mathfrak{p}$-basic system for V. Then a point (ξ) of $\tilde{S}^n$ is contained in $\tilde{V}$ if and only if $\tilde{H}_\alpha(\xi) = 0$ for $1 \leq \alpha \leq s$.*

PROOF. Let (x), (t), (y), F, G be as above; and let (ξ) be a specialization of (x) over $\mathfrak{p}$. As we have $F(t_{ij}, \sum_{j=1}^n t_{ij}x_j) = 0$ and the t_{ij} are independent over $k(x)$, we have $H_\alpha(x) = 0$ for $1 \leq \alpha \leq s$, so that $\tilde{H}_\alpha(\xi) = 0$ for $1 \leq \alpha \leq s$. This proves the "only if" part. Conversely, let (ξ) be a point of $\tilde{S}^n$ such that $\tilde{H}_\alpha(\xi) = 0$ for $1 \leq \alpha \leq s$. Let the τ_{ij} for $0 \leq i \leq r$, $1 \leq j \leq n$ be $(r+1)n$ independent variables over $\tilde{k}(\xi)$. Put $\eta_i = \sum_{j=1}^n \tau_{ij}\xi_j$ for $0 \leq i \leq r$. We have then

$$\tilde{F}(\tau_{ij}, \eta_i) = \tilde{F}\left(\tau_{ij}, \sum_{j=1}^{n} \tau_{ij}\xi_j\right) = 0.$$

Let W be the locus of (y) over $k(t)$ in S^{r+1}. The specialization $(t) \to (\tau)$ ref. $\mathfrak{p}$ gives an extension $\mathfrak{p}'$ of $\mathfrak{p}$ in $k(t)$. By Theorem 21 of [33], we have

$$\mathfrak{p}'(W) = \{(\zeta) \mid (\zeta) \in \tilde{S}^{r+1}, \tilde{F}(\tau_{ij}, \zeta_j) = 0\}.$$

Hence (η) is contained in $\mathfrak{p}'(W)$; so (η) is a specialization of (y) over $\mathfrak{p}'$. Let (ξ') be a specialization of (x) such that

$$(x, y) \to (\xi', \eta) \text{ ref. } \mathfrak{p}'.$$

By Proposition 16 of [33], (ξ') is finite, so that it is a point of $\tilde{S}^n$; and we have $\sum_{j=1}^{n} \tau_{ij}\xi'_j = \eta_i = \sum_{j=1}^{n} \tau_{ij}\xi_j$ for $0 \le i \le r$. Assume that $\xi_j \neq \xi'_j$ for one of the j, say 1; we have then

$$\tau_{i1} = -\sum_{j=2}^{n} \tau_{ij}\left(\frac{\xi_j - \xi'_j}{\xi_1 - \xi'_1}\right) \qquad (0 \le i \le r).$$

As (ξ') is a specialization of (x) over $\mathfrak{p}$, we have $\dim_{\tilde{k}}(\xi') \le r$. Hence we have

$$\dim_{\tilde{k}(\xi)}(\tau) \le \dim_{\tilde{k}(\xi)}(\xi', \tau_{ij}(j > 1)) \le r + (r+1)(n-1) < (r+1)n.$$

This contradicts the assumption that the τ_{ij} are $(r+1)n$ independent variables over $\tilde{k}(\xi)$. Therefore we have $(\xi) = (\xi')$. Hence (ξ) is a point of $\tilde{V}$; this completes the proof.

We will now study the reduction of algebraic varieties modulo infinitely many $\mathfrak{p}$. Let k be a field and $\sum$ a set of discrete places of k. In the rest of this section, we use k and $\sum$ always in this sense. We call a subset σ of $\sum$ an *open set* of $\sum$ if there exist a finite number of elements $a_1, \dots, a_r$, other than 0, in k such that

$$\sigma = \{\mathfrak{p} \mid \mathfrak{p} \in \Sigma, \mathfrak{p}(a_1) \neq 0, \dots, \mathfrak{p}(a_r) \neq 0\}.$$

For any set of elements $\{a_1, \dots, a_r\}$ in k, none of which is 0, the set of $\mathfrak{p}$ in $\sum$ such that the a_i are $\mathfrak{p}$-units, is an open set. In fact, we have

$$\sigma = \{\mathfrak{p} \mid \mathfrak{p} \in \Sigma, \mathfrak{p}(a_1) \neq 0, \mathfrak{p}(a_1^{-1}) \neq 0, \dots, \mathfrak{p}(a_r) \neq 0, \mathfrak{p}(a_r^{-1}) \neq 0\}.$$

We say that a proposition $P(\mathfrak{p})$ concerned with $\mathfrak{p}$ in $\sum$ holds for *almost all* $\mathfrak{p}$ if $P(\mathfrak{p})$ holds for all $\mathfrak{p}$ in an open set of $\sum$.

LEMMA 5. *Given k and $\sum$ as above, let k' be a finitely generated extension of k, of dimension s over k. Let $\sum'$ be a set of discrete places of k' satisfying* *the following conditions:*

(i) *There exists a set of elements* $(t_1, \ldots, t_s)$ *in* k' *such that we have* $\dim_{\mathfrak{p}'(k)}(\mathfrak{p}'(t_1), \ldots, \mathfrak{p}'(t_s)) = s$ *for every* $\mathfrak{p}'$ *in* $\sum'$.
(ii) *For every* $\mathfrak{p}$ *in* $\sum$, *there exists an extension of* $\mathfrak{p}$ *in* $\sum'$.

Let σ' *be an open set of* $\sum'$ *and* σ *the subset of* $\sum$ *consisting of all* $\mathfrak{p}$ *in* $\sum$ *such that* $\mathfrak{p}$ *has at least one extension* $\mathfrak{p}'$ *in* σ'. *Then* σ *contains an open set of* $\sum$.

PROOF. By our definition, there exists a set of elements $(y_1, \ldots, y_r)$ in k' such that $y_i \neq 0$, and if $\mathfrak{p}'(y_1) \neq 0, \ldots, \mathfrak{p}'(y_r) \neq 0$ and if $\mathfrak{p}' \in \sum'$, then $\mathfrak{p}' \in \sigma'$. By our assumption, $t_1, \ldots, t_s$ are independent variables over k, and k' is algebraic over $k(t)$. Let $\sum_\nu b_{i\nu} Y^\nu = 0$ be an irreducible equation for y_i^{-1} over $k(t)$, for each i. We may assume that the $b_{i\nu}$ are polynomials in $k[t_1, \ldots, t_s]$. Let σ_0 be the set of all $\mathfrak{p}$ in $\sum$ such that all non-zero coefficients of the polynomials $b_{i\nu}$ are $\mathfrak{p}$-units. Then σ_0 is an open set of $\sum$. Let $\mathfrak{p}$ be a place in σ_0; by our assumption, there exists in $\sum'$ an extension $\mathfrak{p}'$ of $\mathfrak{p}$, and $\mathfrak{p}'(t_1), \ldots, \mathfrak{p}'(t_s)$ are independent variables over $\mathfrak{p}(k)$, so that the $b_{i\nu}$ are all $\mathfrak{p}'$-integral; moreover, any one of the $b_{i\nu}$, other than 0, is a $\mathfrak{p}'$-unit. It follows that the y_i^{-1} are all $\mathfrak{p}'$-integral, so that $\mathfrak{p}'(y_i) \neq 0$ for every i; namely, $\mathfrak{p}'$ is contained in σ'. Hence σ_0 is contained in σ; this proves our lemma.

12.2. Now, it is easy to verify that the results of [33] §6 can be extended to the present case. Namely, Proposition 29, Lemma 3, Proposition 30, Theorem 26 in that section are all true when we use the terms "for almost all $\mathfrak{p}$" in the sense explained above. In particular, we have

PROPOSITION 17. *Let* V *be a variety in* S^n *defined over* k. *Then, for almost all* $\mathfrak{p}$, $\mathfrak{p}(V)$ *is not empty and* V *is* $\mathfrak{p}$*-simple.*

We have to show that $\mathfrak{p}(V)$ is not empty for almost all $\mathfrak{p}$, as it was not explicitly proved in [33]. Take a point $(a_1, \ldots, a_n)$ in V such that the a_i are contained in a finite algebraic extension k' of k. Let $\sum'$ be the set of all extensions in k' of all $\mathfrak{p}$ in $\sum$ and σ' be the set of $\mathfrak{p}'$ in $\sum'$ such that the a_i are $\mathfrak{p}$-integral. Let σ be the set of all $\mathfrak{p}$ in $\sum$ having an extension $\mathfrak{p}'$ in σ'. Then by Lemma 5, σ is an open set of $\sum$. It is easy to see that $\mathfrak{p}(V)$ is not empty for every $\mathfrak{p}$ in σ.

12.3. We shall now give several properties preserved for almost all $\mathfrak{p}$ in the process of reduction modulo $\mathfrak{p}$.

PROPOSITION 18. *Let* $F_1(X), \ldots, F_r(X)$ *be* r *polynomials in* $k[X_1, \ldots, X_n]$ *and* U *the algebraic set in* S^n *given by*

$$U = \{(x) \mid F_i(x) = 0 \quad (1 \leq i \leq r)\}.$$

Then, we have

$$\mathfrak{p}(U) = \{(\xi) \mid F_{i\mathfrak{p}}(\xi) = 0 \quad (1 \le i \le r)\}$$

for almost all $\mathfrak{p}$*, where the subscript* $\mathfrak{p}$ *means reduction of polynomials modulo* $\mathfrak{p}$.

PROOF. We may consider only those $\mathfrak{p}$ for which the coefficients of the F_i are all $\mathfrak{p}$-integral. It is easy to see

$$\mathfrak{p}(U) \subset \{(\xi) \mid F_{i\mathfrak{p}}(\xi) = 0 \quad (1 \le i \le r)\};$$

so we will now prove that the inverse inclusion holds for almost all $\mathfrak{p}$. Suppose that U is not empty. Let $U_1, \ldots, U_s$ be the components of U and k' be a finite algebraic extension of k such that the U_i are all defined over k'. Let $\sum'$ be the set of all extensions in k' of all $\mathfrak{p}$ in $\sum$. Let $\{H_\alpha^{(i)}(X); 1 \le \alpha \le t_i\}$ be a k'-basic system for U_i, for each i. Then, by the definition of basic system, $\{H_\alpha^{(i)}\}$ is a $\mathfrak{p}'$-basic system for U_i for almost all $\mathfrak{p}'$ in $\sum'$. Put

$$H_{\alpha_1 \cdots \alpha_s}(X) = H_{\alpha_1}^{(1)}(X) \cdots H_{\alpha_s}^{(s)}(X);$$

put $H_1(X) = 1$, if U is empty. Then, by Lemma 4, we see easily

$$\mathfrak{p}'(U) = \{(\xi) \mid H_{\alpha_1 \cdots \alpha_s \mathfrak{p}'}(\xi) = 0 \text{ for every } (\alpha)\}$$

for almost all $\mathfrak{p}'$ in $\sum'$. By Hilbert's theorem, there exists a positive integer ρ such that $H_{\alpha_1, \ldots, \alpha_s}(X)^\rho = \sum Q_{(\alpha)i}(X) F_i(X)$, where the $Q_{(\alpha)i}$ are polynomials in $k'[X]$. The coefficients of $Q_{(\alpha)i}$ are all $\mathfrak{p}'$-integral for almost all $\mathfrak{p}'$ in $\sum'$. For those $\mathfrak{p}'$, we have

$$(H_{(\alpha)\mathfrak{p}'})^\rho = \sum Q_{(\alpha)i\mathfrak{p}'} F_{i\mathfrak{p}'},$$

and hence

$$\mathfrak{p}'(U) \supset \{(\xi) \mid F_{i\mathfrak{p}'}(\xi) = 0 \quad (1 \le i \le r)\}.$$

By virtue of Lemma 5, this proves our proposition.

PROPOSITION 19. *Let* U *and* V *be two algebraic sets in* S^n*, defined over* k*. Then, we have* $\mathfrak{p}(U \cap V) = \mathfrak{p}(U) \cap \mathfrak{p}(V)$ *for almost all* $\mathfrak{p}$.

PROOF. Let $\mathfrak{a}$ and $\mathfrak{b}$ be the ideals of $k[X_1, \ldots, X_n]$ given by

$$\mathfrak{a} = \{F(X) \mid F(x) = 0 \text{ for every } (x) \in U\},$$

$$\mathfrak{b} = \{G(X) \mid G(x) = 0 \text{ for every } (x) \in V\}.$$

Let $\{F_1(X), \ldots, F_r(X)\}$ and $\{G_1(X), \ldots, G_s(X)\}$ be bases for $\mathfrak{a}$ and $\mathfrak{b}$, respectively. Then we have

$$U = \{(x) \mid F_i(x) = 0\,(1 \leq i \leq r)\}, \qquad V = \{(x) \mid G_j(x) = 0\,(1 \leq j \leq s)\},$$

$$U \cap V = \{(x) \mid F_i(x) = 0, G_j(x) = 0\ (1 \leq i \leq r, 1 \leq j \leq s)\}.$$

Therefore, our proposition is an immediate consequence of Proposition 18.

PROPOSITION 20. *Let U be an algebraic set, defined over k, in $S^{n+m} = S^n \times S^m$ and V the projection of U on the first factor S^n. Then, $\mathfrak{p}(V)$ is the projection of $\mathfrak{p}(U)$ on the first factor of $\mathfrak{p}(S^{n+m}) = \mathfrak{p}(S^n) \times \mathfrak{p}(S^m)$, for almost all $\mathfrak{p}$.*

PROOF. It is sufficient to prove our proposition when U is a variety defined over k, since the general case is easily reduced to this special case by means of Lemma 5. Assuming U to be a variety defined over k, let (x, y) be a generic point of U over k with the projection (x) on S^n and (y) on S^m; and let s be the dimension of $(y) = (y_1, \ldots, y_m)$ over $k(x)$. If s is not 0, we may assume that $y_1, \ldots, y_s$ are independent variables over $k(x)$ and (y) is algebraic over $k(x, y_1, \ldots, y_s)$. The locus of $(x, y_1, \ldots, y_s)$ over k is the variety $V \times S^s$. As we have $\mathfrak{p}(V \times S^s) = \mathfrak{p}(V) \times \mathfrak{p}(S^s)$, the projection of $\mathfrak{p}(V \times S^s)$ on the first factor of $\mathfrak{p}(S^n) \times \mathfrak{p}(S^s)$ is $\mathfrak{p}(V)$. Therefore, our proposition is proved if we show that the projection of $\mathfrak{p}(U)$ on the factor $\mathfrak{p}(S^n) \times \mathfrak{p}(S^s)$ of $\mathfrak{p}(S^n) \times \mathfrak{p}(S^s) \times \mathfrak{p}(S^{m-s})$ is $\mathfrak{p}(V \times S^s)$ for almost all $\mathfrak{p}$. Hence it is sufficient to prove our proposition in the case where (y) is algebraic over $k(x)$. Suppose that this is so; let t_{ij}, for $1 \leq i \leq r$, $1 \leq j \leq n$, be rn independent variables over $k(x)$, where r is the dimension of $(x) = (x_1, \ldots, x_n)$ over k. Put $z_i = \sum_{j=1}^n t_{ij} x_j$ for $1 \leq i \leq r$; then (t, z) is $r(n+1)$ independent variables over k and (x) is algebraic over $k(t, z)$. By Proposition 17, $\mathfrak{p}(V)$ is a variety defined over $\mathfrak{p}(k)$ for every $\mathfrak{p}$ in an open set σ of $\sum$. Let $\tau_{ij\mathfrak{p}}$ and $\zeta_{i\mathfrak{p}}$, for $1 \leq i \leq r$, $1 \leq j \leq n$, be $r(n+1)$ independent variables over $\mathfrak{p}(k)$. Take and fix, for each $\mathfrak{p}$ in σ, a point $\xi_\mathfrak{p}$ in the intersection of $\mathfrak{p}(V)$ and the generic linear variety defined by

$$\sum_{j=1}^{n} \tau_{ij\mathfrak{p}} X_j - \zeta_{i\mathfrak{p}} = 0 \qquad (1 \leq i \leq r).$$

Then, $\xi_\mathfrak{p}$ is a generic point of $\mathfrak{p}(V)$ over $\mathfrak{p}(k)$. We see that $(\xi_\mathfrak{p}, \tau_\mathfrak{p}, \zeta_\mathfrak{p})$ is a specialization of (x, t, z) over $\mathfrak{p}$. Let $\sum'$ be the set of all extensions $\mathfrak{p}'$ of $\mathfrak{p}$ in $k(x, t, z, y)$ such that

$$(x, t, z) \to (\xi_\mathfrak{p}, \tau_\mathfrak{p}, \zeta_\mathfrak{p}) \text{ ref. } \mathfrak{p}'.$$

Let σ' be the set of all $\mathfrak{p}'$ in $\sum'$ such that the y_i are all $\mathfrak{p}'$-integral, and let σ_0 be the set of all $\mathfrak{p}$ in σ such that $\mathfrak{p}$ has at least one extension $\mathfrak{p}'$ in σ'. Then, by

Lemma 5, σ_0 contains an open set of $\sum$. Let $\mathfrak{V}_\mathfrak{p}$ denote the projection of $\mathfrak{p}(U)$ on the first factor $\mathfrak{p}(S^n)$ of $\mathfrak{p}(S^n) \times \mathfrak{p}(S^m)$. For every $\mathfrak{p}$ in σ_0, we can find a point $\eta_\mathfrak{p}$ in $\mathfrak{p}(S^m)$ such that $(x, y) \to (\xi_\mathfrak{p}, \eta_\mathfrak{p})$ ref. $\mathfrak{p}$. Hence $\mathfrak{V}_\mathfrak{p}$ contains $\xi_\mathfrak{p}$; this implies $\mathfrak{V}_\mathfrak{p} \supset \mathfrak{p}(V)$. On the other hand, if $\alpha_\mathfrak{p}$ is a generic point of a component of $\mathfrak{V}_\mathfrak{p}$ over the algebraic closure of $\mathfrak{p}(k)$, there exists a point $(\alpha_\mathfrak{p}, \beta_\mathfrak{p})$ in $\mathfrak{p}(U)$. As $(\alpha_\mathfrak{p}, \beta_\mathfrak{p})$ is a specialization of (x, y) over $\mathfrak{p}$, the point $\alpha_\mathfrak{p}$ is contained in $\mathfrak{p}(V)$; so we have $\mathfrak{V}_\mathfrak{p} \subset \mathfrak{p}(V)$. Thus we have proved $\mathfrak{V}_\mathfrak{p} = \mathfrak{p}(V)$ for every $\mathfrak{p}$ in σ_0. This proves our proposition.

PROPOSITION 21. *Let V be a variety in S^n, defined over k, and f a rational mapping of V into S^m, defined over k. Let F be an algebraic set contained in V, defined over k, different from V. Suppose that f is defined at every point in $V - F$. Then, for almost all $\mathfrak{p}$, f is defined at every point in $\mathfrak{p}(V) - \mathfrak{p}(F)$.*

PROOF. It is sufficient to prove the proposition in the case where $m = 1$. Let x be a generic point of V; then we have an expression $f(x) = Q(x)/P(x)$, where P and Q are polynomials in $k[X_1, \dots, X_n]$. Let the $W_{1\beta}$ denote the components of the algebraic set $\{a \mid a \in V, P(a) = 0\}$ which are not contained in F. Then, r being the dimension of V, we have $\dim W_{1\beta} \leq r - 1$. Denote by k_1 the algebraic closure of k; and let $x_{1\beta}$ be a generic point of $W_{1\beta}$ over k_1. As $x_{1\beta}$ is not contained in F, we have expressions $f(x) = Q_{1\beta}(x)/P_{1\beta}(x)$ where $Q_{1\beta}$ and $P_{1\beta}$ are polynomials in $k[X]$ such that $P_{1\beta}(x_{1\beta}) \neq 0$. Let the $W_{2\gamma}$ be the components of the algebraic set

$$\{a \mid a \in V, P(a) = 0, P_{1\beta}(a) = 0 \text{ for every } \beta\},$$

which are not contained in F. We have then $\dim W_{2\gamma} \leq r - 2$. After repeating (at most r times) this procedure, we obtain a set of polynomials $P_{ij}(X)$ in $k[X]$ such that

$$F \supset V \cap \{a \mid P_{ij}(a) = 0 \text{ for every } i \text{ and } j\};$$

and for each (i, j), we have $f(x) = Q_{ij}(x)/P_{ij}(x)$, where Q_{ij} is a polynomial in $k[X]$. By Propositions 18 and 19, we have

$$\mathfrak{p}(F) \supset \mathfrak{p}(V) \cap \{\xi \mid P_{ij\mathfrak{p}}(\xi) = 0 \text{ for every } i \text{ and } j\}$$

for every $\mathfrak{p}$ in an open set σ. If $\mathfrak{p}$ is in σ, we see that, for every point η in $\mathfrak{p}(V) - \mathfrak{p}(F)$, there exists a polynomial P_{ij} such that $P_{ij\mathfrak{p}}(\eta) \neq 0$, namely, f is defined at η. Our proposition is thereby proved.

PROPOSITION 22. *Let V be a variety in S^n, defined over k, and F an algebraic set contained in V, defined over k. Suppose that every point in $V - F$ is simple on V. Then, for almost all $\mathfrak{p}$, every point in $\mathfrak{p}(V) - \mathfrak{p}(F)$ is simple on V.*

PROOF. Let $\{G_1(X), \ldots, G_s(X)\}$ be a basis for the ideal $\mathfrak{a}$ of $k[X]$ given by

$$\mathfrak{a} = \{G(X) \mid G(x) = 0 \text{ for every } x \in V\}.$$

Let the $H_\nu(X)$ denote the determinants of degree $n - r$ belonging to the matrix $(\partial G_i/\partial X_j)$, where r is the dimension of V. Our assumption implies

$$F \supset \{x \mid G_i(x) = 0,\ H_\nu(x) = 0 \text{ for every } i \text{ and every } \nu\}.$$

By Proposition 18, we have, for almost all $\mathfrak{p}$,

$$\mathfrak{p}(F) \supset \{\xi \mid G_{i\mathfrak{p}}(\xi) = 0,\ H_{\nu\mathfrak{p}}(\xi) = 0 \text{ for every } i \text{ and every } \nu\}.$$

Hence, if η is a point in $\mathfrak{p}(V) - \mathfrak{p}(F)$, there exists a polynomial H_ν such that $H_{\nu\mathfrak{p}}(\eta) \neq 0$, namely, we have rank $(\partial G_{i\mathfrak{p}}/\partial X_j(\eta)) \geq n - r$. This proves our proposition.

PROPOSITION 23. *Let $V = [V_\alpha;\ F_\alpha;\ T_{\beta\alpha}]$ be an abstract variety defined over k. Then, for almost all $\mathfrak{p}$, the system $[V_\alpha;\ F_\alpha;\ \mathfrak{p}(F_\alpha);\ T_{\beta\alpha}]$ defines a $\mathfrak{p}$-variety. If V is complete, this $\mathfrak{p}$-variety is $\mathfrak{p}$-complete for almost all $\mathfrak{p}$. Moreover, let H be an algebraic set in V such that every point in $V - H$ is simple on V. Then, for almost all $\mathfrak{p}$, every point in $\mathfrak{p}(V) - \mathfrak{p}(H)$ is simple on V.*

PROOF. Let $B_{\beta\alpha}$ be the set of points in V_α such that the projection from $T_{\beta\alpha}$ to V_α is regular at x if and only if x is not contained in $B_{\beta\alpha}$. Then, $B_{\beta\alpha}$ is an algebraic set in V_α defined over k. This fact is well-known and is proved by the same argument as in the proof of Proposition 21. By Proposition 21, for almost all $\mathfrak{p}$, $T_{\beta\alpha}$ is regular at every point in $\mathfrak{p}(V_\alpha) - \mathfrak{p}(B_{\beta\alpha})$ for every (α, β). By the definition of abstract variety, we have

$$T_{\beta\alpha} \cap [(B_{\beta\alpha} \times V_\beta) \cup (V_\alpha \times B_{\alpha\beta})] \subset T_{\beta\alpha} \cap [(F_\alpha \times V_\beta) \cup (V_\alpha \times F_\beta)].$$

By Proposition 19, for almost all $\mathfrak{p}$,

$$\begin{aligned}\mathfrak{p}(T_{\beta\alpha}) \ \cap\ &[(\mathfrak{p}(B_{\beta\alpha}) \times \mathfrak{p}(V_\beta)) \cup (\mathfrak{p}(V_\alpha) \times \mathfrak{p}(B_{\alpha\beta}))]\\ &\subset\ \mathfrak{p}(T_{\beta\alpha}) \cap [(\mathfrak{p}(F_\alpha) \times \mathfrak{p}(V_\beta)) \cup (\mathfrak{p}(V_\alpha) \times \mathfrak{p}(F_\beta))].\end{aligned}$$

This shows that the system $[V_\alpha;\ F_\alpha;\ \mathfrak{p}(F_\alpha);\ T_{\beta\alpha}]$ is a $\mathfrak{p}$-variety for almost all $\mathfrak{p}$. Let the $(x_{\alpha 1}, \ldots, x_{\alpha n_\alpha})$, for $1 \leq \alpha \leq h$, be corresponding generic points of the V_α by $T_{\beta\alpha}$. Let $\{\varepsilon(\alpha i)\}$ be a set of integers which are equal to 1 or -1. Let W_ε denote the locus of

$$(x_{11}^{\varepsilon(11)}, \ldots, x_{1n_1}^{\varepsilon(1n_1)}, \ldots, x_{h1}^{\varepsilon(h1)}, \ldots, x_{hn_h}^{\varepsilon(hn_h)})$$

over k, where we omit those $x_{\alpha i}$ which are equal to 0. Let $\{G_{\alpha\nu}(X)\}$ be a basis of the ideal

$$\{G(X) \mid G(X) \in k[X], G(x) = 0 \text{ for every } x \in F_\alpha\},$$

for each α. Then, we have, for almost all $\mathfrak{p}$,

$$\mathfrak{p}(F_\alpha) = \{\xi \mid G_{\alpha\nu\mathfrak{p}}(\xi) = 0 \text{ for every } \nu\},$$

for every α. Suppose that V is complete. Then, we have, for every ε,

$$\emptyset = W_\varepsilon \cap \left\{ (u_{1i}) \times \cdots \times (u_{hi}) \,\middle|\, \frac{1-\varepsilon(\alpha i)}{2} u_{\alpha i} = 0 \text{ for every } \alpha, i, \right.$$
$$\left. \text{and } G_{\alpha\nu}(u_{\alpha i}) \prod_i \frac{1+\varepsilon(\alpha i)}{2} = 0 \text{ for every } \alpha, \nu \right\}.$$

By Propositions 18 and 19, we have, for almost all $\mathfrak{p}$,

$$\emptyset = \mathfrak{p}(W_\varepsilon) \cap \left\{ (\xi_{1i}) \times \cdots \times (\xi_{hi}) \,\middle|\, \frac{1-\varepsilon(\alpha i)}{2} \xi_{\alpha i} = 0 \text{ for every } \alpha, i, \right.$$
$$\left. \text{and } G_{\alpha\nu\mathfrak{p}}(\xi_{\alpha i}) \prod_i \frac{1+\varepsilon(\alpha i)}{2} = 0 \text{ for every } \alpha, \nu \right\}$$

for every ε. This shows that V is $\mathfrak{p}$-complete for almost all $\mathfrak{p}$. The algebraic set H has an expression $H = \cup H^{(\alpha)}$, where, for each α, $H^{(\alpha)}$ is the union of the components of H having representatives in V_α. Let H_α be the union of the representatives in V_α for the components of $H^{(\alpha)}$. Then, H_α is an algebraic set defined over k. By our assumption, every point in $V_\alpha - (H_\alpha \cup F_\alpha)$ is simple on V_α. Then, by Proposition 22, every point in $\mathfrak{p}(V_\alpha) - (\mathfrak{p}(H_\alpha) \cup \mathfrak{p}(F_\alpha))$ is simple on V_α for almost all $\mathfrak{p}$. This proves the last assertion of our proposition.

Let $V = [V_\alpha; F_\alpha; T_{\beta\alpha}]$ be an abstract variety defined over k. Then, by Proposition 23, the system $[V_\alpha; F_\alpha; \mathfrak{p}(F_\alpha); T_{\beta\alpha}]$ defines a $\mathfrak{p}$-variety for almost all $\mathfrak{p}$; and, by Proposition 17, the $\mathfrak{p}$-variety is $\mathfrak{p}$-simple for almost all $\mathfrak{p}$. Thus we obtain, for almost all $\mathfrak{p}$, an abstract variety $\mathfrak{p}(V) = [\mathfrak{p}(V_\alpha) : \mathfrak{p}(F_\alpha); \mathfrak{p}(T_{\beta\alpha})]$, defined over $\mathfrak{p}(k)$. Proposition 23 shows that, if V is complete, then, for almost all $\mathfrak{p}$, $\mathfrak{p}(V)$ is complete, and, if V has no multiple point, then, for almost all $\mathfrak{p}$, $\mathfrak{p}(V)$ has no multiple point.

PROPOSITION 24. *Let $V = [V_\alpha; F_\alpha; T_{\beta\alpha}]$ and $W = [W_\lambda; G_\lambda; S_{\mu\lambda}]$ be two abstract varieties defined over k; let f be a rational mapping of V into W and H an algebraic set in V, defined over k, such that f is defined at every point in $V - H$. Then, for almost all $\mathfrak{p}$, f is defined at every point in $\mathfrak{p}(V) - \mathfrak{p}(H)$.*

PROOF. We may consider only those $\mathfrak{p}$ for which $[V_\alpha; F_\alpha; \mathfrak{p}(F_\alpha); T_{\beta\alpha}]$ and $[W_\lambda; G_\lambda; \mathfrak{p}(G_\lambda); S_{\mu\lambda}]$ define $\mathfrak{p}$-varieties. Let H_α be an algebraic set in V_α defined for H in the same manner as in the last part of the proof of Proposition 23. Let Z be the graph of f and $Z_{\alpha\lambda}$ the representative of Z in $V_\alpha \times W_\lambda$; in the following we shall consider only those pairs (α, λ) for which Z has its representative in $V_\alpha \times W_\lambda$. Let $B_{\alpha\lambda}$ be the algebraic set in V_α such that the projection from $Z_{\alpha\lambda}$ to V_α is regular at a if and only if a is not contained in $B_{\alpha\lambda}$. Let $C_{\alpha\lambda}$ be the projection of $Z_{\alpha\lambda} \cap (V_\alpha \times G_\lambda)$ on V_α. We will now prove

$$(1) \qquad \cap_\lambda (B_{\alpha\lambda} \cup C_{\alpha\lambda}) \subset H_\alpha \cup F_\alpha.$$

If a point x_α in V_α is not contained in $H_\alpha \cup F_\alpha$, f is defined at the point x having x_α as its representative in V_α, so that there exists a suffix λ such that $Z_{\alpha\lambda}$ contains a point $x_\alpha \times y_\lambda$ with the projection x_α on V_α, y_λ on $W_\lambda - G_\lambda$ and $Z_{\alpha\lambda}$ is regular at x_α. Suppose that x_α is contained in $B_{\alpha\lambda} \cup C_{\alpha\lambda}$; then, by the definition of $B_{\alpha\lambda}$, we have $x_\alpha \in C_{\alpha\lambda}$. This implies that there exists a point $x'_\alpha \times y'_\lambda$ in $Z_{\alpha\lambda} \cap (V_\alpha \times G_\lambda)$ such that $x'_\alpha \to x_\alpha$ ref. k. As $Z_{\alpha\lambda}$ is regular at x_α, we must have $x'_\alpha \times y'_\lambda \to x_\alpha \times y_\lambda$ ref. k, so that we have $y_\lambda \in G_\lambda$; this is a contradiction; so x_α is not contained in $B_{\alpha\lambda} \cup C_{\alpha\lambda}$. We have thus proved the above inclusion (1). By Propositions 19, 20, 21, there exists an open set σ of $\sum$ such that, if $\mathfrak{p} \in \sigma$, then we have

$$(2) \qquad \cap_\lambda (\mathfrak{p}(B_{\alpha\lambda}) \cup \mathfrak{p}(C_{\alpha\lambda})) \subset \mathfrak{p}(H_\alpha) \cup \mathfrak{p}(F_\alpha)$$

for every α, $Z_{\alpha\lambda}$ is regular at every point in $\mathfrak{p}(V_\alpha) - \mathfrak{p}(B_{\alpha\lambda})$ and $\mathfrak{p}(C_{\alpha\lambda})$ is the projection of $\mathfrak{p}(Z_{\alpha\lambda}) \cap (\mathfrak{p}(V_\alpha) \times \mathfrak{p}(G_\lambda))$ on $\mathfrak{p}(V_\alpha)$. Now $\mathfrak{p}$ being in σ, let ξ be a point in $\mathfrak{p}(V) - \mathfrak{p}(H)$. Then there exists a representative ξ_α of ξ in $\mathfrak{p}(V_\alpha)$ such that $\xi_\alpha \notin \mathfrak{p}(H_\alpha) \cup \mathfrak{p}(F_\alpha)$. By (2), ξ_α is not contained in $\mathfrak{p}(B_{\alpha\lambda}) \cup \mathfrak{p}(C_{\alpha\lambda})$ for some λ. For such a λ, $\mathfrak{p}(Z_{\alpha\lambda})$ contains a point $\xi_\alpha \times \eta_\lambda$ and $Z_{\alpha\lambda}$ is regular at ξ_α. As ξ_α is not contained in $\mathfrak{p}(C_{\alpha\lambda})$, η_λ is not contained in $\mathfrak{p}(G_\lambda)$; so there exists a point η in $\mathfrak{p}(W)$ having η_λ as its representative in W_λ. This shows that f is defined at ξ and $f(\xi) = \eta$. Thus we have proved that, for every $\mathfrak{p}$ in σ, f is defined at every point in $\mathfrak{p}(V) - \mathfrak{p}(H)$.

PROPOSITION 25. *Let A be an abelian variety defined over k. Then, A has no defect for almost all* $\mathfrak{p}$.

This is an immediate consequence of Propositions 17, 23, 24.

12.4. We shall now consider the case where k and $\sum$ are given as follows. Given a field k_0, we take as k a finitely generated extension of k_0 and as $\sum$ the set of discrete places $\mathfrak{p}$ of k, taking values in the universal domain over k, such that $\mathfrak{p}(a) = a$ for every $a \in k_0$. Let A be an abelian variety defined over

k. By Proposition 25, there exists a set of non-zero elements $\{x_1, \ldots, x_r\}$ in k such that, if $\mathfrak{p} \in \sum$ and $\mathfrak{p}(x_i) \neq 0$ for every i, then A has no defect for $\mathfrak{p}$. Take elements $x_{r+1}, \ldots, x_s$ so that $k = k_0(x_1, \ldots, x_r, x_{r+1}, \ldots, x_s)$ and denote by V the locus of $(x_1, \ldots, x_s)$ over k_0. V may not be absolutely irreducible. As we have $x_i \neq 0$ for $1 \leq i \leq r$, V carries a point $(a_1, \ldots, a_s)$ such that $a_i \neq 0$ for $1 \leq i \leq r$ and all the a_i are algebraic over k_0. (Cf. [44] Chap. IV, Proposition 3). By Lemma 2, we can find a place $\mathfrak{p}$ in $\sum$ such that $\mathfrak{p}(x_i) = a_i$ for every i. Then, A has no defect for $\mathfrak{p}$; and $\mathfrak{p}(A)$ is defined over $\mathfrak{p}(k)$. Suppose that $\dim_{k_0} k > 0$. Then, one of the x_i is not algebraic over k_0. Since the a_i are algebraic over k_0, we see that $\dim_{k_0} \mathfrak{p}(k) < \dim_{k_0} k$. We shall use this result in the proof of the following proposition.

PROPOSITION 26. *Let $(F; \{\varphi_i\})$ be a CM-type and (A, ι) an abelian variety of type $(F; \{\varphi_i\})$. Then, there exists an abelian variety of type $(F; \{\varphi_i\})$, isomorphic to (A, ι), defined over an algebraic number field of finite degree.*

PROOF. Take an abelian variety (A_1, ι_1) of type $(F; \{\varphi_i\})$ and a field k of definition for (A_1, ι_1). Let k_0 be the composite of the fields F^{φ_i}. We may assume that k is a finitely generated extension of $\mathbf{Q}$ containing k_0. If $\dim_{k_0} k > 0$, we obtain, by means of the above argument, an abelian variety (A_2, ι_2) of type (F), defined over an extension k_1 of k_0 such that $\dim_{k_0} k_1 < \dim_{k_0} k$. By the result of §11.3, (A_2, ι_2) is of type $(F; \{\varphi_i\})$. Repeating this procedure, we get an abelian variety (A_0, ι_0) of type $(F; \{\varphi_i\})$, defined over an algebraic number field. Now let (A, ι) be an arbitrary abelian variety of type $(F; \{\varphi_i\})$. Then, by Corollary of Theorem 2 of §6 and Remark below it, there exists a homomorphism λ of (A_0, ι_0) into (A, ι). We can find a finite algebraic extension k_0' of k_0, over which (A_0, ι_0) is defined and every point of $\mathrm{Ker}(\lambda)$ is rational. Taking k_0' in place of k_0, apply the above argument to (A, ι). Then we obtain from (A, ι), after several times of reduction, an abelian variety (A', ι') of type $(F; \{\varphi_i\})$, defined over a finite algebraic extension of k_0'; moreover we obtain, at the same time, a homomorphism λ' of (A_0, ι_0) onto (A', ι') as the result of reduction of the homomorphism λ. We observe that A_0 and $\mathrm{Ker}(\lambda)$ never change in the reduction process; so $\mathrm{Ker}(\lambda')$ coincides with $\mathrm{Ker}(\lambda)$. It follows that (A, ι) is isomorphic to (A', ι'). This proves our proposition.

13. The Prime Ideal Decomposition of an $N(\mathfrak{p})$-th Power Homomorphism

13.1. We now prove a fundamental congruence relation for an abelian variety with complex multiplication, which is a generalization of Kronecker's congruence formula for elliptic functions with singular moduli. The relation is described in terms of the prime ideal decomposition of an $N(\mathfrak{p})$-th power homomorphism of the reduction modulo $\mathfrak{p}$ of an abelian variety belonging to a given CM-type.

Let $(F; \{\varphi_i\})$ be a CM-type and $(K^*; \{\psi_\alpha\})$ the reflex of $(F; \{\varphi_i\})$. Let (A, ι) be an abelian variety of type $(F; \{\varphi_i\})$ defined over an algebraic number field k of finite degree. We assume that (A, ι) is *principal*. By Proposition 30 of §8.5, we know that k contains K^*. We extend the ψ_α to isomorphisms of k onto subfields of $\overline{\mathbf{Q}}$, which we denote again by ψ_α. In Proposition 29 of §8.3 we showed that for every fractional ideal $\mathfrak{a}$ in K^* there is an ideal $\mathfrak{b}$ in F such that $\mathfrak{o}_L\mathfrak{b} = \mathfrak{o}_L \prod_\alpha \mathfrak{a}^{\psi_\alpha}$, where $\mathfrak{o}_L$ is the maximal order of the Galois closure L of F over $\mathbf{Q}$. We put now $\mathfrak{b} = g(\mathfrak{a})$. Thus

$$\mathfrak{o}_L g(\mathfrak{a}) = \mathfrak{o}_L \prod_\alpha \mathfrak{a}^{\psi_\alpha}. \tag{1}$$

Similarly we put $g(\xi) = \prod_\alpha \xi^{\psi_\alpha}$ for $\xi \in K^*$. We have seen in the same proposition that $g(\xi) \in F$ and

$$g(\mathfrak{b})g(\mathfrak{b})^\rho = N(\mathfrak{a})\mathfrak{o}_F, \qquad g(\xi)g(\xi)^\rho = N_{K^*/\mathbf{Q}}(\xi). \tag{2}$$

We now take a prime ideal $\mathfrak{p}$ of K^* and a prime factor $\mathfrak{P}$ of $\mathfrak{p}$ in k, and put $q = N(\mathfrak{p})$. We assume that A has no defect for $\mathfrak{P}$, and denote by $(\widetilde{A}, \widetilde{\iota})$ the reduction of (A, ι) modulo $\mathfrak{P}$. Then $(\widetilde{A}^q, \widetilde{\iota}^q)$ is meaningful, as noted in §1.5 and §7.6.

THEOREM 1. *The notation being as above, let π_1 be the $N(\mathfrak{p})$-th power homomorphism of $\widetilde{A}$ onto $\widetilde{A}^q$, and π the $N(\mathfrak{P})$-th power endomorphism of $\widetilde{A}$. Then the following assertions hold:*

(i) *$(\widetilde{A}^q, \widetilde{\iota}^q; \pi_1)$ is a $g(\mathfrak{p})$-transform of $(\widetilde{A}, \widetilde{\iota})$.*
(ii) *There exists an element π_0 of F such that $\widetilde{\iota}(\pi_0) = \pi$. Moreover, with this π_0 one has $g\big(N_{k/K^*}(\mathfrak{P})\big) = \pi_0\mathfrak{o}$, where $\mathfrak{o}$ is the maximal order of F.*

PROOF. Let n be the dimension of A. We take a finite Galois extension k_1 of $\mathbf{Q}$ containing k and F, and take also a prime factor $\mathfrak{P}_1$ of $\mathfrak{P}$ in k_1. We then denote reduction modulo $\mathfrak{P}_1$ by putting tildes, and denote by $\widetilde{k}$ the residue field modulo $\mathfrak{P}$. By Propositions 6 and 9 of §10.4, we can find n invariant differential forms $\eta_1, \ldots, \eta_n$ on A rational over k which form a basis for $\mathfrak{P}$-finite invariant differential forms over the valuation ring at $\mathfrak{P}$. Then $\widetilde{\eta}_1, \ldots, \widetilde{\eta}_n$ are linearly independent over $\widetilde{k}$. By Proposition 7 of §10.4, for any $\gamma \in \mathfrak{o}$ we can put $\delta\iota(\gamma)\eta_i = \sum_j c_{ij}\eta_j$ with $\mathfrak{P}$-integers c_{ij} in k. Then $\det\big(X - (c_{ij})\big) = \prod_i (X - \gamma^{\varphi_i})$. Taking the equality modulo $\mathfrak{P}_1$, we find that $\big\{\widetilde{\gamma^{\varphi_i}}\big\}$ is the set of eigenvalues of $\delta\widetilde{\iota}(\gamma)$. Now, as seen in §11.2, $(\widetilde{A}, \widetilde{\iota})$ is an abelian variety of type (F) defined over the finite field with $N(\mathfrak{P})$ elements, and hence $\pi = \widetilde{\iota}(\pi_0)$ with an element π_0 of $\mathfrak{o}$ as noted at the end of §7.6. By Proposition 2 of §5.1 we have $N_{F/\mathbf{Q}}(\pi_0) = \nu(\pi) = N(\mathfrak{P})^n$. Put $\pi_0\mathfrak{o} = \mathfrak{p}_1^{e_1} \cdots \mathfrak{p}_s^{e_s}$ with different prime ideals $\mathfrak{p}_i$ in F. For each i choose an automorphism σ_i of k_1 that coincides with φ_i^{-1} on F, and denote by d_t the number of i such that $\mathfrak{P}_1^{\sigma_i}$ divides $\mathfrak{p}_t$. Let h be the class number of F. Then $\mathfrak{p}_t^{he_t} = \gamma_t\mathfrak{o}$ with $\gamma_t \in \mathfrak{o}$. It is easy to see that d_t is the number of i such that $\gamma_t^{\varphi_i} \in \mathfrak{P}_1$. Taking γ_t to be the above γ, we find that $\delta\widetilde{\iota}(\gamma_t)$ has rank at least $n - d_t$. Let $\widetilde{x}$ be a generic point of $\widetilde{A}$ over $\widetilde{k}$. Then $\widetilde{k}(\pi^h\widetilde{x}) \subset \widetilde{k}(\widetilde{\iota}(\gamma_t)\widetilde{x})$ since $\pi_0^h \in \mathfrak{p}_t^{he_t} = \gamma_t\mathfrak{o}$. By Theorem 1 of §2.8 we have

$$N(\mathfrak{p}_t^{he_t}) = N(\gamma_t) = \nu(\widetilde{\iota}(\gamma_t)) \leq N(\mathfrak{P})^{hd_t},$$

and hence

$$N(\mathfrak{p}_t)^{e_t} \leq N(\mathfrak{P})^{d_t}. \tag{3}$$

Putting $N_{k_1/k}(\mathfrak{P}_1) = \mathfrak{P}^r$, we obtain

$$N(\mathfrak{p}_t)^{e_t r} \leq N(\mathfrak{P})^{d_t r} = \prod{}' N(\mathfrak{P}_1^{\sigma_i}), \tag{4}$$

where the product is taken over all i such that $\mathfrak{P}_1^{\sigma_i}$ divides $\mathfrak{p}_t$. This shows in particular that $d_t \neq 0$, so that every $\mathfrak{p}_t$ is divisible by at least one of the $\mathfrak{P}_1^{\sigma_i}$. Each $\mathfrak{P}_1^{\sigma_i}$ divides at most one of the $\mathfrak{p}_t$. Hence, from (4) and the decomposition $\pi_0\mathfrak{o} = \mathfrak{p}_1^{e_1} \cdots \mathfrak{p}_s^{e_s}$ we obtain

$$N(\pi_0)^r \leq \prod_{i=1}^{n} N(\mathfrak{P}_1^{\sigma_i}).$$

We note that both sides are equal to $N(\mathfrak{P})^{nr}$. Therefore, (3) and (4) must be equalities, and every $\mathfrak{P}_1^{\sigma_i}$ must divide exactly one of the $\mathfrak{p}_t$. Put $N_{k_1/F}(\mathfrak{P}_1^{\sigma_i}) = \mathfrak{p}_t^{u_t}$ with $u_t \in \mathbf{Z}$ for every $\mathfrak{p}_t$ divisible by $\mathfrak{P}_1^{\sigma_i}$. Since k_1 is a Galois extension of F, u_t does not depend on the choice of i. We have then $N(\mathfrak{P}_1^{\sigma_i}) = N(\mathfrak{p}_t)^{u_t}$, and $N(\mathfrak{p}_t)^{re_t} = N(\mathfrak{p}_t)^{u_t d_t}$ by equality (4). Hence we have $re_t = u_t d_t$, so that

$$\mathfrak{p}_t^{re_t} = (\mathfrak{p}_t^{u_t})^{d_t} = \prod{}' N_{k_1/F}(\mathfrak{P}_1^{\sigma_i}),$$

where the product is taken over all i such that $\mathfrak{P}_1^{\sigma_i}/\mathfrak{p}_t$. Therefore we obtain

$$(\pi_0)^r = \mathfrak{p}_1^{re_1} \cdots \mathfrak{p}_s^{re_s} = N_{k_1/F}\left(\prod_{i=1}^{n} \mathfrak{P}_1^{\sigma_i}\right). \tag{5}$$

Now take a Frobenius automorphism σ for $\mathfrak{P}_1$ over K^*, that is, an automorphism of k_1 over K^* such that $\mathfrak{P}_1^{\sigma} = \mathfrak{P}_1$ and $z^{\sigma} \equiv z^{N(\mathfrak{p})} \pmod{\mathfrak{P}_1}$ for every algebraic integer z in k_1. Put $N(\mathfrak{p}) = q$. Since $\mathfrak{P}_1^{\sigma} = \mathfrak{P}_1$, A^{σ} has no defect for $\mathfrak{P}_1$ and the reduction of $(A^{\sigma}, \iota^{\sigma})$ modulo $\mathfrak{P}_1$ can be identified with $(\widetilde{A}^q, \widetilde{\iota}^q)$. By

Proposition 31 of §8.5 and Proposition 16 (or Proposition 23) of §7, (A^σ, ι^σ) is a c-transform of (A, ι) for an ideal-class c of F; so by Proposition 15 of §11.2, $(\widetilde{A}^q, \widetilde{\iota}^q)$ is a c-transform of $(\widetilde{A}, \widetilde{\iota})$. By the result of §7.6, $\pi_\mathfrak{p}$ is a homomorphism of $(\widetilde{A}, \widetilde{\iota})$ onto $(\widetilde{A}^q, \widetilde{\iota}^q)$, and so by Proposition 13 of §7.2, $\pi_\mathfrak{p}$ is an $\mathfrak{a}$-multiplication for an ideal $\mathfrak{a}$ of $\mathfrak{o}$. Put $N_{k/K^*}(\mathfrak{P}) = \mathfrak{p}^v$; then $N(\mathfrak{p}) = q^v$. Now let π_a be the q-th power homomorphism of $\widetilde{A}^{q^{a-1}}$ onto $\widetilde{A}^{q^a}$ for each positive integer a. Since $(\widetilde{A}^{q^{a-1}}, \widetilde{A}^{q^a}, \pi_a)$ is an isomorphic image of $(\widetilde{A}, \widetilde{A}^q, \pi_\mathfrak{p})$, we observe that $(\widetilde{A}^{q^a}, \widetilde{\iota}^{q^a}; \pi_a)$ is an $\mathfrak{a}$-transform of $(\widetilde{A}^{q^{a-1}}, \widetilde{\iota}^{q^{a-1}})$. Hence $\pi_v \cdots \pi_2\pi_1$ is an $\mathfrak{a}^v$-multiplication of $\widetilde{A}$ onto $\widetilde{A}^{N(\mathfrak{P})} = \widetilde{A}$. On the other hand, we have

$$(\pi_v \cdots \pi_1)\widetilde{x} = \widetilde{x}^{q^v} = \widetilde{x}^{N(\mathfrak{P})} = \pi\widetilde{x},$$

so that $\pi = \pi_v \cdots \pi_1$. Since π is a (π_0)-multiplication, we have $(\pi_0) = \mathfrak{a}^v$. Therefore, by (5) we obtain

$$\mathfrak{a}^{vr} = (\pi_0)^r = \prod_{i=1}^{n} N_{k_1/F}(\mathfrak{P}_1^{\sigma_i}). \tag{6}$$

With $G = \mathrm{Gal}(k_1/\mathbf{Q})$ let S^* be the set of elements of G which coincide with some ψ_α on K^*. Let H resp. H^* be the subgroup of G corresponding to F resp. K^*. The definition of the reflex of a CM-type implies that $S^* = \bigcup_i \sigma_i H = \bigcup_\alpha H^*\psi_\alpha$. From this and (6) we obtain

$$\mathfrak{a}^{vr} \sim \prod_{\tau\in S^*} \mathfrak{P}_1^\tau \sim \prod_\alpha \prod_{\gamma\in H^*} \mathfrak{P}_1^{\gamma\psi_\alpha} \sim \prod_\alpha N_{k_1/K^*}(\mathfrak{P}_1)^{\psi_\alpha} \sim \prod_\alpha (\mathfrak{p}^{rv})^{\psi_\alpha}, \tag{7}$$

where we write $\mathfrak{x} \sim \mathfrak{y}$ if two ideals $\mathfrak{x}$ and $\mathfrak{y}$ of subfields of k_1 coincide in k_1. This proves $\mathfrak{a} = g(\mathfrak{p})$, which is $(\pi 1)$ of our theorem. Since $(\pi_0) = \mathfrak{a}^v$ and $\mathfrak{p}^v = N_{k/K^*}(\mathfrak{P})$, we obtain $(\pi 2)$, which completes the proof.

It should be noted that the above theorem is essential not only for our later treatment, but also for the existence theorem of an abelian variety over a finite field with a given type of Frobenius element, due to T. Honda. For details, the reader is referred to [H].

13.2. We insert here a simple fact which will be employed later. For A as above, there exist n invariant differential forms ω_i on A such that $\delta\iota(\mu)\omega_i = \mu^{\varphi_i}\omega_i$ for $1 \le i \le n$ and every $\mu \in \mathfrak{o}$. In view of the results of Section 2, we may assume that the ω_i are defined over the field k_1 in the above proof. As shown in (iii) of §10.4, changing ω_i for its suitable multiple, we may assume that ω_i is $\mathfrak{P}_1$-finite and $\widetilde{\omega}_i \neq 0$ for every i. Let us now prove that *the $\widetilde{\omega}_i$ form a basis of $\mathfrak{D}_0(\widetilde{A})$ provided $\mathfrak{p} \cap \mathbf{Q}$ is unramified in F.* Take a basis $\{\beta_1, \dots, \beta_{2n}\}$

of $\mathfrak{o}$ over $\mathbf{Z}$ and put $\theta_i = \rho\varphi_i$. Then the determinant of the matrix

$$(8) \qquad \begin{bmatrix} \beta_1^{\varphi_1} & \cdots & \beta_1^{\varphi_n} & \beta_1^{\theta_1} & \cdots & \beta_1^{\theta_n} \\ \beta_2^{\varphi_1} & \cdots & \beta_2^{\varphi_n} & \beta_2^{\theta_1} & \cdots & \beta_2^{\theta_n} \\ \cdots & \cdots & \cdots & \cdots & \cdots & \cdots \\ \beta_{2n}^{\varphi_1} & \cdots & \beta_{2n}^{\varphi_n} & \beta_{2n}^{\theta_1} & \cdots & \beta_{2n}^{\theta_n} \end{bmatrix}$$

is not divisible by $\mathfrak{P}_1$ under our assumption. Hence the submatrix composed of the first n columns of (8) modulo $\mathfrak{P}_1$ has rank n, and so suitable n rows of the submatrix are linearly independent modulo $\mathfrak{P}_1$. Changing the ordering, we may now assume that the upper left $n \times n$-matrix of (8) modulo $\mathfrak{P}_1$ is invertible. Suppose now $\sum_{i=1}^{n} c_i \widetilde{\omega}_i = 0$ with c_i in the residue field of $\mathfrak{P}_1$. Applying $\delta\widetilde{\iota}(\beta_j)$ to this relation, we easily see that $c_i = 0$ for every i, which proves the expected fact.

THEOREM 2. *The notation being as in Theorem 1, suppose that A is simple, and $\mathfrak{p}$ is of absolute degree 1 and unramified over $\mathbf{Q}$. Then $\widetilde{A}$ is simple and* $\mathrm{End}_{\mathbf{Q}}(\widetilde{A}) = \widetilde{\iota}(F)$.

PROOF. Using the same notation as in the proof of Theorem 1, we take a positive integer e so that all the elements of $\mathrm{End}_{\mathbf{Q}}(\widetilde{A})$ are defined over the finite field with $N(\mathfrak{P})^e$ elements. Then π^e is contained in the center of $\mathrm{End}_{\mathbf{Q}}(\widetilde{A})$. Put $\xi = \pi_0^e$; then $\widetilde{\iota}(\xi) = \pi^e$. Let $\sigma \in G$. If $\xi^\sigma = \xi$, we have $\mathfrak{a}^{vr} = (\mathfrak{a}^{vr})^\sigma$, and hence from (7) we obtain

$$(9) \qquad \prod_{\tau \in S^*} \mathfrak{P}_1^{\tau} = \prod_{\tau \in S^*} \mathfrak{P}_1^{\tau\sigma}.$$

Put $Z = \{\tau \in G \mid \mathfrak{P}_1^\tau = \mathfrak{P}_1\}$. Our assumption on $\mathfrak{p}$ implies that $Z \subset H^*$. Therefore from (9) we see that $S^*\sigma \subset ZS^* \subset H^*S^* = S^*$. Since $(F; \{\varphi_i\})$ is primitive, we have $\sigma \in H$ by Proposition 26 of §8.2. We have thus proved that $\xi^\sigma = \xi$ only if $\sigma \in H$, that is, $F = \mathbf{Q}(\xi)$. Consequently every element of $\mathrm{End}_{\mathbf{Q}}(\widetilde{A})$ commutes with $\widetilde{\iota}(F)$. By Proposition 3 of §5.1, $\widetilde{\iota}(F)$ is its commutor in $\mathrm{End}_{\mathbf{Q}}(\widetilde{A})$, and hence $\mathrm{End}_{\mathbf{Q}}(\widetilde{A}) = \widetilde{\iota}(F)$. Then $\widetilde{A}$ must be simple, since $\mathrm{End}_{\mathbf{Q}}(\widetilde{A})$ is a field.

CHAPTER IV

Construction of Class Fields

Throughout this chapter, we shall denote by $\bar{\alpha}$ the complex conjugate of a complex number α; $\mathfrak{a}$ being an ideal of an algebraic number field, $\bar{\mathfrak{a}}$ will denote the ideal consisting of the elements $\bar{\alpha}$ for $\alpha \in \mathfrak{a}$.

14. Polarized Abelian Varieties of Type $(K; \{\varphi_i\})$

14.1. Let $(K; \{\varphi_i\})$ be a primitive CM-type, and $(K^*; \{\psi_\alpha\})$ the reflex of $(K; \{\varphi_i\})$. As is seen in §8.2, K must be a totally imaginary quadratic extension of a totally real field K_0; put $n = [K_0 : \mathbf{Q}]$. The automorphism of K over K_0 other than the identity is given by $\alpha \to \bar{\alpha}$. Let $\mathfrak{o}$ denote the ring of integers in K. In §§14–16, the symbols $(K; \{\varphi_i\})$, $(K^*; \{\psi_\alpha\})$, K_0, $\mathfrak{o}$ and n will always be used in this sense.

Let (A, ι) be an abelian variety of type $(K; \{\varphi_i\})$. Then, by Proposition 6 of §5.1, $\mathrm{End}_{\mathbf{Q}}(A) = \iota(K)$.

PROPOSITION 1. *Given a primitive CM-type* $(K; \{\varphi_i\})$, *let* (A, ι) *and* (A', ι') *be abelian varieties of type* $(K; \{\varphi_i\})$. *Then, every homomorphism of* A *into* A' *is a homomorphism of* (A, ι) *into* (A', ι').

PROOF. By Corollary of Theorem 2 of §6.1 and by Remark below it, there exists a homomorphism λ of (A', ι') onto (A, ι). Let μ be a homomorphism of A into A'. Then $\lambda\mu$ is an element of $\mathrm{End}_{\mathbf{Q}}(A)$. As we have $\mathrm{End}_{\mathbf{Q}}(A) = \iota(K)$, there exists an element ξ of K such that $\lambda\mu = \iota(\xi)$; so we have $\mu = \lambda^{-1}\iota(\xi)$. Hence μ commutes with the operation of K; this proves our proposition.

14.2. We shall now consider polarizations. For an element β of K, we denote by $v(\beta)$ the vector of $\mathbf{C}^n$ with the components $\beta^{\varphi_1}, \dots, \beta^{\varphi_n}$ and by $T(\beta)$ the diagonal matrix with the diagonal elements $\beta^{\varphi_1}, \dots, \beta^{\varphi_n}$. If $\mathfrak{m}$ is a free $\mathbf{Z}$-submodule of K of rank $2n$, we denote by $D(\mathfrak{m})$ the set of all vectors $v(\beta)$ for $\beta \in \mathfrak{m}$. Let (A, ι) be an abelian variety of type $(K; \{\varphi_i\})$. By Theorem 2 of §6.1, (A, ι) is represented by a complex torus $\mathbf{C}^n/D(\mathfrak{m})$ for a suitable $\mathfrak{m}$. Let $E(u, v)$ be a non-degenerate Riemann form on $\mathbf{C}^n/D(\mathfrak{m})$. By Theorem 4 of

§6.2, there exists an element ζ of K such that

$$E(v(\xi), v(\eta)) = \mathrm{Tr}_{K/\mathbf{Q}}(\zeta\xi\bar{\eta}) \tag{1}$$

for every $\xi \in K$, $\eta \in K$; the element ζ satisfies

$$\bar{\zeta} = -\zeta, \qquad \mathrm{Im}(\zeta^{\varphi_i}) > 0 \qquad (1 \leq i \leq n). \tag{2}$$

Conversely, any such element ζ of K determines a non-degenerate Riemann form on $\mathbf{C}^n/D(\mathfrak{m})$ by (1).

Let A^* be the Picard variety of A; put, for every $\alpha \in K$,

$$\iota^*(\alpha) = {}^t\iota(\bar{\alpha}).$$

Then, we have seen in §6.3 that (A^*, ι^*) is of type $(K; \{\varphi_i\})$; and (A^*, ι^*) is analytically represented by the complex torus $\mathbf{C}^n/D(\mathfrak{m}^*)$, where $\mathfrak{m}^*$ is given by

$$\mathfrak{m}^* = \{\beta \mid \beta \in K, \mathrm{Tr}_{K/\mathbf{Q}}(\beta\bar{\mathfrak{m}}) \subset \mathbf{Z}\}. \tag{3}$$

Furthermore, if X is a divisor on A corresponding to the Riemann form defined by (1), then the homomorphism φ_X of A onto A^* is represented by the matrix $T(\zeta)$. The following proposition is an easy consequence of this fact and relation (2).

PROPOSITION 2. *Given (A, ι) and $\mathbf{C}^n/D(\mathfrak{m})$ as above, let X and Y be two non-degenerate divisors on A; and let E_1, E_2 be the Riemann forms on $\mathbf{C}^n/D(\mathfrak{m})$ defined by X, Y, respectively. Let ζ_i, for $i = 1, 2$ be the elements of K determined by the relation*

$$E_i(v(\xi), v(\eta)) = Tr_{K/\mathbf{Q}}(\zeta_i\xi\bar{\eta}).$$

Then, $\zeta_1^{-1}\zeta_2$ is a totally positive element of K_0 and we have

$$\varphi_X^{-1}\varphi_Y = \iota(\zeta_1^{-1}\zeta_2).$$

Now put

$$\mathfrak{r} = \iota^{-1}(\mathrm{End}(A));$$

then $\mathfrak{r}$ is an order in K.

PROPOSITION 3. *The notation being as in Proposition 2, the polarized abelian varieties $(A, \mathcal{C}(X))$ and $(A, \mathcal{C}(Y))$ are isomorphic if and only if there exist a unit ε of $\mathfrak{r}$ and a positive rational number s such that $\zeta_1^{-1}\zeta_2 = s\varepsilon\bar{\varepsilon}$.*

PROOF. As we have $\mathrm{End}(A) = \iota(\mathfrak{r})$, every automorphism of A is given by $\iota(\varepsilon)$ for a unit ε of $\mathfrak{r}$. Suppose that $\iota(\epsilon)^{-1}$ maps $\mathcal{C}(X)$ onto $\mathcal{C}(Y)$. Then, by the definition of $\mathcal{C}(X)$, there exist two positive integers m and m' such that $m\iota(\varepsilon)^{-1}(X)$ is algebraically equivalent to $m'Y$. Consider the homomorphisms φ_X and φ_Y of A onto the Picard variety A^* of A; we have then by (7) of §1.3, $m\,{}^t\iota(\varepsilon)\varphi_X\iota(\varepsilon) = m'\varphi_Y$. Put $s = m/m'$. By relations (5) and (6) of §6.3, and by Proposition 2, we have $s\iota(\varepsilon\bar{\varepsilon}) = \varphi_X^{-1}\varphi_Y = \iota(\zeta_1^{-1}\zeta_2)$. This proves the "only if" part. The "if" part is proved by following up the above argument in the opposite direction.

COROLLARY. *Given (A, ι) and $\mathfrak{r}$ as above, let $\mathcal{C}$ be a polarization of A. Then, for every root of unity ε contained in $\mathfrak{r}$, $\iota(\varepsilon)$ gives an automorphism of $(A, \mathcal{C})$. Conversely, every automorphism of $(A, \mathcal{C})$ is given by $\iota(\varepsilon)$ for a root of unity ε contained in $\mathfrak{r}$.*

PROOF. Put $\mathcal{C} = \mathcal{C}(X) = \mathcal{C}(Y)$ and $\zeta_1 = \zeta_2$ in the above proposition. We see then that if ε is a unit of $\mathfrak{r}$, $\iota(\varepsilon)$ gives an automorphism of $(A, \mathcal{C})$ if and only if $\varepsilon\bar{\varepsilon} = 1$. Since $(\bar{\varepsilon})^{\varphi_i}$ is the complex conjugate of ε^{φ_i} (cf. Lemma 2 of §5.1), the condition $\varepsilon\bar{\varepsilon} = 1$ is equivalent to that the ε^{φ_i} are all of absolute value 1. It is well-known that an algebraic integer α is a root of unity if and only if every conjugate of α over $\mathbf{Q}$ is of absolute value 1. This proves the assertion of our corollary.

14.3. Given $(K; \{\varphi_i\})$ as before, by a *polarized abelian variety of type* $(K; \{\varphi_i\})$, we shall understand a triplet $\mathcal{P} = (A, \iota, \mathcal{C})$ formed by an abelian variety (A, ι) of type $(K; \{\varphi_i\})$ and a polarization $\mathcal{C}$ of A; $\mathcal{P} = (A, \iota, \mathcal{C})$ and $\mathcal{P}_1 = (A_1, \iota_1, \mathcal{C}_1)$ are said to be *isomorphic* if there exists an isomorphism of (A, ι) onto (A_1, ι_1) which sends $\mathcal{C}$ onto $\mathcal{C}_1$. As we have restricted ourselves to primitive CM-types, every isomorphism of $(A, \mathcal{C})$ onto $(A_1, \mathcal{C}_1)$ gives an isomorphism of $\mathcal{P}$ onto $\mathcal{P}_1$ on account of Proposition 1.

Now we impose one more condition on our abelian varieties (A, ι). From now on, until the end of §16, *by an abelian variety (A, ι) of type $(K; \{\varphi_i\})$, we shall understand a principal one*; namely, we assume

$$\iota(\mathfrak{o}) = \mathrm{End}(A),$$

where $\mathfrak{o}$ is the ring of integers in K.

Let (A, ι) be an abelian variety of type $(K; \{\varphi_i\})$. If (A^*, ι^*), $\mathbf{C}^n/D(\mathfrak{m})$, and $\mathbf{C}^n/D(\mathfrak{m}^*)$ are determined as above, we observe that (A^*, ι^*) is also principal and $\mathfrak{m}$, $\mathfrak{m}^*$ are ideals of K. Moreover, if we denote by $\mathfrak{d}$ the different of K relative to $\mathbf{Q}$, relation (3) of §14.2 shows that

$$\mathfrak{m}^* = (\mathfrak{d}\bar{\mathfrak{m}})^{-1}.$$

Let X be a non-degenerate divisor of A, corresponding to the Riemann form given by (1) of §14.2. The homomorphism φ_X of A onto A^* is represented by the matrix $T(\zeta)$. Hence, by Proposition 15 of §7.4, φ_X is a $(\zeta\mathfrak{d}\mathfrak{m}\bar{\mathfrak{m}})$-multiplication of (A, ι) onto (A^*, ι^*). Put

$$\mathfrak{f} = \zeta\mathfrak{d}\mathfrak{m}\bar{\mathfrak{m}}.$$

We shall now prove that $\mathfrak{f}$ is an "ideal of K_0", namely, there exists an ideal $\mathfrak{f}_0$ of K_0 such that $\mathfrak{f} = \mathfrak{o}\mathfrak{f}_0$. Let $\mathfrak{d}_0$ be the different of K_0 relative to $\mathbf{Q}$ and $\mathfrak{d}_1$ the different of K relative to K_0; we have then $\mathfrak{d} = \mathfrak{d}_0\mathfrak{d}_1$. The ideal $\mathfrak{d}_1$ is generated by the elements $\theta - \bar{\theta}$ for $\theta \in \mathfrak{o}$. By property (2) of the element ζ, we see that $\zeta(\theta - \bar{\theta})$ is contained in K_0 for every $\theta \in \mathfrak{o}$, so that $\zeta\mathfrak{d}_1$ is an ideal of K_0. It is obvious that $\mathfrak{d}_0\mathfrak{m}\bar{\mathfrak{m}}$ is an ideal of K_0. Hence $\mathfrak{f} = (\zeta\mathfrak{d}_1)(\mathfrak{d}_0\mathfrak{m}\bar{\mathfrak{m}})$ is an ideal of K_0.

Let $\mathcal{C}$ be a polarization of A. Take a basic polar divisor Y of $(A, \mathcal{C})$ (cf. §4.2). Then, the above argument shows that there exists an ideal $\mathfrak{f}_0$ of K_0 such that φ_Y is an $(\mathfrak{o}\mathfrak{f}_0)$-multiplication of (A, ι) onto (A^*, ι^*). We observe that the ideal $\mathfrak{f}_0$ is determined by $\mathcal{P} = (A, \iota, \mathcal{C})$ and does not depend on the choice of the basic polar divisor Y; so we say that the polarized abelian variety $\mathcal{P}$ is of type $(K; \{\varphi_i\}; \mathfrak{f}_0)$.

PROPOSITION 4. *Let $\mathfrak{f}_0$ be an ideal of K_0. Then, there exists a polarized abelian variety of type $(K; \{\varphi_i\}; s\mathfrak{f}_0)$ for some rational number s if and only if there exists an ideal $\mathfrak{m}$ of K and an element ζ of K such that*

$$\mathfrak{o}\mathfrak{f}_0 = \zeta\mathfrak{d}\mathfrak{m}\bar{\mathfrak{m}},$$

$$\text{(2)} \qquad \bar{\zeta} = -\zeta, \qquad \operatorname{Im}(\zeta^{\varphi_i}) > 0 \qquad (1 \leq i \leq n),$$

where $\mathfrak{d}$ denotes the different of K relative to $\mathbf{Q}$.

PROOF. We have already proved the "only if" part; so we shall now prove the "if" part. We may assume that $\mathfrak{f}_0$ is an integral ideal. If $\mathfrak{m}$ and ζ are given as above, we obtain an abelian variety (A, ι) of type $(K; \{\varphi_i\})$ by means of the complex torus $\mathbf{C}^n/D(\mathfrak{m})$; and ζ defines a bilinear form $E(u, v)$ by relation (1) of §14.2. As $\zeta\mathfrak{d}\mathfrak{m}\bar{\mathfrak{m}} = \mathfrak{o}\mathfrak{f}_0$ is integral, the values of E on $D(\mathfrak{m}) \times D(\mathfrak{m})$ are rational integers; hence by Theorem 4 of §6.2, E is a non-degenerate Riemann form on $\mathbf{C}^n/D(\mathfrak{m})$. If X is a divisor on A corresponding to the form E, φ_X is an $(\mathfrak{o}\mathfrak{f}_0)$-multiplication; so $(A, \iota, \mathcal{C}(X))$ is of type $(K; \{\varphi_i\}; s\mathfrak{f}_0)$ for a suitable rational number s; this completes the proof.

Given (A, ι) as above, let $\mathcal{C}_1$ and $\mathcal{C}_2$ be two polarizations of A; suppose that $(A, \iota, \mathcal{C}_1)$ and $(A, \iota, \mathcal{C}_2)$ are of the same type $(K; \{\varphi_i\}; \mathfrak{f}_0)$. Then $\mathcal{C}_i$ contains a divisor Y_i such that φ_{Y_i} is an $(\mathfrak{o}\mathfrak{f}_0)$-multiplication, for each i. Let ζ_i be the element of K for which the bilinear form $\mathrm{Tr}_{K/\mathbf{Q}}(\zeta_i \xi \bar{\eta})$ defines the Riemann form defined by Y_i. By Proposition 2, we have $(\varphi_{Y_1})^{-1}\varphi_{Y_2} = \iota(\zeta_1^{-1}\zeta_2)$; and $\zeta_1^{-1}\zeta_2$ is a totally positive element of K_0. Since both φ_{Y_i} are $(\mathfrak{o}\mathfrak{f}_0)$-multiplications, $\zeta_1^{-1}\zeta_2$ is a unit of K_0. Conversely, take a totally positive unit ε of K_0 and put $\zeta = \varepsilon\zeta_1$. Then ζ satisfies relation (2), and hence there exists a divisor X on A corresponding to the form $\mathrm{Tr}(\zeta\xi\bar{\eta})$. Then, we see easily that $(A, \iota, \mathcal{C}(X))$ is of type $(K; \{\varphi_i\}; \mathfrak{f}_0)$. By Proposition 3, $(A, \iota, \mathcal{C}_1)$ is isomorphic to $(A, \iota, \mathcal{C}_2)$ if and only if $\zeta_1^{-1}\zeta_2$ is of the form $\alpha\bar{\alpha}$ for a unit α of K. The following proposition is a consequence of these observations.

PROPOSITION 5. *Let U be the group of all totally positive units of K_0, and U_1 the subgroup of U consisting of the elements $N_{K/K_0}(\varepsilon)$ for all units ε of K. Suppose that $(A, \iota, \mathcal{C})$ of type $(K; \{\varphi_i\}; \mathfrak{f}_0)$ exists. Then there exist exactly $[U : U_1]$ isomorphism classes of such polarized abelian varieties with the same (A, ι).*

We note that the index $[U : U_1]$ is finite.

14.4. Let $(A, \iota, \mathcal{C})$ and $(A_1, \iota_1, \mathcal{C}_1)$ be two polarized abelian varieties of the same type $(K; \{\varphi_i\}; \mathfrak{f}_0)$. Put $\mathfrak{f} = \mathfrak{o}\mathfrak{f}_0$. Let X and Y be respectively a basic polar divisor in $\mathcal{C}$ and a basic polar divisor in $\mathcal{C}_1$; then both φ_X and φ_Y are $\mathfrak{f}$-multiplications. Let λ be a homomorphism of (A, ι) onto (A_1, ι_1). By Proposition 23 of §7.5, λ is an $\mathfrak{a}$-multiplication of (A, ι) onto (A_1, ι_1) for an ideal $\mathfrak{a}$ of $\mathfrak{o}$. Put $Z = \lambda^{-1}(Y)$. Then $\varphi_X^{-1}\varphi_Z$ is an element of $\mathrm{End}_{\mathbf{Q}}(A)$; so there exists an element α of K such that $\iota(\alpha) = \varphi_X^{-1}\varphi_Z$. We observe that α is determined only by λ and does not depend on the choice of X and Y; we write

$$\alpha = f(\lambda).$$

Let (A^*, ι^*) and (A_1^*, ι_1^*) be respectively the duals of (A, ι) and (A_1, ι_1). Now we need the following fact.

PROPOSITION 6. *The notation being as above, if λ is an $\mathfrak{a}$-multiplication of (A, ι) onto (A_1, ι_1), then ${}^t\lambda$ is an $\bar{\mathfrak{a}}$-multiplication of (A_1^*, ι_1^*) onto (A^*, ι^*).*

PROOF. By Proposition 1, ${}^t\lambda$ is a homomorphism of (A_1^*, ι_1^*) onto (A^*, ι^*). By Proposition 13 of §7.2 and by Proposition 1, we have $\mathrm{Hom}(A, A_1) = \lambda\iota(\mathfrak{a}^{-1})$. Now consider the mapping $\mu \to {}^t\mu$ which gives an isomorphism of $\mathrm{Hom}(A, A_1)$ onto $\mathrm{Hom}(A_1^*, A^*)$. As we have ${}^t(\lambda\iota(\xi)) = {}^t(\iota_1(\xi)\lambda) = {}^t\lambda{}^t\iota_1(\xi) = {}^t\lambda\iota_1^*(\bar{\xi})$, we obtain $\mathrm{Hom}(A_1^*, A^*) = {}^t\lambda\iota_1^*((\bar{\mathfrak{a}})^{-1})$. This proves our proposition. We can also prove the proposition by means of Proposition 15 of §7.4.

Now, since $Z = \lambda^{-1}(Y)$, we have $\varphi_Z = {}^t\lambda\varphi_Y\lambda$, and hence

$$\iota(f(\lambda)) = \varphi_X^{-1} \cdot {}^t\lambda\varphi_Y\lambda.$$

As both φ_X and φ_Y are $\mathfrak{f}$-multiplications, we see, by the above proposition, that $\iota(f(\lambda))$ is an $\mathfrak{a}\bar{\mathfrak{a}}$-multiplication. On the other hand, by Proposition 2, $f(\lambda)$ is a totally positive element of K_0. We have thus proved that $f(\lambda)$ is a totally positive element of K_0 such that

$$\mathfrak{a}\bar{\mathfrak{a}} = (f(\lambda)).$$

Conversely, let $\mathfrak{b}$ be an ideal of $\mathfrak{o}$ such that there exists a totally positive element β of K_0 for which we have $\mathfrak{b}\bar{\mathfrak{b}} = (\beta)$. Let $(A_2, \iota_2; \mu)$ be a $\mathfrak{b}$-transform of (A, ι). By the results of §7.4, we can take complex tori $\mathbf{C}^n/D(\mathfrak{m})$ and $\mathbf{C}^n/D(\mathfrak{b}^{-1}\mathfrak{m})$ as analytic representations of (A, ι) and (A_2, ι_2). Let ζ be the element of K corresponding to the basic polar divisor X. We have then

$$\mathfrak{f} = \zeta\mathfrak{d}\mathfrak{m}\bar{\mathfrak{m}} = (\beta\zeta)\mathfrak{d}(\mathfrak{b}^{-1}\mathfrak{m})\overline{(\mathfrak{b}^{-1}\mathfrak{m})}; \tag{4}$$

and we see that $\beta\zeta$ satisfies condition (2) of Proposition 4. Then, it can be easily verified, by means of the same argument as in the proof of that proposition, that A_2 has a polarization $\mathcal{C}_2$ for which $(A_2, \iota_2, \mathcal{C}_2)$ is of type $(K; \{\varphi_i\}; \mathfrak{f}_0)$. Thus we have proved:

PROPOSITION 7. *Let $(A, \iota, \mathcal{C})$ be a polarized abelian variety of type $(K; \{\varphi_i\}; \mathfrak{f}_0)$; let $\mathfrak{a}$ be an ideal of $\mathfrak{o}$ and (A_1, ι_1) be an $\mathfrak{a}$-transform of (A, ι). Then, A_1 has a polarization $\mathcal{C}_1$ such that $(A_1, \iota_1, \mathcal{C}_1)$ is of type $(K; \{\varphi_i\}; \mathfrak{f}_0)$ if and only if there exists a totally positive element α of K_0 such that $\mathfrak{a}\bar{\mathfrak{a}} = (\alpha)$.*

For λ as above, we have $\mathrm{Hom}(A, A_1) = \lambda\iota(\mathfrak{a}^{-1})$ by virtue of Proposition 13 of §7.2; we shall now prove that for every $\xi \in \mathfrak{a}^{-1}$,

$$f(\lambda\iota(\xi)) = f(\lambda)\xi\bar{\xi}. \tag{5}$$

Put $\mu = \lambda\iota(\xi)$. We have then $\iota(f(\mu)) = \varphi_X^{-1} \cdot {}^t\mu\varphi_Y\mu = \varphi_X^{-1} \cdot {}^t\iota(\xi){}^t\lambda\varphi_Y\lambda\iota(\xi) = \iota(\bar{\xi})\varphi_X^{-1} \cdot {}^t\lambda\varphi_Y\lambda\iota(\xi) = \iota(\bar{\xi})\iota(f(\lambda))\iota(\xi) = \iota(f(\lambda)\xi\bar{\xi})$. This proves (5). Therefore, f is considered as a hermitian form defined on the module $\mathfrak{a}^{-1}$.

14.5. Class of hermitian forms. Before proceeding further, we give a definition concerning a certain class of hermitian forms on K. Let $\mathfrak{a}$ be an ideal of K such that there exists a totally positive element ρ of K_0 for which we have $\mathfrak{a}\bar{\mathfrak{a}} = (\rho)$. Then the form $\rho\xi\bar{\xi}$ for $\xi \in K$ is a positive hermitian form on K taking algebraic integral values on $\mathfrak{a}^{-1}$. Now let $\{\mathfrak{a}_1, \rho_1\}$ be another pair such that $\mathfrak{a}_1\bar{\mathfrak{a}}_1 = (\rho_1)$ and ρ_1 is totally positive. We say that $\{\mathfrak{a}_1, \rho_1\}$ and $\{\mathfrak{a}, \rho\}$ are *equivalent* if there exists an element $\mu \in K$ for which we have $\mu\mathfrak{a} = \mathfrak{a}_1$ and $\rho_1 = \rho\mu\bar{\mu}$, that is, if the module $\mathfrak{a}^{-1}$ with the form $\rho\xi\bar{\xi}$ is isomorphic to $\mathfrak{a}_1^{-1}$ with $\rho_1\xi\bar{\xi}$ under the mapping $\xi \to \mu^{-1}\xi$. The class determined by this equivalence relation will be denoted by $(\mathfrak{a}, \rho)$; we call it a *class of positive hermitian forms in* K. Define the multiplication of two classes $(\mathfrak{a}, \rho)$ and $(\mathfrak{a}_1, \rho_1)$ by

$$(\mathfrak{a}, \rho)(\mathfrak{a}_1, \rho_1) = (\mathfrak{a}\mathfrak{a}_1, \rho\rho_1).$$

Then the set of classes becomes a group with the identity element given by $(\mathfrak{o}, 1)$. We denote this group by $\mathfrak{C}(K)$.

14.6. Let $\mathcal{P} = (A, \iota, \mathcal{C})$ and $\mathcal{P}_1 = (A_1, \iota_1, \mathcal{C}_1)$ be two polarized abelian varieties of the same type $(K; \{\varphi_i\}; \mathfrak{f}_0)$. Take a homomorphism λ of (A, ι) onto (A_1, ι_1); there exists an ideal $\mathfrak{a}$ of $\mathfrak{o}$ such that λ is an $\mathfrak{a}$-multiplication. We have shown that $f(\lambda)$ is totally positive and $\mathfrak{a}\bar{\mathfrak{a}} = (f(\lambda))$; so we obtain an element $(\mathfrak{a}, f(\lambda))$ of $\mathfrak{C}(K)$. By means of relation (5), we see that the class $(\mathfrak{a}, f(\lambda))$ does not depend on the choice of λ. We write

$$\{\mathcal{P}_1 : \mathcal{P}\} = (\mathfrak{a}, f(\lambda)).$$

PROPOSITION 8. *If $\mathcal{P}$, $\mathcal{P}_1$, $\mathcal{P}_2$ are three polarized abelian varieties of the same type $(K; \{\varphi_i\}; \mathfrak{f}_0)$, we have*

$$\{\mathcal{P}_2 : \mathcal{P}_1\}\{\mathcal{P}_1 : \mathcal{P}\} = \{\mathcal{P}_2 : \mathcal{P}\}.$$

PROOF. By our definition, there exist ideals $\mathfrak{a}$, $\mathfrak{b}$ of $\mathfrak{o}$ and an $\mathfrak{a}$-multiplication λ of $\mathcal{P}$ onto $\mathcal{P}_1$ and a $\mathfrak{b}$-multiplication μ of $\mathcal{P}_1$ onto $\mathcal{P}_2$ such that

$$\{\mathcal{P}_1 : \mathcal{P}\} = (\mathfrak{a}, f(\lambda)), \qquad \{\mathcal{P}_2 : \mathcal{P}_1\} = (\mathfrak{b}, f(\mu)).$$

Then $\mu\lambda$ is an $\mathfrak{a}\mathfrak{b}$-multiplication of $\mathcal{P}$ onto $\mathcal{P}_2$. Let X, Y, Z be respectively basic polar divisors of $\mathcal{P}$, $\mathcal{P}_1$, $\mathcal{P}_2$. We have then

$$\begin{aligned}
\varphi_X^{-1} \cdot {}^t(\mu\lambda)\varphi_Z(\mu\lambda) &= \varphi_X^{-1} \cdot {}^t\lambda^t\mu\varphi_Z\mu\lambda = \varphi_X^{-1} \cdot {}^t\lambda\varphi_Y(\varphi_Y^{-1} \cdot {}^t\mu\varphi_Z\mu)\lambda \\
&= \varphi_X^{-1} \cdot {}^t\lambda\varphi_Y\iota_1(f(\mu))\lambda = \varphi_X^{-1} \cdot {}^t\lambda\varphi_Y\lambda\iota(f(\mu)) \\
&= \iota(f(\lambda))\iota(f(\mu)) = \iota(f(\lambda)f(\mu)).
\end{aligned}$$

Hence we have $\{\mathcal{P}_2 : \mathcal{P}\} = (\mathfrak{a}\mathfrak{b}, f(\lambda)f(\mu))$; this proves the proposition.

PROPOSITION 9. *Let $\mathcal{P} = (A, \iota, \mathcal{C})$ and $\mathcal{P}_1 = (A_1, \iota_1, \mathcal{C}_1)$ be two polarized abelian varieties of the same type $(K; \{\varphi_i\}; \mathfrak{f}_0)$; let η be an isomorphism of A onto A_1. Then, η is an isomorphism of $\mathcal{P}$ onto $\mathcal{P}_1$ if and only if $f(\eta) = 1$.*

PROOF. Let X and Y be basic polar divisors of $\mathcal{C}$ and $\mathcal{C}_1$, respectively. Put $Z = \eta^{-1}(Y)$. If η is an isomorphism of $\mathcal{P}$ onto $\mathcal{P}_1$, there exist two positive integers m and m' such that mZ is algebraically equivalent to $m'X$, namely, $m\varphi_Z = m'\varphi_X$; we have then $f(\eta) = m'/m$. Since η is an $\mathfrak{o}$-multiplication, we have $\mathfrak{o} = (f(\eta))$. It follows that $f(\eta) = 1$; this proves the "only if" part. Conversely, if $f(\eta) = 1$, we obtain $\varphi_X^{-1}\varphi_Z = 1_A$ and hence $\varphi_X = \varphi_Z$, so that $\eta^{-1}(Y) = Z$ is algebraically equivalent to X; hence η is an isomorphism of $\mathcal{P}$ onto $\mathcal{P}_1$. This completes the proof.

PROPOSITION 10. *Let $\mathcal{P}$ and $\mathcal{P}_1$ be two polarized abelian varieties of the same type. Then, $\mathcal{P}$ and $\mathcal{P}_1$ are isomorphic if and only if* $\{\mathcal{P}_1 : \mathcal{P}\} = (\mathfrak{o}, 1)$.

PROOF. The "only if" part follows directly from Proposition 9; so we prove the "if" part. If $\{\mathcal{P}_1 : \mathcal{P}\} = (\mathfrak{o}, 1)$, there exist an ideal $\mathfrak{a}$ of $\mathfrak{o}$ and an $\mathfrak{a}$-multiplication λ of A onto A_1 such that $(\mathfrak{o}, 1) = (\mathfrak{a}, f(\lambda))$. By our definition, we can find an element γ of K such that $\mathfrak{a} = (\gamma)$ and $f(\lambda) = \gamma\bar{\gamma}$. Then, as $\iota(\gamma)$ is an $\mathfrak{a}$-multiplication of A onto itself, there exists an isomorphism η of (A, ι) onto (A_1, ι_1) such that $\eta\iota(\gamma) = \lambda$. By relation (5) of §14.4, we have $f(\eta) = 1$, so that by Proposition 9, $\mathcal{P}$ is isomorphic to $\mathcal{P}_1$. This completes the proof.

COROLLARY. *Let $\mathcal{P}$, $\mathcal{P}_1$, $\mathcal{P}_2$ be three polarized abelian varieties of the same type. Then, $\mathcal{P}_1$ and $\mathcal{P}_2$ are isomorphic if and only if* $\{\mathcal{P}_1 : \mathcal{P}\} = \{\mathcal{P}_2 : \mathcal{P}\}$.

This is an immediate consequence of Propositions 8 and 10.

14.7. Let $\mathcal{P} = (A, \iota, \mathcal{C})$ be a polarized abelian variety of type $(K; \{\varphi_i\}; \mathfrak{f}_0)$. We say that $\mathcal{P}$ is defined over k if (A, ι) is defined over k and $\mathcal{C}$ is defined over k. If that is so, by Proposition 30 of §8.5, k must contain the field K^*. Let τ be an isomorphism of k onto a field k' which leaves the elements of K^* invariant. Then we obtain naturally a system $\mathcal{P}^\tau = (A^\tau, \iota^\tau, \mathcal{C}^\tau)$. By Proposition 31 of §8.5, (A^τ, ι^τ) is of type $(K; \{\varphi_i\})$. Let (A^*, ι^*) be the dual of (A, ι) and k_1 a field of definition for (A^*, ι^*) containing k. We extend τ to an isomorphism of k_1, which we denote again by τ. It is easy to see that $(A^{*\tau}, \iota^{*\tau})$ gives the dual of (A^τ, ι^τ). Let X be a basic polar divisor in $\mathcal{C}$; we take k_1 so large that X is rational over k_1; φ_X is an $(\mathfrak{o}\mathfrak{f}_0)$-multiplication. As is seen in §7.6, $\varphi_{(X^\tau)} = (\varphi_X)^\tau$ is an $(\mathfrak{o}\mathfrak{f}_0)$-multiplication. Therefore, $\mathcal{P}^\tau = (A^\tau, \iota^\tau, \mathcal{C}^\tau)$ is of the same type as $\mathcal{P}$.

Let $\mathcal{P}_1 = (A_1, \iota_1, \mathcal{C}_1)$ be another abelian variety of the same type as $\mathcal{P}$ and λ a homomorphism of (A, ι) onto (A_1, ι_1) onto (A_1, ι_1); let Y be a basic polar divisor in $\mathcal{C}_1$. We may assume that (A_1, ι_1) and λ is defined over k_1 and Y is rational over k_1; if this is not so, we take a suitable extension of k_1 instead of k_1. Now if λ is an $\mathfrak{a}$-multiplication, we have $\mathfrak{a}\bar{\mathfrak{a}} = (f(\lambda))$ and

$\iota(f(\lambda)) = \varphi_X^{-1} \cdot {}^t\lambda\varphi_Y\lambda$. We see easily that $\iota^\tau(f(\lambda)) = \varphi_{(X^\tau)}^{-1} \cdot {}^t(\lambda^\tau)\varphi_{(Y^\tau)}\lambda^\tau$ and λ^τ is an $\mathfrak{a}$-multiplication. We have thus proved the following proposition.

PROPOSITION 11. *Let $\mathcal{P}$ and $\mathcal{P}_1$ be two polarized abelian varieties of the same type $(K; \{\varphi_i\}; \mathfrak{f}_0)$, defined over k; let τ be an isomorphism of k onto a field k′, which leaves the elements of K^* invariant. Then, $\mathcal{P}^\tau$ and $\mathcal{P}_1^\tau$ are of the same type $(K; \{\varphi_i\}; \mathfrak{f}_0)$ as $\mathcal{P}$ and $\mathcal{P}_1$; and we have*

$$\{\mathcal{P}_1^\tau : \mathcal{P}^\tau\} = \{\mathcal{P}_1 : \mathcal{P}\}.$$

15. The Unramified Class Field Obtained from the Field of Moduli

15.1. We shall now proceed to the theory of construction of class fields. First we introduce some symbols concerning ideal groups. Let k be an algebraic number field of finite degree; let $\mathfrak{m}$ be an integral ideal of k. For $\alpha \in k$ we write $\alpha \equiv 1 \operatorname{mod}^\times \mathfrak{m}$ if $(\alpha - 1)\mathfrak{x} = \mathfrak{m}\mathfrak{y}$ with integral ideals $\mathfrak{x}$ and $\mathfrak{y}$ prime to $\mathfrak{m}$. We denote by $I_k(\mathfrak{m})$ the group of ideals of k which are prime to $\mathfrak{m}$, and by $P_k(\mathfrak{m})$ the subgroup of $I_k(\mathfrak{m})$ consisting of all principal ideals (α) such that $\alpha \in k$, $\alpha \equiv 1 \operatorname{mod}^\times \mathfrak{m}$. The factor group $I_k(\mathfrak{m})/P_k(\mathfrak{m})$ is then the group of ideal-classes modulo $\mathfrak{m}$. For our purpose in this section, it is not necessary to consider infinite primes. If k' is a Galois extension of k, we denote by $\mathrm{Gal}(k'/k)$ the Galois group of k' over k.

As before, let $(K; \{\varphi_i\})$ be a primitive CM-type, $[K : \mathbf{Q}] = 2n$, $(K^*; \{\psi_\alpha\})$ the reflex of $(K; \{\varphi_i\})$, K_0 the totally real subfield of K of degree n, and $\mathfrak{o}$ the ring of integers in K. Let (A, ι) be an abelian variety of type $(K; \{\varphi_i\})$ which is principal. We assume in the sequel that (A, ι) is defined over an algebraic number field. Let $\mathcal{C}$ be a polarization of A and Y a basic polar divisor in $\mathcal{C}$. We may assume that Y is algebraic over $\mathbf{Q}$; in fact, if Y is not algebraic over $\mathbf{Q}$, we can find a specialization Y' of Y algebraic over $\mathbf{Q}$; we see easily that Y' is also a basic polar divisor in $\mathcal{C}$. By the result of §14, there exists an integral ideal $\mathfrak{f}_0$ of K_0 such that φ_Y is an $(\mathfrak{o}\mathfrak{f}_0)$-multiplication of (A, ι) onto its dual. Put $\mathcal{P} = (A, \iota, \mathcal{C})$ and $\mathfrak{f} = \mathfrak{o}\mathfrak{f}_0$. Let k be an algebraic number field of finite degree satisfying the following conditions:

(i) *k is normal over K^*.*
(ii) *A is defined over k.*
(iii) *For every $\sigma \in \mathrm{Gal}(k/K^*)$, all the elements of* $\mathrm{Hom}(A, A^\sigma)$ *are defined over k.*
(iv) *Y is rational over k.*

Such a field k really exists, since there are only finitely many transforms A^σ of A over K^*. As $(K; \{\varphi_i\})$ is the reflex of $(K^*; \{\psi_\alpha\})$, k contains K. By

Proposition 30 of §8.5, every element of $\mathrm{End}(A^\sigma)$ is defined over k. Let k_0 be the field of moduli of $(A, \mathcal{C})$. Then, k_0 is contained in k. Put $k_0^* = k_0 K^*$. These symbols and assumptions will be retained until the end of §16.

15.2. Our purpose is to describe the extension k_0^* of K^* in terms of class field theory. Let σ be an element of $\mathrm{Gal}(k/K^*)$. By Proposition 11, $\mathcal{P}^\sigma$ is of the same type as $\mathcal{P}$; so $\{\mathcal{P}^\sigma, \mathcal{P}\}$ has a meaning. Put

$$[\sigma] = \{\mathcal{P}^\sigma : \mathcal{P}\}.$$

By Propositions 8 and 11, we have, for every $\sigma, \tau \in \mathrm{Gal}(k/K^*)$,

$$\begin{aligned}[\sigma\tau] = \{\mathcal{P}^{\sigma\tau} : \mathcal{P}\} &= \{\mathcal{P}^{\sigma\tau} : \mathcal{P}^\tau\}\{\mathcal{P}^\tau : \mathcal{P}\} \\ &= \{\mathcal{P}^\sigma : \mathcal{P}\}\{\mathcal{P}^\tau : \mathcal{P}\} = [\sigma][\tau];\end{aligned}$$

namely, $\sigma \to [\sigma]$ gives a homomorphism of $\mathrm{Gal}(k/K^*)$ into the group $\mathfrak{C}(K)$. Let H be the kernel of this homomorphism. By Proposition 10, we have $\sigma \in H$ if and only if $\mathcal{P}^\sigma$ is isomorphic to $\mathcal{P}$. On the other hand, by Proposition 1 and by the definition of field of moduli, $\mathcal{P}^\sigma$ is isomorphic to $\mathcal{P}$ if and only if σ leaves invariant the elements of the field of moduli k_0. Therefore, H is the set of elements of $\mathrm{Gal}(k/K^*)$ which leave invariant the elements of k_0^*. It follows that $\sigma \to [\sigma]$ gives an injective homomorphism of $\mathrm{Gal}(k_0^*/K^*)$ into $\mathfrak{C}(K)$. As $\mathfrak{C}(K)$ is an abelian group, k_0^* must be an abelian extension of K^*.

15.3. Now we consider reduction of (A, ι) modulo prime ideals of k. By Proposition 25 of §12.3, A has no defect for almost all prime ideals of k. Here and in the following, the terms "almost all" mean "all except a finite number of". Let $\mathfrak{m}$ be the product of prime ideals $\mathfrak{p}$ of K^* satisfying at least one of the following conditions:

(i) There exists a prime ideal of k dividing $\mathfrak{p}$, for which A has defect.
(ii) $\mathfrak{p}$ is ramified in k_0^*.

Let $\mathfrak{p}$ be a prime ideal of K^* which does not divide $\mathfrak{m}$, and $\mathfrak{P}$ a prime ideal of k dividing $\mathfrak{p}$; then, by our definition of $\mathfrak{m}$, for every $\sigma \in \mathrm{Gal}(k/K^*)$, A^σ has no defect for $\mathfrak{P}$. In the following treatment, we denote reduction modulo $\mathfrak{P}$ by putting tildes. Put $N(\mathfrak{p}) = q$. Let σ be a Frobenius automorphism of k for $\mathfrak{P}/\mathfrak{p}$. We can identify $(\tilde{A}^q, \tilde{\iota}^q)$ with the reduction of (A^σ, ι^σ) modulo $\mathfrak{P}$. Let π be the q-th power homomorphism of $\tilde{A}$ onto $\tilde{A}^q$. Put

$$\mathfrak{q} = g(\mathfrak{p})$$

with g defined by (1) of §13.1. Then, by Theorem 1 of §13, π is a $\mathfrak{q}$-multiplication of $(\tilde{A}, \tilde{\iota})$ onto $(\tilde{A}^q, \tilde{\iota}^q)$. If c denotes the ideal-class of $\mathfrak{q}$, $(\tilde{A}^q, \tilde{\iota}^q)$

is a c-transform of $(\tilde{A}, \tilde{\iota})$, so that, by Proposition 15 of §11.2, $(A^{\sigma}, \iota^{\sigma})$ is a c-transform of (A, ι). Hence there exists a $\mathfrak{q}$-multiplication μ of (A, ι) onto $(A^{\sigma}, \iota^{\sigma})$. Then $\tilde{\mu}$ is a $\mathfrak{q}$-multiplication of $(\tilde{A}, \tilde{\iota})$ onto $(\tilde{A}^{q}, \tilde{\iota}^{q})$; so by Proposition 7 of §7.1, there exists an automorphism η of $(\tilde{A}^{q}, \tilde{\iota}^{q})$ such that $\pi = \eta\tilde{\mu}$. By Proposition 3 of §5.1, η must be of the form $\tilde{\iota}^{q}(\varepsilon)$ for $\varepsilon \in \mathfrak{o}$. Put $\lambda = \mu \cdot \iota(\varepsilon)$. We have then $\tilde{\lambda} = \pi$; as ε is a unit of $\mathfrak{o}$, λ is a $\mathfrak{q}$-multiplication. We have thus proved the existence of a $\mathfrak{q}$-multiplication λ of (A, ι) onto $(A^{\sigma}, \iota^{\sigma})$ whose reduction modulo $\mathfrak{P}$ is π.

Put $Z = \lambda^{-1}(Y^{\sigma})$; we have then $\tilde{Z} = \pi^{-1}(\tilde{Y}^{q})$, and hence $\tilde{Z} = q\tilde{Y}$. Take a prime l which is not divisible by $\mathfrak{p}$ and consider l-adic representations of the divisors $Y, Z, \tilde{Y}, \tilde{Z}$. By Proposition 14 of §11.1, with respect to suitable l-adic coordinate-systems, we obtain $E_l(Z) = E_{\ell}(\tilde{Z})$ and $E_l(Y) = E_l(\tilde{Y})$, so that $E_l(Z) = qE_l(Y)$; this implies that Z is algebraically equivalent to qY. Hence we have $\varphi_Y^{-1}\varphi_Z = q \cdot 1_A$ so that

$$f(\lambda) = q,$$

f being determined as in §14.4. We have thus arrived at an important conclusion

$$[\sigma] = (g(\mathfrak{p}), N(\mathfrak{p})). \tag{1}$$

As explained at the end of §15.2, $[\sigma]$ is meaningful for $\sigma \in \mathrm{Gal}(k_0^*/K^*)$. Now, $\mathfrak{p}$ being a prime ideal of K^* which is prime to $\mathfrak{m}$, let $\sigma(\mathfrak{p})$ denote a Frobenius automorphism of k_0^*/K^* for $\mathfrak{p}$. For every ideal $\mathfrak{a} \in I_{K^*}(\mathfrak{m})$, consider the prime ideal decomposition $\mathfrak{a} = \prod \mathfrak{p}^{e(\mathfrak{p})}$ and put

$$\sigma(\mathfrak{a}) = \prod_{\mathfrak{p}} \sigma(\mathfrak{p})^{e(\mathfrak{p})}.$$

This is usually written $\left(\frac{k_0^*/K^*}{\mathfrak{a}}\right)$, but we employ $\sigma(\mathfrak{a})$ for simplicity. Then, by means of the relation (1), we obtain, for every $\mathfrak{a} \in I_{K^*}(\mathfrak{m})$,

$$[\sigma(\mathfrak{a})] = (g(\mathfrak{a}), N(\mathfrak{a})).$$

Let H_1 denote the kernel of the homomorphism $\mathfrak{a} \to \sigma(\mathfrak{a})$ of $I_{K^*}(\mathfrak{m})$ into $\mathrm{Gal}(k_0^*/K^*)$. Since $\sigma \to [\sigma]$ is an isomorphism of $\mathrm{Gal}(k_0^*/K^*)$ into $\mathcal{C}(K)$, H_1 is the set of ideals $\mathfrak{a}$ of $I_{K^*}(\mathfrak{m})$ such that

$$(g(\mathfrak{a}), N(\mathfrak{a})) = (\mathfrak{o}, 1).$$

Let H_0 be the subgroup of $I_{K^*}((1))$ consisting of the ideals $\mathfrak{a}$ such that there exists an element $\mu \in K$ for which we have $g(\mathfrak{a}) = (\mu)$ and $N(\mathfrak{a}) = \mu\bar{\mu}$. Then, we see easily that $H_0 \cap I_{K^*}(\mathfrak{m}) = H_1$. Let $\mathfrak{b} = (\beta)$ be a principal ideal in K^*. Put $\gamma = \prod_{\alpha} \beta^{\psi_{\alpha}}$; then, by Proposition 29 of §8.3, γ is contained in K

and $\gamma\bar{\gamma} = N(\beta)$. As $\gamma\bar{\gamma} > 0$, we obtain $\gamma\bar{\gamma} = N(\mathfrak{b})$. This shows that H_0 contains $P_{K^*}((1))$. Therefore, according to the results of class field theory, k_0^* is the unramified class field over K^* corresponding to the ideal group H_0. We have thus established our first main theorem.

MAIN THEOREM 1. *Let $(K^*; \{\psi_\alpha\})$ be a primitive CM-type and $(K; \{\varphi_i\})$ the reflex of $(K^*; \{\psi_\alpha\})$. Let H_0 be the group of all ideals $\mathfrak{a}$ of K^* such that there exists an element $\mu \in K$ for which we have*

$$g(\mathfrak{a}) = (\mu), \qquad N(\mathfrak{a}) = \mu\bar{\mu},$$

where g is defined by (1) of §13.1 and $\bar{\mu}$ denotes the complex conjugate of μ. Let (A, ι) be an abelian variety of type $(K; \{\varphi_i\})$ and $\mathcal{C}$ a polarization of A. Let k_0 be the field of moduli of $(A, \mathcal{C})$. Then, H_0 is an ideal group of K^* defined modulo (1); and the composite k_0^* of the fields k_0 and K^* is the unramified class field over K^* corresponding to the ideal-group H_0.

NOTE 1. We have to explain why we may drop the condition that (A, ι) is defined over an algebraic number field. Let (A, ι) be an abelian variety of type $(K; \{\varphi_i\})$ and $\mathcal{C}$ a polarization of A. By Proposition 26 of §12.4, there can be found an abelian variety (A_1, ι_1), defined over an algebraic number field, which is isomorphic to (A, ι). If we denote by $\mathcal{C}_1$ the image of $\mathcal{C}$ by an isomorphism of (A, ι) onto (A_1, ι_1), the fields of moduli of $(A, \mathcal{C})$ and of $(A_1, \mathcal{C}_1)$ are the same. Therefore we obtain the results of our theorem, applying the above discussion to (A_1, ι_1).

NOTE 2. The field $k_0^* = k_0K^*$ depends only upon $(K^*; \{\psi_\alpha\})$ and is independent of the choice of (A, ι) and $\mathcal{C}$. A similar fact holds for another type of abelian varieties, which is related to automorphic functions (cf. [36] $n^\circ 23$).

NOTE 3. In the classical case where the dimension of A is equal to 1, every abelian variety of the same type as A is isomorphic to some conjugate A^σ of A over K^*; this implies the so-called "irreducibility of the class equation". We shall now consider this problem in the general case. $(A, \iota, \mathcal{C})$ being as above, let (A_1, ι_1) be another abelian variety of type $(K; \{\varphi_i\})$. By Proposition 16 of §7.4, there exists an ideal $\mathfrak{b}$ of K such that (A_1, ι_1) is a $\mathfrak{b}$-multiplication of (A, ι). By Proposition 7, A_1 has a polarization $\mathcal{C}_1$ such that $(A_1, \iota_1, \mathcal{C}_1)$ is of the same type as $(A, \iota, \mathcal{C})$ if and only if there exists a totally positive element ρ of K_0 such that $\mathfrak{b}\bar{\mathfrak{b}} = (\rho)$. Let h' be the number of ideal-classes of K whose members $\mathfrak{b}$ have this property. Then there exist exactly h' distinct abelian varieties (A_1, ι_1) of type $(K; \{\varphi_i\})$ on which we can find a polarization $\mathcal{C}_1$ such that $(A_1, \iota_1, \mathcal{C}_1)$ is of the given type $(K; \{\varphi_i\}; \mathfrak{f}_0)$. On the other hand, by Proposition 5, there exist $[U : U_1]$ polarized abelian varieties, of which no two are isomorphic, having the same (A_1, ι_1) as the underlying abelian variety, U and U_1 being as in that proposition. Put $[U : U_1] = d$. Thus we observe that there exist exactly dh' polarized abelian varieties of type $(K; \{\varphi_i\}; \mathfrak{f}_0)$, not isomorphic to each other. Let W be the subgroup of the ideal class group of K consisting of the classes of $g(\mathfrak{a})$ with ideals $\mathfrak{a}$ in K^*, and h'_0 the order of W. Then g gives a homomorphism of $I_{K^*}((1))$ onto W. Let H_1 be the kernel of this homomorphism. Then $H_0 \subset H_1$ and $h'_0 = [I_{K^*}((1)) : H_1]$. For $\mathfrak{a} \in H_1$ we can put $g(\mathfrak{a}) = \mu\mathfrak{o}$ with $\mu \in F^\times$. Then $\mu\bar{\mu}N(\mathfrak{a})^{-1} \in \mathfrak{o}^\times$. Let U_0 be the set of all units of the form $\mu\bar{\mu}N(\mathfrak{a})^{-1}$ with such $\mathfrak{a}$ and μ. Then we see that $U_1 \subset U_0 \subset U$ and $[H_1 : H_0] = [U_0 : U_1]$. Putting $d_0 = [U_0 : U_1]$, we thus find that $d_0h'_0 = [I_{K^*}((1)) : H_0] = [k_0^* : K^*]$. In other words, there exist exactly $d_0h'_0$ nonisomorphic conjugates of $\mathcal{P} = (A, \iota, \mathcal{C})$ over K^*. If we have $d_0h'_0 = dh'$, the analogy of the irreducibility of the class equation holds for $(A, \iota, \mathcal{C})$. It is not easy, however, to see in which case the equality $d_0h'_0 = dh'$ holds.

15.4. Examples. (1) We shall first consider the "classical case" where $n = 1$. In this case, K is an imaginary quadratic field and $(K; \varphi) = (K^*; \psi)$. The ideal group H_0 coincides with $P_K((1))$. Hence k_0K is the absolute class field over K. As A is of dimension 1, A has only one polarization $\mathcal{C}$. It is easy to see that k_0 is generated over $\mathbf{Q}$ by the value of the classical modular function $j(\tau)$ (cf. [35] $n^\circ 7$).

(2) Let l be an odd prime and C the plane algebraic curve defined by the equation $y^2 = 1 - x^l$. Then the genus g of C is equal to $(l-1)/2$; and $\omega_\nu = x^\nu dx/y$ for $\nu = 0, 1, \ldots, g-1$ give a basis for the differential forms of the first kind. If ζ denotes a primitive l-th root of unity, then $(x, y) \to (\zeta x, y)$ gives a birational correspondence of C onto itself, which is represented with respect to the basis $\{\omega_\nu\}$ by the diagonal matrix having $\zeta, \zeta^2, \ldots, \zeta^g$ as the diagonal elements. Let C_0 be a complete non-singular curve, defined over $\mathbf{Q}$, birationally equivalent to C over $\mathbf{Q}$; we can find a Jacobian variety J of C_0, defined over $\mathbf{Q}$ (cf. Chow [5], Weil [51], [52]). Now denote by $\iota(\zeta)$ the endomorphism of J corresponding to the above birational correspondence $(x, y) \to (\zeta x, y)$ of C. Then we easily verify that $\zeta \to \iota(\zeta)$ can be extended to an isomorphism ι of $\mathbf{Q}(\zeta)$ into $\mathrm{End}_{\mathbf{Q}}(J)$. Let φ_ν denote the automorphism $\zeta \to \zeta^\nu$ of $\mathbf{Q}(\zeta)$ for $\nu = 1, 2, \ldots, g$. Then (J, ι) is of type $(\mathbf{Q}(\zeta); \{\varphi_\nu\})$. As is seen in §8.4, $(\mathbf{Q}(\zeta); \{\varphi_\nu\})$ is a primitive CM-type and its reflex is given by $(\mathbf{Q}(\zeta); \{\varphi_\nu^{-1}\})$. Therefore, the abelian variety J must be simple; and we have $\mathrm{End}_{\mathbf{Q}}(J) = \mathbf{Q}(\zeta)$. Since $\mathbf{Z}[\zeta]$ is the ring of integers in $\mathbf{Q}(\zeta)$ and $\iota(\zeta) \in \mathrm{End}(J)$, we have $\mathrm{End}(J) = \iota(\mathbf{Z}[\zeta])$, so that (J, ι) is principal. Hence we can apply our theorem to this case; so, for every polarization $\mathcal{C}$ of J, the field of moduli k_0 of $(J, \mathcal{C})$ generates a class field $k_0(\zeta)$ over $\mathbf{Q}(\zeta)$, which corresponds to the ideal

group H_0. Since J is defined over $\mathbf{Q}$, we can find a polarization $\mathcal{C}$ of J, defined over $\mathbf{Q}$; then k_0 must coincide with $\mathbf{Q}$. Therefore in this case, the class field $k_0 K^*$ is not a proper extension of K^*. We can conclude from this the following interesting fact. By class field theory, H_0 must coincide with the whole ideal group $I_K((1))$. Hence, putting $\psi_\nu = \varphi_\nu^{-1}$, we see that, for every ideal $\mathfrak{a}$ of $\mathbf{Q}(\zeta)$, there exists an element μ of K such that $g(\mathfrak{a}) = (\mu)$ and $N(\mathfrak{a}) = \mu\bar{\mu}$. This is a special case of "Stickelberger's relation" (cf. [39], [6]); our result gives a proof of this relation. Finally, we note that this example shows the invalidity of "irreducibility of the class equation" in the general case.

(3) Let us consider example (C) of §8.4. Let the notation be the same as in §8.4 and Note 3 above; put $\psi = \sigma\tau$. We have then $g(\mathfrak{a}) = \mathfrak{a}\mathfrak{a}^{\psi} = N_{L/K}(\mathfrak{a}\mathfrak{o}_L)$ for every ideal $\mathfrak{a}$ of K^*. Now assume that the class-number h of K is an odd number. Let H denote the subgroup of $I_K((1))/P_K((1))$ consisting of the classes which contain $N_{L/K}(\mathfrak{A})$ for an ideal $\mathfrak{A}$ of L. Then H is of index $\leq [L : K] = 2$; as h is odd, the index must be 1. Hence, every ideal-class of K contains an ideal of the form $N_{L/K}(\mathfrak{A})$ for some ideal $\mathfrak{A}$ of L. Let H' be the subgroup of $I_K((1))/P_K((1))$ consisting of the classes whose members $\mathfrak{b}$ have the property $\mathfrak{b}\bar{\mathfrak{b}} = (\xi)$ for a totally positive element ξ of K_0. As the order h' of H' is odd, for every $c \in H'$, there exists an element c' of H' such that $c = c'^2$. Take an ideal $\mathfrak{A}$ of L such that $N_{L/K}(\mathfrak{A}) \in c'$. Put $\mathfrak{b} = \mathfrak{A}\mathfrak{A}^{\tau}\mathfrak{A}^{\psi}\mathfrak{A}^{\psi\tau}$; then $\mathfrak{b}$ is an ideal of K^*. Since $N_{L/K}(\mathfrak{A})$ is contained in c', by the definition of H', there exists a totally positive element ξ of K_0 such that we have $(\xi) = N_{L/K}(\mathfrak{A})\overline{N_{L/K}(\mathfrak{A})}$. We can easily verify that

$$g(\mathfrak{b}) = N_{L/K}(\mathfrak{b}) = N_{L/K}(\mathfrak{A})^2(\xi^{\tau}).$$

This shows that the class c contains $g(\mathfrak{b})$. It follows from this that $h' = h'_o$. Since K_0 is a real quadratic field, the group of units in K_0 is the direct product of $\{1, -1\}$ and the free cyclic group $\{\varepsilon^n \mid n \in \mathbf{Z}\}$ generated by a unit ε, which we call a fundamental unit of K_0. Assume that $N_{K_0/\mathbf{Q}}(\varepsilon) = -1$. Then the group of totally positive units in K_0 is the free cyclic group generated by ε^2; and, since $N_{K/K_0}(\varepsilon) = \varepsilon^2$, we have $d = d_0 = 1$. Therefore, if the class-number h of K is odd and if the norm $N_{K_0/\mathbf{Q}}(\varepsilon)$ of a fundamental unit ε of K_0 is -1, we have $dh' = d_0 h'_0$, so that the analogy of the irreducibility of class equation holds in this case. This result is due to Hecke [20]. See also the Appendix of [S75b] for the discussion of related problems in a more general case.

16. The Class Fields Generated by Ideal-Section Points

16.1. The notation and assumptions being as in the preceding section, let (V, F) be a normalized Kummer variety of $(A, \mathcal{C})$ (cf. §4.4). Let $\mathfrak{b}$ be an integral ideal of K and t a proper $\mathfrak{b}$-section point of A; then we say that a system

$(\mathcal{P}, t) = (A, \iota, \mathcal{C}, t)$ is of type $(K; \{\varphi_i\}; \mathfrak{f}_0; \mathfrak{b})$. We shall denote $(K; \{\varphi_i\}; \mathfrak{f}_0; \mathfrak{b})$ briefly by $\mathfrak{K}(\mathfrak{b})$. Our purpose is to show that the field $k_0^*(F(t))$ is a class field over K^* and to determine the corresponding ideal group of K^*.

Let $(\mathcal{P}_1, t_1) = (A_1, \iota_1, \mathcal{C}_1, t_1)$ be another system of type $\mathfrak{K}(\mathfrak{b})$. By Proposition 24 of §7.5, there exists a homomorphism λ of A onto A_1 such that $\lambda t = t_1$; and λ is an $\mathfrak{a}$-multiplication for an ideal $\mathfrak{a}$ of K which is prime to $\mathfrak{b}$. Now consider the function $f(\lambda)$ defined in §14.4; $f(\lambda)$ is a totally positive element of K_0, and $f(\lambda\iota(\xi)) = f(\lambda)\xi\bar{\xi}$ for every $\xi \in \mathfrak{a}^{-1}$. Let λ_0 be another homomorphism of A onto A_1 such that $\lambda_0 t = t_1$. As we have $\mathrm{Hom}(A, A_1) = \lambda\iota(\mathfrak{a}^{-1})$, there exists an element $\mu \in \mathfrak{a}^{-1}$ such that $\lambda_0 = \lambda\iota(\mu)$. Then, λ_0 is a $(\mu\mathfrak{a})$-multiplication and $f(\lambda_0) = f(\lambda)\mu\bar{\mu}$. Moreover, we see that $\lambda - \lambda_0 = \lambda\iota(1-\mu)$ is a $(1-\mu)\mathfrak{a}$-multiplication. Hence, as $(\lambda - \lambda_0)t = 0$, we have $(1-\mu)\mathfrak{a} \subset \mathfrak{b}$, by virtue of Proposition 24 of §7.5. Since $\mathfrak{a}$ is prime to $\mathfrak{b}$, we get $\mu \equiv 1 \bmod^{\times} \mathfrak{b}$. Thus we are led to the following definition.

For a fixed integral ideal $\mathfrak{b}$ of K consider a pair $\{\mathfrak{a}, \rho\}$ formed by an ideal $\mathfrak{a}$ of K which is *prime to* $\mathfrak{b}$ and a totally positive element ρ of K_0 such that

$$\mathfrak{a}\bar{\mathfrak{a}} = (\rho).$$

Given another pair $\{\mathfrak{a}_1, \rho_1\}$ satisfying these conditions, we say that $\{\mathfrak{a}, \rho\}$ and $\{\mathfrak{a}_1, \rho_1\}$ are *equivalent modulo* $\mathfrak{b}$, if there exists an element $\mu \in K$ for which we have

$$\mathfrak{a}_1 = \mu\mathfrak{a}, \qquad \rho_1 = \rho\mu\bar{\mu}, \qquad \mu \equiv 1 \bmod^{\times} \mathfrak{b}.$$

The class determined by this equivalence relation will be denoted by $(\mathfrak{a}, \rho)_{\mathfrak{b}}$. Define the multiplication of two classes $(\mathfrak{a}, \rho)_{\mathfrak{b}}$ and $(\mathfrak{a}_1, \rho_1)_{\mathfrak{b}}$ by

$$(\mathfrak{a}, \rho)_{\mathfrak{b}}(\mathfrak{a}_1, \rho_1)_{\mathfrak{b}} = (\mathfrak{a}\mathfrak{a}_1, \rho\rho_1)_{\mathfrak{b}}.$$

It can be easily verified that the classes $(\mathfrak{a}, \rho)_{\mathfrak{b}}$ form a group by this law of multiplication; we denote this group by $\mathfrak{C}(K; \mathfrak{b})$; the identity element of $\mathfrak{C}(K; \mathfrak{b})$ is given by $(\mathfrak{o}, 1)_{\mathfrak{b}}$.

Now consider the pair $\{\mathfrak{a}, f(\lambda)\}$ determined by $(\mathcal{P}, t)$ and $(\mathcal{P}_1, t_1)$; by the above observations, the class $(\mathfrak{a}, f(\lambda))_{\mathfrak{b}}$ does not depend on the choice of λ; so we write

$$\{(\mathcal{P}_1, t_1) : (\mathcal{P}, t)\} = (\mathfrak{a}, \rho)_{\mathfrak{b}}.$$

If furthermore $(\mathcal{P}_2, t_2)$ is of type $\mathfrak{K}(\mathfrak{b})$, we have

$$\{(\mathcal{P}_2, t_2) : (\mathcal{P}_1, t_1)\}\{(\mathcal{P}_1, t_1) : (\mathcal{P}, t)\} = \{\mathcal{P}_2, t_2) : (\mathcal{P}, t)\}.$$

This is proved in a straightforward way, using the proof of Proposition 8.

An isomorphism η of $\mathcal{P}$ onto $\mathcal{P}_1$ is called an *isomorphism* of $(\mathcal{P}, t)$ onto $(\mathcal{P}_1, t_1)$ if we have $\eta t = t_1$.

PROPOSITION 12. *Given* $(\mathcal{P}, t)$ *and* $(\mathcal{P}_1, t_1)$ *of type* $\mathfrak{K}(\mathfrak{b})$, *we have* $\{(\mathcal{P}_1, t_1) : (\mathcal{P}, t)\} = (\mathfrak{o}, 1)_{\mathfrak{b}}$ *if and only if there exists an isomorphism of* $(\mathcal{P}, t)$ *onto* $(\mathcal{P}_1, t_1)$.

PROOF. The "if" part is an easy consequence of our definition and Proposition 9. Suppose that $\{(\mathcal{P}_1, t_1) : (\mathcal{P}, t)\} = (\mathfrak{o}, 1)_{\mathfrak{b}}$. Take an $\mathfrak{a}$-multiplication λ of A onto A_1 such that $\lambda t = t_1$. By our definition, there exists an element μ of K such that $\mathfrak{a} = (\mu)$, $f(\lambda) = \mu\bar{\mu}$, and $\mu \equiv 1 \bmod^{\times} \mathfrak{b}$. Then, we can easily verify that $\lambda\iota(\mu^{-1})$ is an isomorphism of $(\mathcal{P}, t)$ onto $(\mathcal{P}_1, t_1)$. This proves the "only if" part.

PROPOSITION 13. *Given* $(\mathcal{P}, t)$ *and* $(\mathcal{P}_1, t_1)$ *of type* $\mathfrak{K}(\mathfrak{b})$, *let* k *be a field of definition for* $\mathcal{P}$ *and* $\mathcal{P}_1$, *over which both* t *and* t_1 *are rational. Let* σ *be an isomorphism of* k *into a field* k', *which leaves invariant the elements of* K^*. *Then* $(\mathcal{P}^\sigma, t^\sigma)$ *and* $(\mathcal{P}_1^\sigma, t_1^\sigma)$ *are of type* $\mathfrak{K}(\mathfrak{b})$ *and*

$$\{(\mathcal{P}_1^\sigma, t_1^\sigma) : (\mathcal{P}^\sigma, t^\sigma)\} = \{(\mathcal{P}_1, t_1) : (\mathcal{P}, t)\}.$$

This is an easy consequence of Proposition 11 and the above definition.

16.2. Now we fix a system $(\mathcal{P}, t) = (A, \iota, \mathcal{C}, t)$ of type $\mathfrak{K}(\mathfrak{b})$. Let (V, F) be a normalized Kummer variety of $(A, \mathcal{C})$. Let k be an algebraic number field of finite degree satisfying the conditions (i)–(iv) of §15.1 and the following condition:

(v) *t is rational over k.*

Then, for every $\sigma \in \mathrm{Gal}(k/K^*)$, we obtain, by Proposition 13, an element $\{(\mathcal{P}^\sigma, t^\sigma) : (\mathcal{P}, t)\}$ of $\mathfrak{C}(K; \mathfrak{b})$; put

$$[\sigma]_t = \{(\mathcal{P}^\sigma, t^\sigma) : (\mathcal{P}, t)\}.$$

We can prove, in the same way as for $[\sigma]$, that $\sigma \to [\sigma]_t$ gives a homomorphism of $\mathrm{Gal}(k/K^*)$ into $\mathfrak{C}(K; \mathfrak{b})$. Let $\mathfrak{H}$ be the kernel of this homomorphism. If $\sigma \in \mathfrak{H}$, there exists, by Proposition 12, an isomorphism η of $\mathcal{P}$ onto $\mathcal{P}^\sigma$ such that $\eta t = t^\sigma$. By the definition of field of moduli, σ leaves the elements of k_0^* invariant. Since V is defined over k_0, we have $V^\sigma = V$; moreover, by property (N3) of Theorem 3 of §4.4, we have $F = F^\sigma \circ \eta$. Hence we have

$$F(t)^\sigma = F^\sigma(t^\sigma) = F^\sigma(\eta t) = F(t).$$

This shows that σ is the identity map on $k_0^*(F(t))$. Conversely, suppose that an element σ of $\mathrm{Gal}(k/K^*)$ fixes the elements of $k_0^*(F(t))$. Then, by the result of §15.2, there exists an isomorphism ε of $\mathcal{P}$ onto $\mathcal{P}^\sigma$; and again by property

(N3), we have $F = F^\sigma \circ \varepsilon$. Hence, we obtain $F(t) = F(t)^\sigma = F(\varepsilon^{-1}t^\sigma)$. By property (K2) of Proposition 16 of §4.3, there exists an automorphism μ of $\mathcal{P}$ such that $\varepsilon^{-1}t^\sigma = \mu t$. Then $\varepsilon\mu$ is an isomorphism of $\mathcal{P}$ onto $\mathcal{P}^\sigma$ satisfying $(\varepsilon\mu)t = t^\sigma$; so, by Proposition 12, we have $[\sigma]_t = (\mathfrak{o}, 1)_\mathfrak{b}$, and hence $\sigma \in \mathfrak{H}$. We have thus proved that $k_0^*(F(t))$ is the subfield of k corresponding to $\mathfrak{H}$; in other words, $\sigma \to [\sigma]_t$ induces an isomorphism of $\mathrm{Gal}(k_0^*(F(t))/K^*)$ into $\mathfrak{C}(K; \mathfrak{b})$; we shall also denote this isomorphism by the same notation $[\sigma]_t$. Since $\mathfrak{C}(K; \mathfrak{b})$ is an abelian group, $k_0^*(F(t))$ is an abelian extension of K^*.

16.3. The ideal $\mathfrak{m}$ being defined for the field k as in §15.3, let $\mathfrak{p}$ be a prime ideal of K^* which does not divide $\mathfrak{m}$, and $\mathfrak{P}$ a prime ideal of k dividing $\mathfrak{p}$. Put $N(\mathfrak{p}) = q$. Let σ be a Frobenius automorphism of k/K^* for $\mathfrak{P}/\mathfrak{p}$. Consider now reduction modulo $\mathfrak{P}$. By the result of §15.3, there exists a $g(\mathfrak{p})$-multiplication λ of A onto A^σ such that the reduction $\tilde{\lambda}$ of λ modulo $\mathfrak{P}$ is the q-th power homomorphism π of $\tilde{A}$ onto $\tilde{A}^q$; and we have $f(\lambda) = q$. We see easily

$$(1) \qquad \widetilde{t^\sigma} = \tilde{t}^q = \pi\tilde{t} = \tilde{\lambda}\tilde{t} = \widetilde{(\lambda t)}.$$

Assume that $\mathfrak{p}$ is prime to $N(\mathfrak{b})$; then, by Proposition 16 of §11.2, reduction modulo $\mathfrak{P}$ gives an isomorphism of $\mathfrak{g}(\mathfrak{b}, A^\sigma)$ onto $\mathfrak{g}(\mathfrak{b}, \tilde{A}^q)$. Therefore, (1) implies $t^\sigma = \lambda t$, since both t^σ and λt are contained in $\mathfrak{g}(\mathfrak{b}, A^\sigma)$. Using this homomorphism λ, we obtain

$$(2) \qquad [\sigma]_t = (g(\mathfrak{p}), N(\mathfrak{p}))_\mathfrak{b}.$$

Put $\mathfrak{n} = \mathfrak{m}N(\mathfrak{b})$. For every prime ideal $\mathfrak{p} \in I_{K^*}(\mathfrak{n})$, let $\sigma(\mathfrak{p})$ denote the Frobenius automorphism of $k_0^*(F(t))/K^*$ for $\mathfrak{p}$; here recall that $k_0^*(F(t))$ is an abelian extension of K^*. For every ideal $\mathfrak{a} = \prod_\mathfrak{p} \mathfrak{p}^{e(\mathfrak{p})}$ in $I_{K^*}(\mathfrak{n})$, put

$$\sigma(\mathfrak{a}) = \prod_\mathfrak{p} \sigma(\mathfrak{p})^{e(\mathfrak{p})}.$$

Then, from (2) we obtain, for every $\mathfrak{a} \in I_{K^*}(\mathfrak{n})$,

$$[\sigma(\mathfrak{a})]_t = (g(\mathfrak{a}), N(\mathfrak{a}))_\mathfrak{b}.$$

Let H_1 be the kernel of the homomorphism $\mathfrak{a} \to \sigma(\mathfrak{a})$ of $I_{K^*}(\mathfrak{n})$ into $\mathrm{Gal}(k_0^*(F(t))/K^*)$. Since $\sigma \to [\sigma]_t$ is an isomorphism of $\mathrm{Gal}(k_0^*(F(t))/K^*)$ into $\mathfrak{C}(K; \mathfrak{b})$, H_1 consists of the ideals $\mathfrak{a} \in I_{K^*}(\mathfrak{n})$ such that

$$(g(\mathfrak{a}), N(\mathfrak{a}))_\mathfrak{b} = (\mathfrak{o}, 1)_\mathfrak{b}.$$

Now let b be the smallest positive integer divisible by $\mathfrak{b}$; and let $H(\mathfrak{b})$ be the subgroup of $I_{K^*}((b))$ consisting of the ideals $\mathfrak{a}$ for which there exists an element μ of K such that $g(\mathfrak{a}) = (\mu)$, $N(\mathfrak{a}) = \mu\bar{\mu}$ and $\mu \equiv 1 \bmod \mathfrak{b}$. Then we see easily that $H(\mathfrak{b}) \cap I_{K^*}(\mathfrak{n}) = H_1$. Let ξ be an element of K^* such that $\xi \equiv 1 \bmod^{\times}(b)$. Put $\gamma = \prod_\alpha \xi^{\psi_\alpha}$; then by Proposition 29 of §8.3, we have $\gamma \in K$, $\gamma\bar{\gamma} = N((\xi))$; moreover, it is obvious that $\gamma \equiv 1 \bmod \mathfrak{b}$. This shows that $P_{K^*}((b))$ is contained in $H(\mathfrak{b})$. Hence, by class field theory, we observe that $k_0^*(F(t))$ is the class field over K^* corresponding to the ideal group $H(\mathfrak{b})$. We have thus arrived at the following conclusion.

MAIN THEOREM 2. *The notation and assumptions being as in Main theorem 1, let (V, F) be a normalized Kummer variety of $(A, \mathcal{C})$. Let $\mathfrak{b}$ be an integral ideal of K and b the smallest positive integer divisible by $\mathfrak{b}$. Let $H(\mathfrak{b})$ be the group of all ideals $\mathfrak{a}$ of K^*, prime to (b), such that there exists an element $\mu \in K$ for which we have*

$$g(\mathfrak{a}) = (\mu), \quad N(\mathfrak{a}) = \mu\bar{\mu}, \quad \mu \equiv 1 \bmod^{\times} \mathfrak{b}.$$

Let t be a proper $\mathfrak{b}$-section point of A. Then, $H(\mathfrak{b})$ is an ideal group of K^* defined modulo (b); and $k_0^*(F(t))$ is the class field over K^* corresponding to the ideal group $H(\mathfrak{b})$.

The reason why we may dispense with the condition that A is defined over an algebraic number field, is the same as in Main theorem 1.

We now consider the case of dimension 1; it is then clear that $H(\mathfrak{b})$ coincides with $P_K(\mathfrak{b}) \cap I_K((b))$; it follows that $k_0^*(F(t))$ is the class field over K corresponding to the ideal group $P_K(\mathfrak{b})$. We can also easily verify that the normalized Kummer variety (V, F) is explicitly given by the Weierstrass $\wp$-function, or more precisely, by Weber's τ-function; then the above theorem implies the main theorem of the classical theory of complex multiplication, which asserts that the abelian extensions of an imaginary quadratic field are generated by the values of certain elliptic or elliptic modular functions with singular moduli. For details of the classical theory and later refinements, the reader is referred to Kronecker [25], Weber [43], Takagi [40], Hasse [17], Deuring [De], and Ramachandra [Ra]. A modern formulation of complex multiplication of elliptic modular functions is given in [S71a].

17. The Field of Moduli in a Generalized Setting

17.1. Let us first generalize the notion of field of moduli introduced in Chapter I, §4. We consider a structure $\mathcal{P} = (A, \mathcal{C}, \iota; \{t_i\}_{i=1}^r)$ formed by a polarized abelian variety $(A, \mathcal{C})$ in the sense of §§4.1, 4.2, a ring-injection ι of a $\mathbf{Q}$-algebra W (with identity element) into $\mathrm{End}_{\mathbf{Q}}(A)$, and an ordered set of points $\{t_i\}_{i=1}^r$ of A of finite order. We always assume that $\iota(1)$ is the identity element of $\mathrm{End}(A)$. We say that $\mathcal{P}$ *is defined (or rational) over a field* k and that k *is a field of definition (or rationality) for* $\mathcal{P}$ if $(A, \mathcal{C})$, every element of $\iota(W) \cap \mathrm{End}(A)$, and every t_i are all rational over k. We always take such a k to be a subfield of $\mathbf{C}$, except when we consider reduction of $\mathcal{P}$ modulo a prime divisor. Given such a k and an isomorphism σ of k onto a field (contained in $\mathbf{C}$ for the moment), we put $\mathcal{P}^\sigma = (A^\sigma, \mathcal{C}^\sigma, \iota^\sigma; \{t_i^\sigma\}_{i=1}^r)$, where $\iota^\sigma(a) = \iota(a)^\sigma$. If $\mathcal{P}' = (A', \mathcal{C}', \iota'; \{t_i'\}_{i=1}^r)$ is another such structure, we understand by an *isomorphism* of $\mathcal{P}$ onto $\mathcal{P}'$ an isomorphism f of $(A, \mathcal{C})$ onto $(A', \mathcal{C}')$ in the sense of §4.2 such that $f \circ \iota(a) = \iota'(a) \circ f$ for every $a \in W$ and $f(t_i) = t_i'$ for every i. We call f an *automorphism* of $\mathcal{P}$ if $\mathcal{P} = \mathcal{P}'$, and denote by $\mathrm{Aut}(\mathcal{P})$ the group of all automorphisms of $\mathcal{P}$.

17.2. Proposition *Given* $\mathcal{P}$*, there exists a subfield* k_1 *of* $\mathbf{C}$ *with the following properties:*

(i) *Every field of definition of* $\mathcal{P}$ *contains* k_1.
(ii) *If* $\mathcal{P}$ *is defined over* k *and* σ *is an isomorphism of* k *onto a subfield of* $\mathbf{C}$*, then* σ *is the identity map on* k_1 *if and only if* $\mathcal{P}^\sigma$ *is isomorphic to* $\mathcal{P}$.
(iii) k_1 *is uniquely determined for* $\mathcal{P}$ *by these two properties.*
(iv) k_1 *is algebraic over the field of moduli of* $(A, \mathcal{C})$ *defined in §4.2.*

This, except for (iv), is a generalization of Theorem 2 in §4.2 and is given in [36] for $\mathcal{P}$ without $\{t_i'\}$, and in [S65, §1.4] for $\mathcal{P}$ with $\{t_i'\}$. As for (iv), see [36, Proposition 8] and [S65, Proposition 1.11]. We call k_1 *the field of moduli of* $\mathcal{P}$.

17.3. Let us now assume the following two conditions:

(17.3a) *W is a commutative semisimple algebra over* $\mathbf{Q}$ *such that* $[W : \mathbf{Q}] = 2\dim(A)$*, or* W *is a division algebra;*
(17.3b) $\iota(W)$ *is stable under the involution* $\alpha \mapsto \alpha'$ *of* $\mathrm{End}_{\mathbf{Q}}(A)$ *determined by a divisor* X *in* $\mathcal{C}$ *as in (5) in §1.3.*

Given $\mathcal{P}$ satisfying these conditions, let d be the dimension of A. Then we can find a lattice Λ in $\mathbf{C}^d$ and an exact sequence

$$0 \longrightarrow \Lambda \longrightarrow \mathbf{C}^d \xrightarrow{\ \xi\ } A \longrightarrow 0 \qquad (17.3c)$$

so that ξ gives a complex analytic isomorphism of $\mathbf{C}^d/\Lambda$ onto A. Then we find a ring-injection Ψ: $W \to \mathbf{C}_d^d$ given by $\iota(a)\xi(u) = \xi(\Psi(a)u)$ for $a \in W$ and $u \in \mathbf{C}^d$. Let $\mathbf{Q}\Lambda$ denote the $\mathbf{Q}$-linear span of Λ in $\mathbf{C}^d$. Then $\mathbf{Q}\Lambda$ is a $2d$-dimensional vector space over $\mathbf{Q}$, and is stable under $\Psi(W)$. Thus it has a structure of W-module. Under (17.3a) $\mathbf{Q}\Lambda$ is isomorphic to W_m^1 with the integer

m such that $2d = m[W : \mathbf{Q}]$. Put $W_{\mathbf{R}} = W \otimes_{\mathbf{Q}} \mathbf{R}$. Then the isomorphism of W_m^1 onto $\mathbf{Q}\Lambda$ can be extended to an $\mathbf{R}$-linear isomorphism q of $(W_{\mathbf{R}})_m^1$ onto $\mathbf{C}^d$ such that $q(ax) = \Psi(a)q(x)$ for $a \in W$ and $x \in W_m^1$. Let L be the inverse image of Λ. Then we obtain a commutative diagram

(17.3d)
$$\begin{array}{ccccccccc} 0 & \longrightarrow & L & \longrightarrow & (W_{\mathbf{R}})_m^1 & \longrightarrow & (W_{\mathbf{R}})_m^1/L & \longrightarrow & 0 \\ & & \downarrow & & \downarrow{\scriptstyle q} & & \downarrow & & \\ 0 & \longrightarrow & \Lambda & \longrightarrow & \mathbf{C}^d & \xrightarrow{\xi} & A & \longrightarrow & 0 \end{array}$$

Let $E_X(x, y)$ be the Riemann form of $\xi^{-1}(X)$ with a divisor X in $\mathcal{C}$, and ρ the involution of W such that $\iota(a^\rho) = \iota(a)'$, where α' is as in (17.3b). For the reason explained at the end of §3.3, we have

(17.3e) $$E_X\big(\Psi(a)u, v\big) = E_X\big(u, \Psi(a^\rho)v\big) \quad \text{for every} \quad a \in W.$$

Put $f(x, y) = E_X\big(q(x), q(y)\big)$. Then $(x, y) \mapsto f(x, y)$ is a $\mathbf{Q}$-valued alternating form on $W_m^1 \times W_m^1$ such that $f(ax, y) = f(x, a^\rho y)$. Let $\mathrm{Tr}_{W/\mathbf{Q}}$ denote the reduced trace map $W \to \mathbf{Q}$. For fixed x, y we consider a $\mathbf{Q}$-linear map $a \mapsto f(ax, y)$ of W to $\mathbf{Q}$, and find an element $g(x, y) \in W$ such that $f(ax, y) = \mathrm{Tr}_{W/\mathbf{Q}}\big(a \cdot g(x, y)\big)$. Then we easily see that $(x, y) \mapsto g(x, y)$ is a skew-hermitian form. Putting $g(x, y) = xT \cdot {}^t y^\rho$ with an element T of $GL_m(W)$, we thus obtain

(17.3f) $${}^t T^\rho = -T,$$

(17.3g) $$E_X\big(q(x), q(y)\big) = \mathrm{Tr}_{W/\mathbf{Q}}\big(xT \cdot {}^t y^\rho\big) \quad \text{for every} \quad x, y \in W_m^1.$$

In this way we obtain a set of data

(17.3h) $$\Omega = \big\{ W, \Psi, \rho, L, T, \{u_i\}_{i=1}^r \big\}.$$

We note two simple facts:

(17.3i) *ρ is a positive involution in the sense that* $\mathrm{Tr}_{W/\mathbf{Q}}(aa^\rho) > 0$ *for every* $a \in W, \neq 0$;

(17.3j) *The direct sum of Ψ and its complex conjugate is equivalent to a $\mathbf{Q}$-rational representation of W.*

The first fact follows from formula (6) in §1.3, and the second one from the statement at the end of §3.2. Given such an Ω, we say that $\mathcal{P} = (A, \mathcal{C}, \iota; \{t_i\}_{i=1}^r)$ *is of type* Ω (with respect to ξ and q) if there is an $\mathbf{R}$-linear isomorphism $q : (W_{\mathbf{R}})_m^1 \to \mathbf{C}^d$ with which we have (17.3d) and such that $q(ax) = \Psi(a)q(x)$ and $\iota(a) \circ \xi = \xi \circ \Psi(a)$ for every $a \in W$, $q(u_i) = t_i$ for every i, and (17.3g) holds with some $X \in \mathcal{C}$. We also say that $\mathcal{P}$ is *strictly* of type Ω if in addition (17.3g) holds with a *basic polar divisor* X of $\mathcal{C}$ defined at the end of §4.2.

17.4. Lemma *Let h be the field of moduli of a structure $\mathcal{P}$ of type Ω. Then h contains* $\mathrm{tr}\,[\Psi(a)]$ *for all $a \in W$.*

PROOF. If $\sigma \in \mathrm{Aut}(\mathbf{C}/h)$, then there exists an isomorphism λ of $\mathcal{P}$ onto $\mathcal{P}^\sigma$ such that $\lambda \cdot \iota(a) = \iota(a)^\sigma \lambda$. With $\delta\lambda$ defined as in §2.8, we have $\delta[\iota(a)]\delta\lambda = \delta\lambda\delta[\iota(a)^\sigma]$. Define $\mathfrak{D}_0(A)$ as in §2.6. Then we see that the trace of $\delta[\iota(a)]$ on $\mathfrak{D}_0(A)$ equals that of $\delta[\iota(a)^\sigma]$ on $\mathfrak{D}_0(A^\sigma)$. Since σ sends $\mathfrak{D}_0(A)$ onto $\mathfrak{D}_0(A^\sigma)$, we see that the latter trace is the image of the former under σ. Now if $z_1, \ldots, z_d$ are the coordinate functions on $\mathbf{C}^d$, then $dz_1, \ldots, dz_d$ form a basis of $\mathfrak{D}_0(A)$, and hence the trace of $\delta[\iota(a)]$ on $\mathfrak{D}_0(A)$ equals $\mathrm{tr}\big[\Psi(a)\big]$. Therefore we obtain $\mathrm{tr}\big[\Psi(a)\big] = \mathrm{tr}\big[\Psi(a)\big]^\sigma$ for every $\sigma \in \mathrm{Aut}(\mathbf{C}/h)$, which proves our lemma, by virtue of (ii) of the following lemma.

17.5. Lemma *Let h and k be subfields of* $\mathbf{C}$ *with countably many elements. Then the following assertions hold:*

(i) *If k is stable under* $\mathrm{Aut}(\mathbf{C}/h)$, *then the composite field hk is a finite or an infinite Galois extension of h.*
(ii) *If every element of* $\mathrm{Aut}(\mathbf{C}/h)$ *gives the identity map on k, then $k \subset h$.*
(iii) *If Y is an algebro-geometric object such as a variety, a divisor, or a rational map, and $Y^\sigma = Y$ for every $\sigma \in$* $\mathrm{Aut}(\mathbf{C}/h)$, *then Y is rational over h.*

The proof, being completely elementary, is left to the reader. In the following sections we shall often make use of these principles, though we shall not explicitly mention them in each instance.

17.6. Let us now state an easy fact which is implicit in our treatment in §3, but not explicitly stated. Let A_i for $i = 1, 2$ be abelian varieties identified with complex tori $\mathbf{C}^n/D_i$ with lattices D_i as in §3.1; let $X_1 = \lambda^{-1}(X_2)$ with an isogeny λ of A_1 to A_2 and a divisor X_2 on A_2. Further let $E_i(x, y)$ be the Riemann form of X_i and Λ the $\mathbf{C}$-linear endomorphism of $\mathbf{C}^n$ that represents λ as in §3.2. Then we have

$$E_1(x, y) = E_2(\Lambda x, \Lambda y). \tag{17.6a}$$

In fact, take a theta function f on $\mathbf{C}^n/D_2$ whose divisor is X_2 as explained in §3.1. Then X_1 is the divisor of $f \circ \Lambda$, and we easily see that $f \circ \Lambda$ determines the right-hand side of (17.6a) as expected.

18. The Main Theorem of Complex Multiplication in the Adelic Language

18.1. By a *CM-field* we mean a totally imaginary quadratic extension of a totally real algebraic number field of finite degree. This is the type of field discussed in §§5 and 8. Given a CM-field K and an absolute equivalence class Φ of representations of K by complex matrices, we call (K, Φ) a *CM-type* if the direct sum of Φ and its complex conjugate is equivalent to the regular representation of K over $\mathbf{Q}$. In §5.2, we considered a more general type of field, but from now on, we speak of a CM-type only with a CM-field. If (K, Φ) is a CM-type and $[K : F] = 2n$, then we can put $\Phi = \{\varphi_1, \ldots, \varphi_n\}$ as we did in §5.2 with n isomorphic embeddings φ_i of K into $\mathbf{C}$ such that $\{\varphi_1, \ldots, \varphi_n, \varphi_1\rho, \ldots, \varphi_n\rho\}$ is exactly the set of all embeddings K into $\mathbf{C}$, where ρ denotes complex conjugation.

18.2. Lemma *Let K be an algebraic number field and J the set of all isomorphic embeddings of K into $\mathbf{C}$. Then the following assertions hold:*

(i) *K is a CM-field if and only if it has an automorphism α of order 2 such that $z^{\alpha\sigma} = z^{\sigma\rho}$ for every $\sigma \in J$ and $z \in K$, where ρ denotes complex conjugation in $\mathbf{C}$.*
(ii) *The composite of finitely many CM-fields is a CM-field.*
(iii) *The Galois closure of a CM-field over $\mathbf{Q}$ is a CM-field.*
(iv) *A subfield of a CM-field is either totally real or a CM-field.*

PROOF. Assuming the existence of such an α, let F be the fixed subfield of α in K. Then $x^\sigma = x^{\sigma\rho}$ for every $\sigma \in J$ and every $x \in F$, so that F is totally real. Take $y \in K^\times$ so that $y^\alpha = -y$. Then $y^{\sigma\rho} = -y^\sigma$ for every $\sigma \in J$, and hence K is totally imaginary. Conversely, let K be a totally imaginary quadratic extension of a totally real algebraic number field F. Let α be the generator of $\mathrm{Gal}(K/F)$. Take $y \in K^\times$ so that $y^\alpha = -y$. Then, for every $\sigma \in J$ we have $K^\sigma = F^\sigma(y^\sigma)$ and $(y^\sigma)^2 \in F^\sigma$, so that $y^{\sigma\rho} = -y^\sigma = y^{\alpha\sigma}$. Since F is totally real we have $x^{\alpha\sigma} = x^\sigma = x^{\sigma\rho}$ for every $x \in F$. Combining both equalities, we obtain $z^{\alpha\sigma} = z^{\sigma\rho}$ for every $z \in K$. This proves (i). Then assertions (ii)–(iv) can be easily verified either directly or by means of (i).

If R is the Galois closure of a CM-field K over $\mathbf{Q}$ and β is the automorphism of R of order 2 with the property of (i) for R, then clearly β belongs to the center of $\mathrm{Gal}(R/\mathbf{Q})$.

18.3. Given an algebraic number field M of finite degree, we denote its maximal abelian extension by M_{ab}. If M is a subfield of $\mathbf{C}$, which is the case with few exceptions, we take M_{ab} to be a subfield of $\mathbf{C}$. We denote also the adele ring and idele group of M by $M_\mathbf{A}$ and $M_\mathbf{A}^\times$. The archimedean and nonarchimedean factors of $M_\mathbf{A}^\times$ are denoted by $M_\mathbf{a}^\times$ and $M_\mathbf{h}^\times$. We view these, as well as the localization $M_v^\times$ of $M^\times$ at a prime v of M, as subgroups of $M_\mathbf{A}^\times$ as usual, and for $x \in M_\mathbf{A}^\times$ we denote its projections to $M_v^\times$, $M_\mathbf{a}^\times$, and

$M_{\mathbf{h}}^{\times}$ by $x_v, x_{\mathbf{a}}$, and $x_{\mathbf{h}}$. Now by class field theory there exists a canonical homomorphism of $M_{\mathbf{A}}^{\times}$ onto $\mathrm{Gal}(M_{\mathrm{ab}}/M)$ whose kernel is the closure of the product of $M^{\times}$ and the identity component of $M_{\mathbf{a}}^{\times}$. We denote by $[a, M]$ the element of $\mathrm{Gal}(M_{\mathrm{ab}}/M)$ which is the image of $a \in M_{\mathbf{A}}^{\times}$. For $t \in M_{\mathbf{A}}^{\times}$ we can speak of its p-component t_p for each rational prime p, which belongs to $M_p = M \otimes_{\mathbf{Q}} \mathbf{Q}_p$. Given a $\mathbf{Z}$-lattice $\mathfrak{a}$ in M, we put $\mathfrak{a}_p = \mathfrak{a} \otimes_{\mathbf{Z}} \mathbf{Z}_p$. Then we denote by $t\mathfrak{a}$ the $\mathbf{Z}$-lattice in M such that $(t\mathfrak{a})_p = t_p\mathfrak{a}_p$ for every p. In particular, if $\mathfrak{a}$ is a fractional ideal in M, so is $t\mathfrak{a}$. We can also associate with t an isomorphism of $M/\mathfrak{a}$ onto $M/t\mathfrak{a}$ as follows. We first observe that $M/\mathfrak{a}$ is canonically isomorphic to the direct sum of $M_p/\mathfrak{a}_p$ for all p, which easily follows from the fact that $\mathbf{Q}/\mathbf{Z}$ is canonically isomorphic to the direct sum of $\mathbf{Q}_p/\mathbf{Z}_p$ for all p. Then multiplication by t_p defines an isomorphism of $M_p/\mathfrak{a}_p$ onto $M_p/t_p\mathfrak{a}_p$. Combining these isomorphisms together we obtain an isomorphism of $M/\mathfrak{a}$ onto $M/t\mathfrak{a}$, and denote by tw the image of $w \in M/\mathfrak{a}$ by this isomorphism. If $w = u \pmod{\mathfrak{a}}$ with $u \in M$, then $tw = v \pmod{t\mathfrak{a}}$ with an element $v \in M$ such that $v \equiv t_p u \pmod{t_p\mathfrak{a}_p}$ for all p. Thus we can write

(18.3a) $$t \cdot \bigl(u \pmod{\mathfrak{a}}\bigr) = tu \pmod{t\mathfrak{a}}$$

if we understand by the right-hand side the element $\{t_p u \pmod{t_p\mathfrak{a}_p}\}_p \in M/\mathfrak{a}$, though tu is an element of $M_{\mathbf{A}}$.

18.4. Let us now consider $\mathcal{P} = (A, \mathcal{C}, \iota)$ as in §17.3, assuming W to be a CM-field K such that $2\dim(A) = [K : \mathbf{Q}]$. This is exactly the setting of §6.2. In this case, putting $[K : \mathbf{Q}] = 2n$, we can write (17.3d) (using $\mathfrak{a}$ instead of L) in the form

(18.4a) $$\begin{array}{ccccccccc} 0 & \longrightarrow & \mathfrak{a} & \longrightarrow & K_{\mathbf{R}} & \longrightarrow & K_{\mathbf{R}}/\mathfrak{a} & \longrightarrow & 0 \\ & & \downarrow & & \downarrow{\scriptstyle q} & & \downarrow & & \\ 0 & \longrightarrow & q(\mathfrak{a}) & \longrightarrow & \mathbf{C}^n & \xrightarrow{\ \xi\ } & A & \longrightarrow & 0 \end{array}$$

with a $\mathbf{Z}$-lattice $\mathfrak{a}$ in K. In order to distinguish the present case from the general case of §17, let us now use the letter Φ instead of Ψ there. Then Φ is a representation $K \to \mathbf{C}_n^n$. As seen in §§5 and 6, we may assume that $\Phi(a)$ for $a \in K$ is the diagonal matrix whose diagonal elements are $a^{\varphi_1}, \dots, a^{\varphi_n}$ with n embeddings φ_i of K into $\mathbf{C}$, and that $q(a)$ is the vector with components $a^{\varphi_1}, \dots, a^{\varphi_n}$. We often identify Φ with the set $\{\varphi_i\}_{i=1}^n$; then (K, Φ) is a CM-type. To make our later reference clearer, let us state condition (17.3b) again in the present case:

(18.4b) *$\iota(K)$ is stable under the involution $\alpha \mapsto \alpha'$ of* $\mathrm{End}_{\mathbf{Q}}(A)$ *determined by a divisor X in $\mathcal{C}$ as in (5) in §1.3.*

Now (17.3g) can be written

(18.4c) $$E_X\bigl(q(x), q(y)\bigr) = \mathrm{Tr}_{K/\mathbf{Q}}\bigl(\zeta x y^\rho\bigr) \quad \text{for every} \quad x, y \in K$$

with an element ζ of K such that $\zeta^\rho = -\zeta$ and $\mathrm{Im}(\zeta^{\varphi_i}) > 0$ for every i, as observed in Theorem 4 of §6.2. Here X is a divisor of $\mathcal{C}$ and ρ is the Galois involution of K over its maximal real subfield. Thus omitting ρ, we say that $\mathcal{P}$ *is of type* Ω *with respect to* ξ, where

(18.4d) $$\Omega = \{\, K, \Phi, \mathfrak{a}, \zeta \,\}.$$

Since we fix K, Φ, and q, we refer only to ξ. Clearly (and as observed in that theorem) the isomorphism class of $\mathcal{P}$ over $\mathbf{C}$ is completely determined by Ω.

18.5. Let (K^*, Φ^*) be the reflex of (K, Φ) defined in §8.3. Taking a Galois extension L of $\mathbf{Q}$ containing K, put $G = \mathrm{Gal}(L/\mathbf{Q})$, $H = \mathrm{Gal}(L/K)$, and $H^* = \mathrm{Gal}(L/K^*)$. Take elements φ_ν of G so that $\Phi = \{\varphi_1, \dots, \varphi_n\}$. By Proposition 28 of §8.3 we have

(18.5a) $$\bigsqcup_{\nu=1}^{n} \varphi_\nu^{-1} H = \bigsqcup_{\mu=1}^{m} H^* \tau_\mu$$

with $\tau_\mu \in G$, where $m = [K^* : \mathbf{Q}]/2$. As explained in that proposition, $\{\tau_\mu\}$ gives Φ^*.

By Proposition 29 of §8.3 we can define a map g: $(K^*)^\times \to K^\times$ by

(18.5b) $$g(a) = \prod_{\mu=1}^{m} a^{\tau_\mu} \qquad (a \in (K^*)^\times).$$

The right-hand side can be written also $\det\bigl(\Phi^*(a)\bigr)$. The map g can be extended naturally to a homomorphism $(K^*)_\mathbf{A}^\times \to K_\mathbf{A}^\times$ as follows.

Take $b \in K^*$ so that $K^* = \mathbf{Q}(b)$. Since the polynomial $P(x) = \prod_{\mu=1}^{m}(x - b^{\tau_\mu})$ has coefficients in K, there is an element $T \in K_m^m$ whose characteristic polynomial is P. Then, putting $\Phi^0\bigl(\sum_k c_k b^k\bigr) = \sum_k c_k T^k$ for $c_k \in \mathbf{Q}$, we obtain a $\mathbf{Q}$-linear ring-injection Φ^0: $K^* \to K_m^m$, which is clearly equivalent to Φ^* and $g(a) = \det[\Phi^o(a)]$. Naturally Φ^0 can be extended $\mathbf{Q}_p$-linearly to K_p^* and $\mathbf{Q}_\mathbf{A}$-linearly to $K_\mathbf{A}^*$. Taking the determinant of this extension, we thus obtain a continuous homomorphism $(K^*)_\mathbf{A}^\times \to K_\mathbf{A}^\times$, which we also write g.

Given a fractional ideal $\mathfrak{x}$ in K^*, we have, as shown in the same proposition, $\prod_{\mu=1}^{m} \mathfrak{x}^{\tau_\mu}\mathfrak{O} = \mathfrak{y}\mathfrak{O}$ with a fractional ideal $\mathfrak{y}$ in K, where $\mathfrak{O}$ is the maximal order of L. We write then $\mathfrak{y} = g(\mathfrak{x})$ as we did in (1) of §13.1. Clearly this notation is consistent with (18.5b). By the same proposition we have

(18.5c) $$g(a)g(a)^\rho = N_{K^*/\mathbf{Q}}(a), \qquad g(\mathfrak{x})g(\mathfrak{x})^\rho = N(\mathfrak{x})\mathfrak{o},$$

where $\mathfrak{o}$ is the maximal order of K.

18.6. Theorem *Let $\mathcal{P} = (A, \mathcal{C}, \iota)$ be a structure strictly of type Ω, and let K^* be as above. Further let σ be an element of $\mathrm{Aut}(\mathbf{C}/K^*)$, and s an element of $(K^*)^\times_{\mathbf{A}}$ such that $\sigma = [s, K^*]$ on K^*_{ab}. Then there is an exact sequence*

$$0 \longrightarrow q\big(g(s)^{-1}\mathfrak{a}\big) \longrightarrow \mathbf{C}^n \xrightarrow{\ \xi'\ } A^\sigma \longrightarrow 0$$

with the following properties:

(1) $\mathcal{P}^\sigma$ is strictly of type $(K, \Phi, g(s)^{-1}\mathfrak{a}, \zeta')$ with $\zeta' = N(s\mathfrak{r})\zeta$ with respect to ξ', where $\mathfrak{r}$ is the maximal order of K^*.

(2) $\xi\big(q(w)\big)^\sigma = \xi'\big(q(g(s)^{-1}w)\big)$ for every $w \in K/\mathfrak{a}$, that is, the following diagram is commutative:

$$\begin{array}{ccc} K/\mathfrak{a} & \xrightarrow{\ \xi\circ q\ } & A \\ {\scriptstyle g(s)^{-1}}\big\downarrow & & \big\downarrow{\scriptstyle \sigma} \\ K/g(s)^{-1}\mathfrak{a} & \xrightarrow{\ \xi'\circ q\ } & A^\sigma \end{array} .$$

Clearly ξ' is unique for σ, once ξ is fixed.

PROOF. In this theorem it is unnecessary to assume that A is defined over an algebraic number field. However, if we can prove our assertion for a $\mathcal{P}$, then it is easy to see that the theorem is true for every structure isomorphic to $\mathcal{P}$ over $\mathbf{C}$. Therefore, in view of Proposition 26 of §12.4 and the observation in §15.1, we may assume that $\mathcal{P}$ is rational over an algebraic number field. Next, we can show that it is sufficient to prove the case $\iota(\mathfrak{o}) \subset \mathrm{End}(A)$ with the maximal order $\mathfrak{o}$ of K as follows. Take a fractional ideal $\mathfrak{b}$ in K contained in $\mathfrak{a}$. Then we can find $\mathcal{P}_1 = (A_1, \mathcal{C}_1, \iota_1)$ strictly of type $\{ K, \Phi, \mathfrak{b}, \zeta_1 \}$ with some ζ_1 and an isomorphism ξ_1: $\mathbf{C}^n/q(\mathfrak{b}) \to A_1$. Let λ: $A_1 \to A$ be the isogeny which makes the diagram

(18.6a)
$$\begin{array}{ccccc} \mathbf{C}^n & \longrightarrow & \mathbf{C}^n/q(\mathfrak{b}) & \xrightarrow{\ \xi_1\ } & A_1 \\ {\scriptstyle \mathrm{id.}}\big\downarrow & & \big\downarrow & & \big\downarrow{\scriptstyle \lambda} \\ \mathbf{C}^n & \longrightarrow & \mathbf{C}^n/q(\mathfrak{a}) & \xrightarrow{\ \xi\ } & A \end{array}$$

commutative. Let us put $S = g(s)^{-1}$ for simplicity. Assuming our assertion to be true for $\mathcal{P}_1$, we obtain an isomorphism ξ_1': $\mathbf{C}^n/q\big(S\mathfrak{b}\big) \to A_1$ and a commutative diagram:

$$\begin{array}{ccc} K/\mathfrak{b} & \xrightarrow{\ \xi_1\circ q\ } & A_1 \\ {\scriptstyle S}\big\downarrow & & \big\downarrow{\scriptstyle \sigma} \\ K/S\mathfrak{b} & \xrightarrow{\ \xi_1'\circ q\ } & A_1^\sigma \end{array} .$$

Now we see that $\mathrm{Ker}(\lambda) = \xi_1\big(q(\mathfrak{a}/\mathfrak{b})\big)$, so that $\mathrm{Ker}(\lambda^\sigma) = \mathrm{Ker}(\lambda)^\sigma = \xi_1\big(q(\mathfrak{a}/\mathfrak{b})\big)^\sigma = \xi_1'\big(q(S\mathfrak{a}/S\mathfrak{b})\big)$. Clearly we can find an abelian variety A' isomorphic to $\mathbf{C}^n/q(S\mathfrak{a})$ and an isogeny λ': $A_1^\sigma \to A'$ such that the diagram

$$\begin{array}{ccccc} \mathbf{C}^n & \longrightarrow & \mathbf{C}^n/q(S\mathfrak{b}) & \xrightarrow{\xi_1'} & A_1^\sigma \\ {\scriptstyle \mathrm{id.}}\downarrow & & \downarrow & & \downarrow{\scriptstyle \lambda'} \\ \mathbf{C}^n & \longrightarrow & \mathbf{C}^n/q(S\mathfrak{a}) & \xrightarrow{\eta} & A' \end{array}$$

is commutative. Then $\mathrm{Ker}(\lambda') = \xi_1'\big(q(S\mathfrak{a}/S\mathfrak{b})\big) = \mathrm{Ker}(\lambda^\sigma)$, and hence we can find an isomorphism ε: $A' \to A^\sigma$ such that $\varepsilon \circ \lambda' = \lambda^\sigma$. Putting $\xi' = \varepsilon \circ \eta$, we obtain a commutative diagram:

$$\text{(18.6b)} \qquad \begin{array}{ccccc} \mathbf{C}^n & \longrightarrow & \mathbf{C}^n/q(S\mathfrak{b}) & \xrightarrow{\xi_1'} & A_1^\sigma \\ {\scriptstyle \mathrm{id.}}\downarrow & & \downarrow & & \downarrow{\scriptstyle \lambda^\sigma} \\ \mathbf{C}^n & \longrightarrow & \mathbf{C}^n/q(S\mathfrak{a}) & \xrightarrow{\xi'} & A^\sigma \end{array} .$$

Then, for $u \in K$ we have

$$\begin{aligned} \xi\big(q(u \ (\mathrm{mod}\,\mathfrak{a}))\big)^\sigma &= \lambda^\sigma\big(\xi_1\big(q(u \ (\mathrm{mod}\,\mathfrak{b}))\big)^\sigma\big) \\ &= \lambda^\sigma\big(\xi_1'\big(q(Su \ (\mathrm{mod}\,S\mathfrak{b}))\big)^\sigma\big) = \xi'\big(q(Su \ (\mathrm{mod}\,S\mathfrak{a}))\big), \end{aligned}$$

which proves assertion (2) for A. To prove (1), let X be determined by (18.4c) with respect to ξ. Put $X_1 = \lambda^{-1}(X)$. By (18.6a) and (17.6a) X_1 corresponds to ζ with respect to ξ_1. Since we are assuming (1) for $\mathcal{P}_1$, $(\lambda^\sigma)^{-1}(X^\sigma) = X_1^\sigma$ corresponds to ζ' as given in (1) with respect to ξ_1'. Then (18.6b) together with (17.6a) shows that X^σ corresponds to ζ', which proves assertion (1) for A^σ.

Thus we may assume that $\mathfrak{a}$ is a fractional ideal in K, so that $\iota(\mathfrak{o}) \subset \mathrm{End}(A)$. Under this condition our theorem is essentially a reformulation of formulas (1) in §15.3 and (2) in §16.3, and in fact we can employ them to give a shorter proof. However, we give here a more direct proof. (In §§14–16 we assumed that the CM-type is primitive, but it can easily be seen that those formulas are valid without that assumption.)

As remarked at the beginning of the proof, we may take $\mathcal{P}$ to be defined over an algebraic number field. Now we choose a complete set of representatives $\{\mathfrak{a}_i\}$ for the ideal classes of K and consider a structure $\mathcal{P}_i = (A_i, C_i, \iota_i)$ strictly of type $(K, \Phi, \mathfrak{a}_i, \zeta_i)$ with some ζ_i. We take an isomorphism ξ_i: $\mathbf{C}^n/q(\mathfrak{a}_i) \to A_i$ and $X_i \in C_i$ corresponding to ζ_i with respect to ξ_i. We may naturally assume that $\mathcal{P}$ is one of these $\mathcal{P}_i$. Take any positive integer $M > 2$ with the property that if ε is a root of unity in K such that $\varepsilon \equiv 1 \ (\mathrm{mod}\, M\mathfrak{o})$, then $\varepsilon = 1$. Let C_M denote the ray-class field over K^* modulo (M). We then take an algebraic number field L of finite degree such that: (i) the A_i, X_i, and the points on A_i

annihilated by M are all rational over L; (ii) L is normal over $\mathbf{Q}$; (iii) $C_M \subset L$; (iv) all homomorphisms of A_i to A_j are rational over L for every i and j. Now, given σ and s as in our theorem, we can find a prime ideal $\mathfrak{P}$ in L with the following properties: (1) $\mathfrak{P}$ does not divide M; (2) if $\mathfrak{p} = K^* \cap \mathfrak{P}$, then $N(\mathfrak{p})$ is a rational prime, and $\mathfrak{p}$ is unramified in L; (3) the restriction of σ to L is a Frobenius element of $\mathrm{Gal}(L/K^*)$ for $\mathfrak{P}$; (4) A_i^γ has no defect for $\mathfrak{P}$ for every i and every $\gamma \in \mathrm{Gal}(L/\mathbf{Q})$. Such a $\mathfrak{P}$ is guaranteed by the Tschebotareff density theorem. Put $p = N(\mathfrak{p})$ and $\mathfrak{q} = g(\mathfrak{p})$. Now $\mathbf{C}^n/q(\mathfrak{q}^{-1}\mathfrak{a})$ is isomorphic to A_i with a unique i. Fix an isomorphism η: $\mathbf{C}^n/q(\mathfrak{q}^{-1}\mathfrak{a}) \to A_i$. Take an integral ideal $\mathfrak{x}$ in K prime to p so that $\mathfrak{x}\mathfrak{q} = \alpha\mathfrak{o}$ with $\alpha \in \mathfrak{o}$. Then we obtain a commutative diagram:

$$\begin{array}{ccccc}
\mathbf{C}^n & \longrightarrow & \mathbf{C}^n/q(\mathfrak{a}) & \xrightarrow{\ \xi\ } & A \\
{\scriptstyle \mathrm{id.}}\downarrow & & \downarrow & & \downarrow{\scriptstyle \lambda} \\
\mathbf{C}^n & \longrightarrow & \mathbf{C}^n/q(\mathfrak{q}^{-1}\mathfrak{a}) & \xrightarrow{\ \eta\ } & A_i \\
{\scriptstyle \Phi(\alpha)}\downarrow & & \downarrow & & \downarrow{\scriptstyle \mu} \\
\mathbf{C}^n & \longrightarrow & \mathbf{C}^n/q(\mathfrak{a}) & \xrightarrow{\ \xi\ } & A
\end{array}$$

with isogenies λ and μ. Then $\mu \circ \lambda = \iota(\alpha)$, $\nu(\lambda) = N(\mathfrak{q})$, and $\nu(\mu) = N(\mathfrak{x})$.

We now consider reduction modulo $\mathfrak{P}$ and indicate by $\tilde{Y}$ or $Y^\sim$ the object obtained from Y by reduction modulo $\mathfrak{P}$. Let $\mathfrak{O}$ be the maximal order of L and f the automorphism $x \mapsto x^p$ of $\mathfrak{O}/\mathfrak{P}$. Then $\tilde{A}^f$ is meaningful, and we can define the p-th power homomorphism π: $\tilde{A} \to \tilde{A}^f$. We are going to show that there is an isomorphism ψ: $\tilde{A}_i \to \tilde{A}^f$ such that $\psi \circ \tilde{\lambda} = \pi$. For this purpose we take a basis $\{\omega_1, \dots, \omega_n\}$ of L-rational invariant differential forms on A such that $\delta\iota(a)\omega_i = a^{\varphi_i}\omega_i$ for every $a \in \mathfrak{o}$ (see §5.2). As shown in §13.2, we can take them so that $\tilde{\omega}_1, \dots, \tilde{\omega}_n$ are well defined and form a basis of invariant differential forms on $\tilde{A}$. Let H, H^*, and τ_μ be as in (18.5a). Since $\mathfrak{q}\mathfrak{O} = \prod_\mu \mathfrak{p}^{\tau_\mu}\mathfrak{O}$, we have $N(\mathfrak{q})^{[L:K]} = N(\mathfrak{p}\mathfrak{O})^m = p^{m[L:K^*]}$, which combined with the equality $n[L:K] = m[L:K^*]$ shows that $N(\mathfrak{q}) = p^n$. Given φ_ν, we have $\varphi_\nu = \tau_\mu^{-1}\gamma$ with some μ and $\gamma \in H^*$, and hence $\mathfrak{q}^{\varphi_\nu} \subset \mathfrak{p}^\gamma\mathfrak{O} \subset \mathfrak{P}$. Now $\delta\lambda\delta\mu(\omega_\nu) = \delta\iota(\alpha)\omega_\nu = \alpha^{\varphi_\nu}\omega_\nu$. Since $\alpha^{\varphi_\nu} \in \mathfrak{q}^{\varphi_\nu} \subset \mathfrak{P}$, we have $\delta\tilde{\lambda}\delta\tilde{\mu}(\tilde{\omega}_\nu) = 0$, which is true for every ν. We have $\nu(\tilde{\mu}) = \nu(\mu) = N(\mathfrak{x})$, and this number is prime to p. Therefore $\delta\tilde{\mu}$ is surjective, and hence $\delta\tilde{\lambda} = 0$. Since $\nu(\tilde{\lambda}) = \nu(\lambda) = N(\mathfrak{q}) = p^n$, Proposition 6 in §2.8 shows that $\pi = \psi \circ \tilde{\lambda}$ with an isomorphism ψ: $\tilde{A}_i \to \tilde{A}^f$ as expected.

By our choice of $\mathfrak{P}$, we see that $(Y^\sigma)^\sim = \tilde{Y}^f$ for every object Y rational over L. Since $\pi \circ \beta = \beta^f \circ \pi$ for every $\beta \in \mathrm{End}(\tilde{A})$ rational over $\mathfrak{O}/\mathfrak{P}$, we see that ψ is an isomorphism of $(A_i, \iota_i)^\sim$ to $(A^\sigma, \iota^\sigma)^\sim$. By Proposition 16 of §7.4, (A^σ, ι^σ) is a c-transform of (A_i, ι_i) with an ideal class c of K. By Proposition 15 of §11.2, c must be the identity class, that is, (A_i, ι_i) must be isomorphic to

(A^σ, ι^σ). Namely, we can put (A^σ, ι^σ) in place of (A_i, ι_i) in the above diagram. Then ψ is an automorphism of $(A^\sigma, \iota^\sigma)^\sim$ and so by Proposition 1 of §5.1 we have $\psi = \iota^\sigma(\varepsilon)^\sim$ with $\varepsilon \in \mathfrak{o}$. Putting $\xi^* = \iota^\sigma(\varepsilon) \circ \eta$ and $\kappa = \iota^\sigma(\varepsilon) \circ \lambda$, we have $\tilde{\kappa} = \pi$ and a commutative diagram:

$$\begin{array}{ccccc} \mathbf{C}^n & \longrightarrow & \mathbf{C}^n/q(\mathfrak{a}) & \xrightarrow{\ \xi\ } & A \\ {\scriptstyle \mathrm{id.}}\downarrow & & \downarrow & & \downarrow{\scriptstyle \kappa} \\ \mathbf{C}^n & \longrightarrow & \mathbf{C}^n/q(\mathfrak{q}^{-1}\mathfrak{a}) & \xrightarrow{\ \xi^*\ } & A^\sigma \end{array} .$$

Let X be a basic polar divisor in $\mathcal{C}$ as in (18.4c). Then $\kappa^{-1}(X^\sigma)^\sim = \pi^{-1}(\tilde{X}^f) = p\tilde{X}$. By Proposition 14 of §11.1 we obtain $E_\ell(\kappa^{-1}(X^\sigma)) = E_\ell(pX)$ with E_ℓ there. This means that $\kappa^{-1}(X^\sigma)$ and pX determine the same Riemann form. Combining this fact with the above diagram, we see from (17.6a) that X^σ corresponds to $p\zeta$ with respect to ξ^*. Thus we have proved that $\mathcal{P}^\sigma$ is strictly of type $(K, \Phi, \mathfrak{q}^{-1}\mathfrak{a}, p\zeta)$ with respect to ξ^* of the above diagram.

Let us now take $t \in A$ such that $Mt = 0$. Then $(t^\sigma)^\sim = \pi(\tilde{t}) = (\kappa t)^\sim$. Since M is prime to p, we have $t^\sigma = \kappa t$ by Proposition 16 of §11.2. Let $w = u \pmod{\mathfrak{a}}$ and $x = u \pmod{\mathfrak{q}^{-1}\mathfrak{a}}$ with $u \in M^{-1}\mathfrak{a}$. Put $r = \xi \circ q$ and $r^* = \xi^* \circ q$. Then $r(w)^\sigma = \kappa(r(w)) = r^*(x)$. Take $c \in (K^*)^\times_{\mathbf{A}}$ so that its $\mathfrak{p}$-component is a prime element of $K^*_{\mathfrak{p}}$ and all other components are 1. Then $[s, K^*] = \sigma = [c, K^*]$ on C_M. Therefore $c = sde$ with $d \in (K^*)^\times$ and an element $e \in (K^*)^\times_{\mathbf{A}}$ such that $e_v \in \mathfrak{r}_v^\times$ and $e_v - 1 \in M\mathfrak{r}_v$ for every finite prime v of K^*, where $\mathfrak{r}$ is the maximal order of K^*. Then $\mathfrak{q}^{-1}\mathfrak{a} = g(c)^{-1}\mathfrak{a} = g(d^{-1})g(s^{-1})\mathfrak{a}$, and we can extend the above diagram to a commutative diagram

$$(*)\qquad \begin{array}{ccccc} \mathbf{C}^n & \longrightarrow & \mathbf{C}^n/q(\mathfrak{a}) & \xrightarrow{\ \xi\ } & A \\ {\scriptstyle \mathrm{id.}}\downarrow & & \downarrow & & \downarrow{\scriptstyle \kappa} \\ \mathbf{C}^n & \longrightarrow & \mathbf{C}^n/q(\mathfrak{q}^{-1}\mathfrak{a}) & \xrightarrow{\ \xi^*\ } & A^\sigma \\ {\scriptstyle \Phi(g(d))}\downarrow & & \downarrow & & \downarrow{\scriptstyle \mathrm{id.}} \\ \mathbf{C}^n & \longrightarrow & \mathbf{C}^n/q(g(s)^{-1}\mathfrak{a}) & \xrightarrow{\ \xi'\ } & A^\sigma \end{array}$$

with a suitable ξ'. Put $r' = \xi' \circ q$. Then $r(w)^\sigma = r^*(x) = r'(g(d)u \bmod g(s)^{-1}\mathfrak{a})$. We are going to show that if $u \in M^{-1}\mathfrak{a}$, then

$$(**)\qquad \big(g(d)u\big)_\ell - \big(g(s)^{-1}u\big)_\ell \in \big(g(s)^{-1}\mathfrak{a}\big)_\ell$$

for every rational prime ℓ. First suppose $\ell \neq p$. Our choice of c implies that $g(s)_\ell g(d)_\ell \equiv 1 \pmod{M\mathfrak{o}_\ell}$, so that $g(d)_\ell - g(s)_\ell^{-1} \in g(s)_\ell^{-1}M\mathfrak{o}_\ell$. Therefore if $y \in M^{-1}\mathfrak{a}_\ell$, then $g(d)_\ell y - g(s)_\ell^{-1}y \in \big(g(s)^{-1}\mathfrak{a}\big)_\ell$. Next, we have $\big(g(s)g(d)g(e)\big)_p = g(c)_p \in \mathfrak{q}_p$. Since $g(e)_p \in \mathfrak{o}_p^\times$, we have $g(d)_p \in g(s)_p^{-1}\mathfrak{q}_p$. Let $y \in M^{-1}\mathfrak{a}_p$. Since $p \nmid M$, we have $y \in \mathfrak{a}_p$, and hence

$g(d)_p y \in g(s)_p^{-1}\mathfrak{q}_p\mathfrak{a}_p \subset g(s)_p^{-1}\mathfrak{a}_p$. This completes the proof of formula (**), which can be written

$$(***) \qquad r(w)^\sigma = r'(g(s)^{-1}w) \quad \textit{for every} \quad w \in M^{-1}\mathfrak{a}/\mathfrak{a}.$$

From (*) we see that X^σ corresponds to $\big(g(d)g(d)^\rho\big)^{-1}p\zeta$ with respect to ξ'. By (18.5c) we have $g(d)g(d)^\rho = N(d\mathfrak{r}) = N(s\mathfrak{r})^{-1}N(c\mathfrak{r}) = N(s\mathfrak{r})^{-1}p$, since $c\mathfrak{r} = \mathfrak{p}$. This shows that X^σ corresponds to ζ' of (2) of our theorem with respect to ξ'.

Now replace M by its multiple N. Then we obtain an isomorphism ξ'': $\mathbf{C}^n/q(g(s)^{-1}\mathfrak{a}) \to A^\sigma$ which has all the properties of ξ' with N in place of M. Then $\xi'' = \iota^\sigma(b) \circ \xi'$ with $b \in \mathfrak{o}^\times$. Since X^σ corresponds to ζ' with respect to both ξ' and ξ'', we see that $bb^\rho = 1$. By Lemma 18.2(i), $|b^\tau| = 1$ for every isomorphic embedding τ of K into $\mathbf{C}$. Therefore b must be a root of unity. Now we have (***) with $r'' = \xi'' \circ q$ in place of r'. Thus, for $w \in M^{-1}\mathfrak{a}/\mathfrak{a}$ we have

$$r(bw)^\sigma = \iota(b)^\sigma r(w)^\sigma = \iota(b)^\sigma r'(g(s)^{-1}w) = r''(g(s)^{-1}w) = r(w)^\sigma,$$

so that $bw = w$. It follows that $b - 1 \in M\mathfrak{o}$, and hence $b = 1$ because of our assumption $M > 2$. Thus $\xi' = \xi''$ and (***) holds for $w \in N^{-1}\mathfrak{a}/\mathfrak{a}$ with any multiple N of M. This proves (2) of our theorem, and our proof is now complete.

18.7. Let us now generalize Theorem 18.6 by taking a CM-algebra instead of a CM-field. Here we understand by a *CM-algebra* a direct sum of finitely many CM-fields. Thus let $Y = K_1 \oplus \cdots \oplus K_t$ with CM-fields K_i. We then consider $\mathcal{P} = (A, \mathcal{C}, \iota)$ as in §17.3 such that $2\dim(A) = [Y : \mathbf{Q}]$ with Y as W. In this case we have (17.3d) with $m = 1$. Let e_i be the identity element of K_i and let $A_i = \iota(m_i e_i)A$ with a positive integer m_i such that $\iota(m_i e_i) \in \mathrm{End}(A)$. We easily see that A is isogenous to $A_1 \times \cdots \times A_t$ and ι embeds K_i into $\mathrm{End}_{\mathbf{Q}}(A_i)$. Denote this embedding by ι_i. Since $[K_i : \mathbf{Q}] \leq 2\dim(A_i)$ by Proposition 1 of §5.1 and $[Y : \mathbf{Q}] = 2\dim(A)$, we obtain $[K_i : \mathbf{Q}] = 2\dim(A_i)$. Thus (A_i, ι_i) determines a CM-type (K_i, Φ_i), and Ψ of (17.3h) is the direct sum of $\Phi_1, \ldots, \Phi_t$ in the sense that we can put

$$(18.7\text{a}) \qquad \Psi\left(\sum_{i=1}^{t} a_i e_i\right) = \mathrm{diag}\,[\Phi_1(a_1), \ldots, \Phi_t(a_t)] \qquad (a_i \in K_i).$$

Similarly we may take the map q: $Y \to \mathbf{C}^d$ to be as follows: $q(\sum_{i=1}^t a_i e_i)$ with $a_i \in K_i$ is the column vector which is the juxtaposition of $q_1(a_1), \ldots, q_t(a_t)$, where q_i is defined for (K_i, Φ_i) as in §18.4. Then (17.3g) can be written

$$(18.7\text{b}) \qquad E_X\big(q(x), q(y)\big) = \mathrm{Tr}_{Y/\mathbf{Q}}\big(\zeta x y^\rho\big) \quad \text{for every} \quad x, y \in Y$$

with $\zeta = (\zeta_1, \ldots, \zeta_t)$, $-\zeta_i^\rho = \zeta_i \in K_i$, where ρ is the automorphism of Y whose restriction to each K_i is complex conjugation.

Let (K_i^*, Φ_i^*) be the reflex of (K_i, Φ_i) and K^* the composite field $K_1^* \cdots K_t^*$. Notice that K^* is generated over $\mathbf{Q}$ by $\mathrm{tr}[\Psi(a)]$ for all $a \in Y$. We then put $Y_\mathbf{A} = \prod_{i=1}^t (K_i)_\mathbf{A}$, $Y_\mathbf{A}^\times = \prod_{i=1}^t (K_i)_\mathbf{A}^\times$, and define a map g: $(K^*)_\mathbf{A}^\times \to Y_\mathbf{A}^\times$ by

(18.7c) $$g(x) = \left(\det \Phi_i^*\left(N_{K^*/K_i^*}(x)\right)\right)_{i=1}^t,$$

where $\det \Phi_i^*$ for each i is defined by (18.5b) (see the line below that formula). For a $\mathbf{Z}$-lattice $\mathfrak{a}$ in Y and $x \in Y_\mathbf{A}^\times$ we can define, generalizing what we did in §18.3, a $\mathbf{Z}$-lattice $x\mathfrak{a}$ in Y and a map of $Y/\mathfrak{a}$ to $Y/x\mathfrak{a}$ given by multiplication by x. Now the generalization of Theorem 18.6 is as follows:

18.8. Theorem *The statement of Theorem 18.6 holds with the following modifications: take Y in place of K so that $\Omega = (Y, \Psi, \mathfrak{a}, \zeta)$ with Ψ as in (18.7a), a $\mathbf{Z}$-lattice $\mathfrak{a}$ in Y, and ζ as in (18.7b); K^*, q, and g are defined as above.*

PROOF. Take a fractional ideal $\mathfrak{b}_i$ in K_i for each i so that $\mathfrak{b}_1 \oplus \cdots \oplus \mathfrak{b}_t \subset \mathfrak{a}$. The same argument as in the first part of the proof of Theorem 18.6 shows that it is sufficient to prove the case $\mathfrak{a} = \mathfrak{b}_1 \oplus \cdots \oplus \mathfrak{b}_t$. In such a case $A = A_1 \times \cdots \times A_t$, and we can apply Theorem 18.6 to $(A_i, \mathcal{C}_i, \iota_i)$ with $\mathcal{C}_i$ determined by (18.4c) with ζ_i as ζ. Combining the results for all i, we obtain the present theorem.

18.9. Corollary *The notation and the assumption being as in Theorem 18.8, let $w_1, \ldots, w_t$ be elements of $Y/\mathfrak{a}$, $r = \xi \circ q$, and let T be the subgroup of $(K^*)_\mathbf{A}^\times$ consisting of the elements s such that*

(18.9a) $bb^\rho N(s\mathfrak{r}) = 1, \quad bg(s)\mathfrak{a} = \mathfrak{a}, \quad bg(s)w_i = w_i \quad (i = 1, \ldots, t)$

for some $b \in Y^\times$. Then the field of moduli of $\mathcal{Q} = \left(A, \mathcal{C}, \iota, r(w_1), \ldots, r(w_t)\right)$ is the subfield of K_{ab}^* corresponding to T.

PROOF. Let h be the field of moduli of $\mathcal{Q}$ and let $\sigma \in \mathrm{Aut}(\mathbf{C}/h)$. Then there exists an isomorphism λ of $\mathcal{Q}$ onto $\mathcal{Q}^\sigma$. By Lemma 17.4 we have $K^* \subset h$, and hence we can find $s \in (K^*)_\mathbf{A}^\times$ such that $\sigma = [s, K^*]$ on K_{ab}^*. Take ξ' as in (2) of Theorem 18.6 (in the setting of Theorem 18.8) and put $\gamma = g(s)$. Then λ corresponds to a $Y_\mathbf{R}$-linear automorphism of $Y_\mathbf{R}$ that maps $\mathfrak{a}$ onto $\gamma^{-1}\mathfrak{a}$. Clearly such a map is given by multiplication with an element of $Y^\times$, say, b. Then $b\mathfrak{a} = \gamma^{-1}\mathfrak{a}$ and $\lambda\left(r(w)\right) = r'(bw)$ for $w \in Y/\mathfrak{a}$, where $r' = \xi' \circ q$. By (2) of Theorem 18.6 we have $r'(bw_i) = \lambda\left(r(w_i)\right) = r(w_i)^\sigma = r'(\gamma^{-1}w_i)$, and hence $bw_i \equiv \gamma^{-1}w_i \pmod{\gamma^{-1}\mathfrak{a}}$, that is, $b\gamma w_i \equiv w_i \pmod{\mathfrak{a}}$ for every i. Since λ

sends $\mathcal{C}^\sigma$ back to $\mathcal{C}$, (1) of Theorem 18.6 shows that $bb^\rho\zeta' = \zeta$. Thus (18.9a) holds. Conversely, if such a $b \in Y$ exists for a given $s \in (K^*)^\times_{\mathbf{A}}$, and $\sigma = [s, k]$ on K^*_{ab}, then we easily see that b corresponds to an isomorphism of $\mathcal{Q}$ onto $\mathcal{Q}^\sigma$, so that $\sigma =$ id. on h. This proves our assertion.

This corollary is a reformulation of Main Theorem 2 of §16.3, since it can easily be shown that the field $k^*\big(F(t)\big)$ of that theorem is the field of moduli of $(A, \mathcal{C}, \iota, t)$ if A is simple. In general $\mathrm{End}_{\mathbf{Q}}(A)$ may be larger than $\iota(Y)$ and so may be noncommutative. Theorem 18.8 can be formulated in terms of such a larger algebra (see [S70, Theorem 4.3]).

CHAPTER V

The Zeta Function of an Abelian Variety with Complex Multiplication

19. The Zeta Function Relative to a Field over which Some Endomorphisms Are Defined

19.1. We start with a somewhat general setting in which A is an abelian variety of dimension n defined over an algebraic number field k of finite degree. Let $\mathfrak{g}_\ell(A)$ denote the set of all points of A that are annihilated by powers of ℓ, and $k^{(\ell)}$ the smallest extension of k over which all the points of $\mathfrak{g}_\ell(A)$ are rational. Clearly $k^{(\ell)}$ is a subfield of the algebraic closure $\overline{\mathbf{Q}}$ of $\mathbf{Q}$. Take an ℓ-adic coordinate system of $\mathfrak{g}_\ell(A)$ as in §1.2. Since every element of $\mathrm{Gal}(\overline{\mathbf{Q}}/k)$ gives an automorphism of the module $\mathfrak{g}_\ell(A)$, we obtain a homomorphism

$$\mathfrak{M}_\ell\colon \mathrm{Gal}(\overline{\mathbf{Q}}/k) \to GL_{2n}(\mathbf{Z}_\ell). \tag{19.1a}$$

Clearly $\mathrm{Ker}(\mathfrak{M}_\ell) = \mathrm{Gal}(\overline{\mathbf{Q}}/k^{(\ell)})$, so that $\mathfrak{M}_\ell$ can be viewed also as a faithful representation of $\mathrm{Gal}(k^{(\ell)}/k)$. Notice that $k^{(\ell)}$ is an infinite extension of k, since every root of unity whose order is a power of ℓ is contained in it. (The infiniteness follows also from the Mordell-Weil theorem.) Given a prime ideal $\mathfrak{p}$ in k, take a prime divisor $\mathfrak{P}$ in $k^{(\ell)}$ that divides $\mathfrak{p}$ and a Frobenius automorphism σ of $k^{(\ell)}$ for $\mathfrak{P}/\mathfrak{p}$. Since $\mathfrak{M}_\ell$ is faithful, once we know the nature of $\mathfrak{M}_\ell(\sigma)$, that may give a significant piece of information about the arithmetic of the extension $k^{(\ell)}/k$, which is our basic philosophy.

19.2. We now consider reduction modulo $\mathfrak{p}$ of A. If A has no defect for $\mathfrak{p}$ in the sense of §11.1, then there is no problem. However, to obtain better results, we need the notion of *Néron model*. Given A over k and a prime ideal $\mathfrak{p}$ of k, Néron showed that there is an abelian variety A' isomorphic to A over k with the "best behavior" modulo $\mathfrak{p}$. For details, we refer the reader to his paper [N]. If A' has no defect for $\mathfrak{p}$, then we say that A has *good reduction modulo* $\mathfrak{p}$, and when we speak of reduction modulo $\mathfrak{p}$ of A, we understand that we are considering this "best model" A' instead of A. In fact, we need this model only in the context of the following fact, due to Serre and Tate [ST], which we quote without proof.

19.3. Lemma. *Let A be an abelian variety defined over k as in §19.1. Then A has good reduction modulo* $\mathfrak{p}$ *if and only if* $\mathfrak{p}$ *is unramified in* $k^{(\ell)}$ *for some* ℓ *prime to* $\mathfrak{p}$, *in which case* $\mathfrak{p}$ *is unramified in* $k^{(\ell)}$ *for every* ℓ *prime to* $\mathfrak{p}$.

In fact, the result holds more generally for k and $\mathfrak{p}$, as in §9.1. The "only if" part can easily be shown as in the proof of Lemma 19.5 below, and so the main point of Lemma 19.3 is the "if" part.

19.4. Suppose now that A in the setting of §19.1 has good reduction modulo $\mathfrak{p}$ and ℓ is prime to $\mathfrak{p}$. Given a $k^{(\ell)}$-rational object X, we denote by $\tilde{X}$ or $X^{\sim}$ the reduction of X modulo $\mathfrak{P}$. We put $\tilde{X} = \mathfrak{P}(X)$ if $\mathfrak{P}$ needs to be specified. As in the proof of Proposition 14 of §11.1, we can choose ℓ-adic coordinate systems of $\mathfrak{g}_\ell(A)$ and $\mathfrak{g}_\ell(\tilde{A})$ so that every $t \in \mathfrak{g}_\ell(A)$ and its reduction modulo $\mathfrak{P}$ have the same coordinates. Let $\varphi_\mathfrak{p}$ be the $N(\mathfrak{p})$-th power endomorphism of $\tilde{A}$. Observing that $(t^\sigma)^{\sim} = \varphi_\mathfrak{p}(\tilde{t})$ for every $t \in \mathfrak{g}_\ell(A)$, we easily see that

$$\mathfrak{M}_\ell(\sigma) = M_\ell(\varphi_\mathfrak{p}), \tag{19.4a}$$

where M_ℓ is the ℓ-adic representation of $\mathrm{End}_\mathbf{Q}(\tilde{A})$. This shows that σ is uniquely determined by $\mathfrak{P}$. Thus we have obtained the following result:

19.5. Lemma. *The notation being as above, suppose that A has good reduction modulo* $\mathfrak{p}$; *let* $f_{\mathfrak{p},\ell}$ *be the characteristic polynomial of* $\mathfrak{M}_\ell(\sigma)$. *Then a prime ideal* $\mathfrak{p}$ *in k prime to* ℓ *is always unramified in* $k^{(\ell)}$. *Moreover,* $f_{\mathfrak{p},\ell} = f_{\mathfrak{p},\ell'}$ *for any two primes* ℓ *and* ℓ'.

19.6. We now define the zeta function of A over k, written $Z(s; A/k)$, by

$$Z(s; A/k) = \prod_\mathfrak{p} \det\left[1 - N(\mathfrak{p})^{-s} M_\ell(\varphi_\mathfrak{p})\right]^{-1}, \tag{19.6a}$$

where $\mathfrak{p}$ runs over all the prime ideals in k modulo which A has good reduction, $\varphi_\mathfrak{p}$ is the $N(\mathfrak{p})$-th power endomorphism of $\mathfrak{p}(A)$, and M_ℓ is the ℓ-adic representation of $\mathrm{End}_\mathbf{Q}\left(\mathfrak{p}(A)\right)$. For each $\mathfrak{p}$ we must take ℓ prime to $\mathfrak{p}$. A well-known fact on the eigenvalues of $\varphi_\mathfrak{p}$ cited in §1.5 guarantees the absolute convergence of the product on the right-hand side for $\mathrm{Re}(s) > 3/2$. It should also be noted that in view of (19.4a) we may consider (19.6a) an analogue of Artin's L-functions.

19.7. Let us now take a CM-type (K, Φ) as in §18.4 and a structure (A, ι) of type (K, Φ) defined over an algebraic number field k of finite degree. We do not need any polarization of A here, though it will appear in the proof; nor do we consider the general case of §18.7. By Proposition 30 of §8.5 we have

$K^* \subset k$. We take $\mathfrak{a}$ so that (18.4a) holds, and define maps $f\colon k_{\mathbf{A}}^{\times} \to K_{\mathbf{A}}^{\times}$ and $r\colon K_{\mathbf{R}} \to A$ by

(19.7a) $$f = g \circ N_{k/K^*}, \qquad r = \xi \circ q$$

with q, ξ of (18.4a) and g of (18.5b). From (18.5c) we obtain

(19.7b) $$f(x)f(x)^{\rho} = N_{k/\mathbf{Q}}(x) \qquad (x \in k_{\mathbf{A}}^{\times}).$$

19.8. Theorem. *The notation and assumptions being as above, every point of finite order on A is rational over k_{ab}. Moreover, there exists a homomorphism $\alpha\colon k_{\mathbf{A}}^{\times} \to K^{\times}$ such that* $\mathrm{Ker}(\alpha)$ *is open in $k_{\mathbf{A}}^{\times}$, $\alpha(x)f(x)^{-1}\mathfrak{a} = \mathfrak{a}$, $\alpha(x)\alpha(x)^{\rho} = N(x\mathfrak{g})$, and $r(w)^{[x,k]} = r\bigl(\alpha(x)f(x)^{-1}w\bigr)$ for every $x \in k_{\mathbf{A}}^{\times}$ and every $w \in K/\mathfrak{a}$, where $\mathfrak{g}$ is the maximal order of k.*

PROOF. Let k' be the smallest extension of k over which all the points of A of finite order are rational. Given $x \in k_{\mathbf{A}}^{\times}$, let σ be an element of $\mathrm{Gal}(k'/k)$ such that $\sigma = [x, k]$ on $k' \cap k_{\mathrm{ab}}$; put $y = N_{k/K^*}(x)$. Then $\sigma = [y, K^*]$ on $k' \cap K^*_{\mathrm{ab}}$. Take ξ' as in Theorem 18.6 for the present σ with y in place of s there. Take any polarization $\mathcal{C}$ of A satisfying (18.4b), which can be obtained by (18.4c) from an element ζ of K. Then $\mathcal{C}$ is rational over k, as will be shown in Proposition 20.17 below. Since $\sigma = \mathrm{id.}$ on k, we have $(A, \mathcal{C}, \iota)^{\sigma} = (A, \mathcal{C}, \iota)$, so that there exists a $\mathbf{C}$-linear automorphism T such that $\xi' = \xi \circ T$. Clearly T commutes with the elements of $\Phi(K)$, $T\bigl(q(g(y)^{-1}\mathfrak{a})\bigr) = q(\mathfrak{a})$, and $E_X(Tu, Tv) = N(y\mathfrak{r})E_X(u, v)$ for E_X of (18.4c) by (17.6a) and (1) of Theorem 18.6. Then we easily see that $T = \Phi(\alpha)$ with an element α of K. Since $g(y) = f(x)$, we have $\alpha\alpha^{\rho} = N(y\mathfrak{r}) = N(x\mathfrak{g})$, $\alpha f(x)^{-1}\mathfrak{a} = \mathfrak{a}$, and

$$r(w)^{\sigma} = \xi\bigl(q(w)\bigr)^{\sigma} = \xi'\bigl(q(f(x)^{-1}w)\bigr) = \xi\bigl(q(\alpha f(x)^{-1}w)\bigr) = r\bigl(\alpha f(x)^{-1}w\bigr)$$

for every $w \in K/\mathfrak{a}$. For a positive integer m put

$$Z_m = \bigl\{t \in K_{\mathbf{A}}^{\times} \mid t\mathfrak{a} = \mathfrak{a}, tw = w \quad \textit{for every} \quad w \in m^{-1}\mathfrak{a}/\mathfrak{a}\bigr\}.$$

Let Z' be the projection of Z_1 to the nonarchimedean part of $K_{\mathbf{A}}^{\times}$ and let z be the nonarchimedean part of $\alpha f(x)^{-1}$. Then $z \in Z'$ and $r(w)^{\sigma} = r(zw)$ for all $w \in K/\mathfrak{a}$. Clearly the element z of Z' is uniquely determined by the last equality. Also the map $\sigma \to z$ defines a homomorphism of $\mathrm{Gal}(k'/k)$, which must be injective because of the definition of k'. It follows that $\mathrm{Gal}(k'/k)$ is abelian. This proves the first assertion. The element α is unique for x since z is unique for σ. Thus taking α to be $\alpha(x)$, we obtain our theorem, except the assertion that $\mathrm{Ker}(\alpha)$ is open. To prove it, take $m > 2$ and denote by k_m the smallest extension of k over which the points of $r(m^{-1}\mathfrak{a}/\mathfrak{a})$ are rational. Since $k_m \subset k_{\mathrm{ab}}$, class field theory assigns to k_m a subgroup Y of $k_{\mathbf{A}}^{\times}$ containing $k^{\times}k_{\mathbf{a}}^{\times}$. Let x be an element of Y such that $f(x) \in Z_m$. Put $\sigma = [x, k]$ and

$\alpha = \alpha(x)$. Then $\mathfrak{a} = \alpha\mathfrak{a}$, $\alpha\alpha^\rho = 1$, and hence α is a root of unity. Moreover, for $w \in m^{-1}\mathfrak{a}/\mathfrak{a}$ we have $r(w) = r(w)^\sigma = r(\alpha f(x)^{-1}w) = r(\alpha w)$, so that $(\alpha - 1)\mathfrak{a} \subset m\mathfrak{a}$. Since $m > 2$ we have $\alpha = 1$, which proves that Ker(α) is open, because Y and Z_m are open. This completes our proof.

19.9. By a *Hecke character of k* we mean a continuous homomorphism of $k_{\mathbf{A}}^\times$ into $\mathbf{C}^\times$ trivial on $k^\times$. For a prime $\mathfrak{p}$ of k let $k_\mathfrak{p}$ and $\mathfrak{g}_\mathfrak{p}$ denote the $\mathfrak{p}$-completions of k and $\mathfrak{g}$, where $\mathfrak{g}$ is the maximal order of k. We say that a Hecke character χ of k is *unramified* at $\mathfrak{p}$ if $\chi(\mathfrak{g}_\mathfrak{p}^\times) = 1$. Then the L-function of χ is defined as usual by

(19.9a) $$L(s, \chi) = \prod_{\mathfrak{p}} \left[1 - N(\mathfrak{p})^{-s}\chi(\pi_\mathfrak{p})\right]^{-1},$$

where $\mathfrak{p}$ runs over all the primes of k where χ is unramified, and $\pi_\mathfrak{p}$ is a prime element of $k_\mathfrak{p}$. For $x \in k_{\mathbf{A}}^\times$ we define as usual its *idele norm* $|x|_{\mathbf{A}}$ to be the product of $|x_v|_v$ for all primes v of k, where $|\ |_v$ is the normalized valuation at v.

Coming back to (K, Φ), we denote by J the set of all isomorphic embeddings of K into $\mathbf{C}$. Fixing an element ε of J, we identify K with K^ε. We can also identify $K_{\mathbf{a}}^\times$ with the group

(19.9b) $$\left\{ (y_\tau)_{\tau\in J} \in (\mathbf{C}^\times)^J \mid y_{\tau\rho} = (y_\tau)^\rho \quad \text{for every} \quad \tau \in J \right\}.$$

Then, for $z \in K_{\mathbf{A}}^\times$ and $\tau \in J$ we define the *τ-component* z_τ to be the τ-component of $z_\mathbf{a}$. We can then view our CM-type Φ as a subset of J.

19.10. Proposition. *For each $\tau \in J$ define χ_τ: $k_{\mathbf{A}}^\times \to \mathbf{C}^\times$ by $\chi_\tau(x) = \big(\alpha(x)/f(x)\big)_\tau$ and put $\chi = \chi_\varepsilon$ for the identity embedding ε of K into $\mathbf{C}$. Then χ_τ is a Hecke character of k satisfying the following conditions:*

(19.10a) $\chi(y) = f(y)^{-1}$ *for every* $y \in k_{\mathbf{a}}^\times$ *with f of (19.7a).*

(19.10b) *If $x \in k_{\mathbf{h}}^\times$, then $\chi(x) \in K^\times$, $\chi(x)\chi(x)^\rho = |x|_{\mathbf{A}}^{-1}$, and $\chi(x)\mathfrak{a} = f(x)\mathfrak{a}$.*

(19.10c) $\chi_\tau(z) = f(z)_\tau^{-1}\chi(z_{\mathbf{h}})^\tau$ *for every* $z \in k_{\mathbf{A}}^\times$.

(19.10d) $\chi_{\tau\rho}(x) = \overline{\chi_\tau(x)}$ *for every* $x \in k_{\mathbf{A}}^\times$.

(19.10e) *χ_τ and χ are ramified at the same prime ideals of k.*

PROOF. Clearly χ_τ is a homomorphism. If $x \in k^\times$, we have $[x, k] = \text{id}$. and $f(x) \in K^\times$, and hence from the relation $r(w)^{[x,k]} = r\big(\alpha(x)f(x)^{-1}w\big)$ we obtain $\alpha(x) = f(x)$, so that $\chi_\tau(x) = 1$. If $x \in k_{\mathbf{a}}^\times$, we have $[x, k] = \text{id}$.

and $f(x) \in K_{\mathbf{a}}^{\times}$, and hence $\alpha(x) = 1$, so that $\chi_\tau(x) = f(x)_\tau^{-1}$, which proves (19.10a). Since $\mathrm{Ker}(\alpha)$ is open, we easily see that χ_τ is continuous. Properties (19.10b, c, d) follow easily from Theorem 19.8 and our definition of χ_τ; (19.10e) follows from (19.10c).

Notice that the map α: $k_{\mathbf{A}}^{\times} \to K^{\times}$ can be obtained from χ by

(19.10f) $$\alpha(x) = \chi(x_{\mathbf{h}}) \qquad (x \in k_{\mathbf{A}}^{\times}).$$

We call χ *the Hecke character determined by* (A, ι) *over* k in the setting of Proposition 19.10, that is, if conditions (19.10a, b) are satisfied and

(19.10g) $$r(w)^{[x,k]} = r\big(\chi(x_{\mathbf{h}})f(x)^{-1}w\big) \quad \text{for every} \quad w \in K/\mathfrak{a} \text{ and } x \in k_{\mathbf{A}}^{\times}$$

with f and r of (19.7a). Notice that χ determines χ_τ for all $\tau \in J$ by (19.10c).

Now the main result of this section is:

19.11. Theorem. *The notation and assumptions being as above, χ is unramified at a prime ideal $\mathfrak{p}$ in k if and only if A has good reduction modulo $\mathfrak{p}$. Moreover, we have*

(19.11a) $$Z(s, A/k) = \prod_{\tau \in J} L(s, \chi_\tau).$$

PROOF. Suppose A has good reduction modulo $\mathfrak{p}$. Let $\sigma = [\pi_{\mathfrak{p}}, k]$ with $\pi_{\mathfrak{p}}$ as in (19.9a). By Proposition 14 in §11.1, we can define ℓ-adic representations M_ℓ of $\mathrm{End}_{\mathbf{Q}}(A)$ and M'_ℓ of $\mathrm{End}_{\mathbf{Q}}(\tilde{A})$ so that $M_\ell(\lambda) = M'_\ell(\tilde{\lambda})$ for every $\lambda \in \mathrm{End}(A)$. Suppose that $\mathfrak{p}$ is prime to ℓ. By Lemma 19.5, $\mathfrak{p}$ is unramified in $k^{(\ell)}$. Since $k^{(\ell)} \subset k_{\mathrm{ab}}$ by Theorem 19.8, we can take the present σ to be that of (19.4a). Put $\beta = \alpha(\pi_{\mathfrak{p}})$ and $H_\ell = \bigcup_{m=1}^{\infty} \ell^{-m}\mathfrak{a}$. Since the ℓ-component of $\pi_{\mathfrak{p}}$ is 1, we have $r(w)^\sigma = r\big(\beta f(\pi_{\mathfrak{p}})^{-1}w\big) = \iota(\beta)r(w)$ for every $w \in H_\ell/\mathfrak{a}$. This combined with (19.4a) shows that the eigenvalues of $M_\ell(\varphi_{\mathfrak{p}})$ are $\{\,\beta_\tau\,\}_{\tau \in J}$. Also, we have $\chi_\tau(\pi_{\mathfrak{p}}) = \beta_\tau$. Thus, with an indeterminate X we have

$$\det\big[1 - M_\ell(\varphi_{\mathfrak{p}})X\big] = \prod_{\tau \in J}\big(1 - \beta_\tau X\big) = \prod_{\tau \in J}\big(1 - \chi_\tau(\pi_{\mathfrak{p}})X\big).$$

Therefore our proof is complete if we prove the first assertion of our theorem. Let $\gamma = \alpha(y)$ with $y \in \mathfrak{o}_{\mathfrak{p}}^{\times}$. Take ℓ prime to $\mathfrak{p}$. Suppose that A has good reduction modulo $\mathfrak{p}$. By Lemma 19.5, $[y, k] = \mathrm{id.}$ on $k^{(\ell)}$, so that $r(w) = r(w)^{[y,k]} = r\big(\gamma f(y)^{-1}w\big) = r(\gamma w)$ for every $w \in H_\ell/\mathfrak{a}$ since $f(y)_\ell = 1$. Thus $\gamma = 1$, which implies that $\chi(y) = 1$. Conversely, if χ is unramified at $\mathfrak{p}$, then $\chi(y) = 1$, so that $\gamma = 1$. Thus $r(w)^{[y,k]} = r\big(\gamma f(y)^{-1}w\big) = r(w)$ for every $w \in H_\ell/\mathfrak{a}$, that is, $[y, k] = \mathrm{id.}$ on $k^{(\ell)}$ for all $y \in \mathfrak{o}_{\mathfrak{p}}^{\times}$. By Lemma 19.3, A has good reduction modulo $\mathfrak{p}$. This completes the proof.

19.12. Lemma. *Let* (A, ι) *and* (A', ι') *be two structures of type* (K, Φ) *defined over an algebraic number field* k, *which determine the same Hecke character of* k. *Then every isogeny of* (A, ι) *to* (A', ι') *is rational over* k.

PROOF. Take $\mathfrak{a}$ and $r \colon K_{\mathbf{R}}/\mathfrak{a} \to A$ as above for (A, ι); define similarly $\mathfrak{a}'$ and $r' \colon K_{\mathbf{R}}/\mathfrak{a}' \to A'$. Let λ be an isogeny of (A, ι) to (A', ι'). Then there exists an element c of K such that $c\mathfrak{a} \subset \mathfrak{a}'$ and $\lambda\big(r(w)\big) = r'(cw)$. If $\sigma \in \mathrm{Aut}(\mathbf{C}/k)$ and $\sigma = [x, k]$ on k_{ab} with $x \in k_{\mathbf{A}}^{\times}$, then our assumption implies that $r(v)^{\sigma} = r\big(\alpha(x) f(x)^{-1} v\big)$ for $v \in K/\mathfrak{a}$ and $r'(w)^{\sigma} = r'\big(\alpha(x) f(x)^{-1} w\big)$ for $w \in K/\mathfrak{a}'$ with the same α and f. Therefore $\lambda^{\sigma}\big(r(v)^{\sigma}\big) = \lambda\big(r(v)^{\sigma}\big)$ for all $v \in K/\mathfrak{a}$, and hence $\lambda^{\sigma} = \lambda$. This shows that λ is k-rational as expected.

If (A, ι), k, and χ are as above, and if (A, ι) is rational over a subfield k' of k, then (A, ι) determines a Hecke character χ' of k', and it is easy to verify that $\chi = \chi' \circ N_{k/k'}$. In general, in the setting of Theorem 19.11, we can ask whether χ can be obtained as $\chi = \psi \circ N_{k/h}$ with a Hecke character ψ of a subfield h of k. This can be answered as follows.

19.13. Theorem. *The notation being the same as in Theorem 19.11, let* h *be a subfield of* k *containing* K^*. *Then the following two conditions are equivalent:*

(1) *There exists a Hecke character* ψ *of* h *such that* $\chi = \psi \circ N_{k/h}$.
(2) *Every point of finite order on* A *is rational over the composite field* $h_{\mathrm{ab}} \cdot k$.

Moreover, if these conditions are satisfied , the number of characters ψ *as in (1) for a fixed* χ *is exactly* $[h_{\mathrm{ab}} \cap k : h]$.

For the proof, which is not so involved, the reader is referred to [S71a, Theorem 7.44]. The result of Theorem 19.11 in a somewhat less precise form was given by Taniyama [42]. In this and the following three sections we took and rearranged the material from [S71a], [S71b], and [S82]. In the next section we shall investigate the zeta function of A over a field which is not a field of rationality for (A, ι).

20. The Zeta Function over Smaller Fields

20.1. Throughout this section we fix a CM-type (K, Φ), denote its reflex by (K^*, Φ^*), and denote the reflex of (K^*, Φ^*) by (Z, Ξ); further we let F denote the maximal real subfield of K. Since Z is a CM-field contained in K, we have $K = FZ$. It should be noted that (K^*, Φ^*) is the reflex of (Z, Ξ) (see the paragraph preceding Proposition 29 in §8.3). We also employ the symbol J as in §19.9.

Given $\mathcal{P} = (A, \mathcal{C}, \iota)$ of type (K, Φ) satisfying (18.4b) and a subfield D of K, we denote by ι_D the restriction of ι to D, and by M_D the field of moduli of $(A, \mathcal{C}, \iota_D)$. In particular, we put $M = M_{\mathbf{Q}}$.

20.2. Lemma. *If (Z, Ξ) is as above, the center of $\mathrm{End}_{\mathbf{Q}}(A)$ coincides with $\iota(Z)$, and $\mathrm{tr}\,[\Phi(a)] = \mathrm{tr}\left[\Xi\big(Tr_{K/Z}(a)\big)\right]$ for every $a \in K$.*

PROOF. We use the same notation as in (18.5a). Let $H' = \mathrm{Gal}(L/Z)$. Then relation (18.5a) applied to (K^*, Φ^*) and (Z, Ξ) gives

$$\bigsqcup_{\mu=1}^{m} \tau_\mu^{-1} H^* = \bigsqcup_{\nu=1}^{\ell} H'\alpha_\nu \tag{20.2a}$$

with elements $\alpha_1, \ldots, \alpha_\ell$ such that $\Xi = \{\alpha_1, \ldots, \alpha_\ell\}$. This combined with (18.5a) shows that $\bigsqcup_{\nu=1}^{n} H\varphi_\nu = \bigsqcup_{\nu=1}^{\ell} H'\alpha_\nu$, which proves the last equality of our lemma. Also the restriction of Φ to Z is the sum of t copies of Ξ, where $t = [K : Z]$. Let (B, κ) be a structure of type (Z, Ξ). Since (Z, Ξ) is the reflex of a CM-type, it is primitive by Proposition 28 of §8.3, and hence B is simple, and $\mathrm{End}_{\mathbf{Q}}(B) = \kappa(Z)$ by Proposition 6 of §5.1. Then we can easily find a ring-injection θ of K into $\mathrm{End}_{\mathbf{Q}}(B^t)$ such that (B^t, θ) is of type (K, Φ). By the Corollary to Theorem 2 in §6.1 (B^t, θ) is isogenous to (A, ι), which proves our assertion.

20.3. Proposition. *Let D be a subfield of K and let $E = DZ$. Then the following assertions hold:*

(i) *K^* is normal over $K^* \cap M$. (This is equivalent to the statement that M contains the fixed subfield of $\mathrm{Aut}(K^*)$ in K^*.)*
(ii) *$K^* M_D = M_E$.*
(iii) *M_E is normal over M_D, and $\mathrm{Gal}(M_E/M_D)$ is isomorphic to a subgroup of $\mathrm{Aut}(E/D)$.*

PROOF. Let $\sigma \in \mathrm{Aut}(\mathbf{C}/M_D)$. Then there is an isomorphism λ of $(A, \mathcal{C}, \iota_D)$ onto $(A, \mathcal{C}, \iota_D)^\sigma$. Since $\iota(Z)$ is the center of $\mathrm{End}_{\mathbf{Q}}(A)$, we have $\lambda\iota(Z)\lambda^{-1} = \iota^\sigma(Z)$, and hence

$$\iota^\sigma(a)\lambda = \lambda \cdot \iota(a^\gamma) \quad \text{for every} \quad a \in E \tag{20.3a}$$

with $\gamma \in \mathrm{Aut}(E/D)$. Since K^* is generated by $\mathrm{tr}\,[\Xi(a)]$ for all $a \in Z$, the equality of Lemma 20.2 together with (20.3a) show that $K^{*\sigma} = K^*$. Taking D to be $\mathbf{Q}$, we see that M contains the fixed subfield of $\mathrm{Aut}(K^*)$ in K^*, which is (i). Next suppose $\gamma = \mathrm{id.}$ on Z. Then we see from (20.3a) that $\sigma = \mathrm{id.}$ on K^*. This shows that $K^* \subset M_Z$. Now let the notation be as in §18.5 and

(20.2a). Take an element of $\mathrm{Gal}(L/\mathbf{Q})$ representing γ and denote it again by γ. Since $Z^\gamma = Z$, we have $\gamma H' = H'\gamma$. If $\sigma = \mathrm{id.}$ on K^*, then (20.3a) shows that $\{\gamma\alpha_1, \ldots, \gamma\alpha_\ell\}$ coincides with $\{\alpha_1, \ldots, \alpha_\ell\}$ as a whole. We have then $\bigcup \gamma H'\alpha_\nu = \bigcup H'\gamma\alpha_\nu = \bigcup H'\alpha_\nu$, and so $\gamma \in H'$ by Proposition 28 of §8.3 (with H' and Z as H^* and K^* there). Thus $\gamma = \mathrm{id.}$ on Z, and hence λ is an isomorphism of $(A, \mathcal{C}, \iota_E)$ onto $(A, \mathcal{C}, \iota_E)^\sigma$. This shows that $M_E \subset K^*M_D$. Since $K^* \subset M_Z \subset M_E$, we obtain (ii). By (i), K^* is normal over $K^* \cap M_D$, and hence K^*M_D (which equals M_E) is normal over M_D. Now, assigning γ to σ, we obtain a homomorphism of $\mathrm{Aut}(\mathbf{C}/M_D)$ into $\mathrm{Aut}(E/D)$. We have seen that $\sigma = \mathrm{id.}$ on K^* if and only if $\gamma = \mathrm{id.}$ on Z, which proves the last statement of (iii).

20.4. Proposition. *The notation being the same as in Proposition 20.3, if (A, ι_D) is rational over a field h, then the following assertions hold:*

(i) *K^*h is normal over h; K^* is normal over $K^* \cap h$.*
(ii) *K^*h is the smallest field of rationality for (A, ι_E) containing h.*
(iii) $\mathrm{Gal}(K^*h/h)$ *is isomorphic to a subgroup of* $\mathrm{Aut}(E/D)$.
(iv) *If $D = F$, the maximal real subfield of K^* is contained in h.*

PROOF. Let $\sigma \in \mathrm{Aut}(\mathbf{C}/h)$. Then $\iota(Z)^\sigma = \iota(Z)$, so that (20.3a) holds with $\lambda = \mathrm{id}_A$. Thus we can repeat the proof of Proposition 20.3 disregarding $\mathcal{C}$, and obtain (i-iii) in the same manner. If $D = F$, we have $E = K$ and $\iota(a)^\sigma = \iota(a)$ for every $a \in K$, or $\iota(a)^\sigma = \iota(a^\rho)$ for every $a \in K$. It follows that $\sigma = \rho$ or $\sigma = \mathrm{id.}$ on K^*. This completes the proof.

20.5. Remark. (1) Since $FZ = K$, we obtain $M_K = K^*M_F$ from Proposition 20.3 (ii). If (A, ι_F) is rational over a field h, then $M_F \subset h$ by Proposition 20.17 below. Therefore, from Proposition 20.3 (iii) we obtain

(20.5a) $$[K^*h : h] \le [M_K : M_F] \le 2.$$

(2) If A is simple, then $Z = K$, and hence $M_K = K^*M$ and $[M_K : M] \le [K : \mathbf{Q}]$.

(3) Proposition 20.3 (i) shows that if K^* is not normal over $\mathbf{Q}$, then $M \neq \mathbf{Q}$. For example, if K is a field of degree 4 discussed in Example (2) (C) in §8.4, then M contains the real quadratic subfield of K^*.

20.6. Lemma. *Let (K, Φ), (A, ι), k, f, and α be as in Theorem 19.8; let γ be an automorphism of K and $\sigma \in \mathrm{Gal}(\overline{\mathbf{Q}}/\mathbf{Q})$. Put $\Phi'(a) = \Phi(a^\gamma)^\sigma$ and $\iota'(a) = \iota(a^\gamma)^\sigma$ for $a \in K$. Further let α' and f' be the maps defined for (A^σ, ι') over k^σ corresponding to α and f. Then (A^σ, ι') is of type (K, Φ'), $f'(x) = f(x^{\sigma^{-1}})^{\gamma^{-1}}$ and $\alpha'(x) = \alpha(x^{\sigma^{-1}})^{\gamma^{-1}}$.*

PROOF. The first assertion is obvious. Extend γ to an automorphism of $\overline{\mathbf{Q}}$ and denote it again by γ. Put $\Psi(a) = \Phi^*(a^{\sigma^{-1}})^{\gamma^{-1}}$ for $a \in K$. It can easily be verified that $(K^{*\sigma}, \Psi)$ is the reflex of (K, Φ'), and hence $f'(x) = f(x^{\sigma^{-1}})^{\gamma^{-1}}$. Take $\mathfrak{a}$ as before, and take $\mathfrak{a}'$ and ξ': $\mathbf{C}^n \to A^\sigma$ so that $\xi' \circ q'$ gives an isomorphism of $K_\mathbf{R}/\mathfrak{a}'$ to A^σ, where q' is defined relative to Φ'. Put $r = \xi \circ q$ and $r' = \xi' \circ q'$. Then the map $u \mapsto (r')^{-1}\big(r(u^\gamma)^\sigma\big)$ is an isomorphism of $K/\mathfrak{a}^{\gamma^{-1}}$ onto $K/\mathfrak{a}'$ commuting with the action of elements of K. Therefore $\mathfrak{a}' = t\mathfrak{a}^{\gamma^{-1}}$ and $(r')^{-1}\big(r(u^\gamma)^\sigma\big) = tu$ with $t \in K_\mathbf{A}^\times$. Now, given $x \in K_\mathbf{A}^\times$, put $\tau = [x^\sigma, k^\sigma]$. Then we have $r'(tu)^\tau = r(u^\gamma)^{\sigma\tau} = r(u^\gamma)^{[x,k]\sigma} = r\big(\alpha(x)f(x)^{-1}u^\gamma\big)^\sigma = r'\big(\alpha(x)^{\gamma^{-1}}\big(f(x)^{\gamma^{-1}}\big)^{-1}tu\big)$, so that $\alpha'(x^\sigma) = \alpha(x)^{\gamma^{-1}}$, which completes the proof.

20.7. Theorem. *Suppose that (A, ι_F) is rational over an algebraic number field h that does not contain K^*; let $k = hK^*$. Then $[k : h] = 2$ and (A, ι) is rational over k. Define χ_τ for (A, ι) over k as in Proposition 19.10 (which is meaningful in view of (ii) of Proposition 20.4). Then A has good reduction modulo a prime $\mathfrak{q}$ of h if and only if $\mathfrak{q}$ has a prime factor in k where χ is unramified, in which case χ is unramified at every prime factor of $\mathfrak{q}$ in k and $\mathfrak{q}$ is unramified in k. Moreover we have*

$$Z(s, A/h) = \prod_{\tau \in \Phi} L(s, \chi_\tau).$$

By Proposition 20.4 (ii), (A, ι) is rational over k, and $[k : h] = 2$ by (20.5a). We shall prove the remaining part after the proof of Theorem 20.9 below, which concerns a more general situation.

20.8. Let (A, ι) be of type (K, Φ); let k_0 be an algebraic number field over which A is defined. We assume

(20.8a) $\qquad \iota(K)$ is stable under $\mathrm{Gal}(\overline{\mathbf{Q}}/k_0)$.

This is obviously so if A is simple. (Thus Theorem 20.9 below is applicable to any simple A of type (K, Φ) with no extra condition.) Now let k be the smallest field of rationality for (A, ι) containing k_0. We easily see that under (20.8a) k is a finite Galois extension of k_0, and there is an injective homomorphism π of $\mathrm{Gal}(k/k_0)$ into $\mathrm{Aut}(K)$ such that

(20.8b) $\qquad \iota(a)^\sigma = \iota\big(a^{\pi(\sigma)}\big) \qquad (a \in K, \sigma \in \mathrm{Gal}(k/k_0)).$

By Proposition 30 of §8.5 we have $k \supset K^*k_0$ in general, and $k = k_0K^*$ if A is simple.

20.9. Theorem. *The notation being as above, let χ be defined as in Proposition 19.10. Then, for a prime ideal $\mathfrak{q}$ of k_0, the following conditions are equivalent:*

(i) *A has good reduction modulo $\mathfrak{q}$.*
(ii) *$\mathfrak{q}$ is unramified in k and has a prime factor in k where χ is unramified.*
(iii) *$\mathfrak{q}$ is unramified in k and χ is unramified at every prime factor of $\mathfrak{q}$ in k.*

Moreover, if $\mathfrak{q}$ is such a prime, then the Euler $\mathfrak{q}$-factor of $Z(s, A/k_0)$ coincides with that of $\prod_{\tau\in T} L(s, \chi_\tau)$, where T is a complete set of representatives for $\pi\big(\mathrm{Gal}(k/k_0)\big)\backslash J$ with J of §19.9. Consequently one has

$$P(s)Z(s, A/k_0) = \prod_{\tau\in T} L(s, \chi_\tau), \tag{20.9a}$$

where P is the product of the Euler $\mathfrak{p}$-factors on the right-hand side for all the primes $\mathfrak{p}$ of k ramified over k_0.

PROOF. Put $G = \mathrm{Gal}(k/k_0)$. Define $k^{(\ell)}$ as in §19.1 and define similarly $k_0^{(\ell)}$. We easily see that every element of $\mathrm{End}(A)$ is defined over $k_0^{(\ell)}$, and hence $K^* \subset k_0^{(\ell)}$ by Proposition 30 of §8.5. This shows that $k_0^{(\ell)} = k^{(\ell)}$. Therefore from Theorem 19.11 and Lemma 19.3 we see that (i) is equivalent to (iii). If a prime factor $\mathfrak{p}$ of $\mathfrak{q}$ in k is unramified in $k_0^{(\ell)}$, then clearly $\mathfrak{p}^\sigma$ is unramified in $k_0^{(\ell)}$ for every $\sigma \in G$, and hence we obtain our first assertion. Now, taking $\pi(\sigma)^{-1}$ to be γ in Lemma 20.6, we obtain $\iota' = \iota$, so that $\alpha(x^\sigma) = \alpha(x)^{\pi(\sigma)}$ and $f(x^\sigma) = f(x)^{\pi(\sigma)}$ for $x \in k_{\mathbf{A}}^\times$, and hence $\chi(x^\sigma) = \chi_{\pi(\sigma)}(x)$ for every $\sigma \in G$. Let $\mathfrak{p}$ be a prime ideal in k unramified over k_0, and let $\mathfrak{q} = k_0 \cap \mathfrak{p}$; suppose that A has good reduction modulo $\mathfrak{q}$. Then χ is unramified at $\mathfrak{p}^\sigma$ for all $\sigma \in G$. Denote by $\tilde{X}$ the reduction of an object X modulo $\mathfrak{p}$. Let $\varphi_\mathfrak{p}$ (resp. $\varphi_\mathfrak{q}$) be the $N(\mathfrak{p})$-th (resp. $N(\mathfrak{q})$-th) power endomorphism of $\tilde{A}$. Take a prime element c of $k_\mathfrak{p}$ and consider it an element of $k_{\mathbf{A}}^\times$; put $\beta = \alpha(c)$. The proof of Theorem 19.11 shows that $\varphi_\mathfrak{p} = \tilde{\iota}(\beta)$. Let γ denote the Frobenius element of G for $\mathfrak{p}$. From the equality $\iota(a)^\gamma = \iota(a^{\pi(\gamma)})$ we obtain $\varphi_\mathfrak{q} \circ \tilde{\iota}(a) = \tilde{\iota}\big(a^{\pi(\gamma)}\big) \circ \varphi_\mathfrak{q}$ for every $a \in K$. Therefore $\chi(c^\sigma) = \alpha(c^\sigma) = \alpha(c)^{\pi(\sigma)} = \beta^{\pi(\sigma)}$ for every $\sigma \in G$, so that $\beta^{\pi(\gamma)} = \beta$. (Notice that $\chi(c)$ does not depend on the choice of c, since χ is unramified at $\mathfrak{p}$.) Let S be the subalgebra of $\mathrm{End}_\mathbf{Q}(\tilde{A})$ generated by $\tilde{\iota}(K)$ and $\varphi_\mathfrak{q}$; put $Y = \{\sigma \in G \mid \mathfrak{p}^\sigma = \mathfrak{p}\}$ and $E = \big\{a \in K \mid a^{\pi(\gamma)} = a\big\}$. Then γ is a generator of Y, and S is a central simple algebra of degree ν^2 over E, where $\nu = [K : E] = [Y : 1]$. Consider the ℓ-adic representation M_ℓ of $\mathrm{End}_\mathbf{Q}(\tilde{A})$ for an ℓ prime to $N(\mathfrak{q})$. Since $\nu[E : \mathbf{Q}] = [K : \mathbf{Q}] = 2n$, from Lemma 1 of §5.1 we see that the restriction of M_ℓ to S is equivalent to the reduced representation of S over $\mathbf{Q}$. Since $\varphi_\mathfrak{q}^\nu = \varphi_\mathfrak{p} = \tilde{\iota}(\beta)$ and $\beta \in E$, the characteristic polynomial of $M_\ell\big(\varphi_\mathfrak{q}\big)$ equals $\prod_\delta (X^\nu - \beta^\delta)$, where X is an indeterminate and δ runs over all the isomorphic embeddings of E into $\mathbf{Q}$, which are clearly represented by $\pi(Y)\backslash J$. Therefore

$$\det\left[1 - M_\ell(\varphi_\mathfrak{q})X\right] = \prod_{\delta\in\pi(Y)\backslash J}\left(1 - \beta^\delta X^\nu\right).$$

On the other hand, consider all prime factors $\mathfrak{p}'$ of $\mathfrak{q}$ in k, and take a prime element $c(\mathfrak{p}')$ at $\mathfrak{p}'$. Then, with T as in our theorem, we obtain

$$\begin{aligned}\prod_{\mathfrak{p}'|\mathfrak{q}}\prod_{\tau\in T}\left[1 - \chi_\tau(c(\mathfrak{p}'))X\right] &= \prod_{\sigma\in Y\backslash G}\prod_{\tau\in T}\left[1 - \chi_\tau(c^\sigma)X\right] \\ &= \prod_{\sigma\in Y\backslash G}\prod_{\tau\in T}\left(1 - \beta^{\pi(\sigma)\tau}X^\nu\right) \\ &= \prod_{\delta\in\pi(Y)\backslash J}\left(1 - \beta^\delta X^\nu\right).\end{aligned}$$

This completes the proof.

20.10. We can ask a natural question:

(20.10a) *Can one take $P = 1$ in (20.9a)?*

In fact, we can find an example of A/k_0 for which the answer to (20.10a) is negative; see [S71b, p. 532, Example 5]. To obtain a sufficient condition for an affirmative answer, we first prove two lemmas.

20.11. Lemma. *Let Y be an abelian variety of dimension n defined over a field of any characteristic, where $n = [F : \mathbf{Q}]$. Suppose that* $\mathrm{End}_\mathbf{Q}(Y)$ *has a subalgebra W that is either a field of degree $2n$ or an algebra of degree 2 over F with identity element. Then W is a totally imaginary field.*

PROOF. First suppose that W is a field. By Propositions 1 and 2 of §5.1, W and $\mathrm{End}_\mathbf{Q}(Y)$ have the same identity element, and for every $x \in W$ we have $N_{W/\mathbf{Q}}(x) = \nu(x) \geq 0$. Such a number field W is obviously totally imaginary. Next suppose that W is not a field. Then $W \cap \mathrm{End}(Y)$ has an element α such that $0 < \dim(\alpha Y) < n$. Clearly F has an isomorphic image in $\mathrm{End}_\mathbf{Q}(\alpha Y)$. By Proposition 2 of §5.1, n divides $2\dim(\alpha Y)$, and hence $\dim(\alpha Y) = n/2$. Taking F and αY to be W and Y, we find that F is totally imaginary, which is a contradiction. This completes the proof.

20.12. Lemma. *The notation being as in Theorem 20.9, let $\mathfrak{p}$ be a prime ideal in k ramified over k_0, I the inertia group of $\mathfrak{p}$, and W the fixed subfield of K by $\pi(I)$. If W is not totally imaginary, then every abelian variety B isomorphic to A over k does not have good reduction modulo $\mathfrak{p}$.*

PROOF. Suppose B has good reduction modulo $\mathfrak{p}$. Let μ be a k-rational isomorphism of A to B. Put $\lambda_\sigma = \mu^\sigma \circ \mu^{-1}$ for $\sigma \in \mathrm{Gal}(k/k_0)$. Then $\lambda_{\sigma\tau} = \lambda_\sigma^\tau \lambda_\tau$. Define $\kappa\colon K \to \mathrm{End}_{\mathbf{Q}}(B)$ by $\kappa(a) = \mu \cdot \iota(a)\mu^{-1}$. Then we obtain the same $(\tilde{B}, \tilde{\kappa})$ from $(B, \kappa)^\sigma$ for all $\sigma \in I$ by reduction modulo $\mathfrak{p}$. Now, $\tilde{\lambda}_\sigma$ is an automorphism of $\tilde{B}$ and $\tilde{\lambda}_{\sigma\tau} = \tilde{\lambda}_\sigma^\tau \tilde{\lambda}_\tau$. Since $\kappa(a)^\sigma \lambda_\sigma = \lambda_\sigma \cdot \kappa(a^{\pi(\sigma)})$, we have $\tilde{\kappa}(a)\tilde{\lambda}_\sigma = \tilde{\lambda}_\sigma \cdot \tilde{\kappa}(a^{\pi(\sigma)})$ for every $a \in K$. We then easily see that $\tilde{\kappa}(K)$ and the $\tilde{\lambda}_\sigma$ for all $\sigma \in I$ generate a subalgebra of $\mathrm{End}_{\mathbf{Q}}(\tilde{B})$ isomorphic to a total matrix algebra of degree d over W, where $d = [K : W] = [I : 1]$. Then $\tilde{B}$ must be isomorphic to a product of d copies of an abelian variety P such that $\mathrm{End}_{\mathbf{Q}}(P)$ contains a subalgebra isomorphic to W. Since $\dim(P) = n/d = [W : \mathbf{Q}]/2$, W must be totally imaginary by Lemma 20.11. This proves our lemma.

20.13. Proposition. *In the setting of Theorem 20.9 let S be the subfield of K such that* $\mathrm{Gal}(K/S) = \pi\big(\mathrm{Gal}(k/k_0)\big)$. *If S is not totally imaginary and every prime ideal of k ramified over k_0 is completely ramified over k_0, then $P = 1$ in (20.9a).*

This is an immediate consequence of Lemma 20.12 and Theorem 20.9.

20.14. Proof of Theorem 20.7. If h is as in Theorem 20.7, then of course $\iota(F)$ is stable under $\mathrm{Gal}(\overline{\mathbf{Q}}/h)$. The same is true for $\iota(Z)$ by Lemma 20.2. Since $K = FZ$, (20.8a) is true in the present case, and so Theorem 20.9 is applicable. Clearly S of Proposition 20.13 is F. Therefore by that proposition we obtain our assertion.

20.15. Theorem. *The notation being as in Theorem 20.7, let χ be the Hecke character of $k_{\mathbf{A}}^\times$ determined by (A, ι), and ψ the quadratic Hecke character of $h_{\mathbf{A}}^\times$ corresponding to the extension k/h. Then*

$$\chi(z) = \psi(z)|z|^{-1} \quad \text{for every} \quad z \in h_{\mathbf{A}}^\times, \tag{20.15a}$$

where $|z|$ is the idele norm of z defined on $h_{\mathbf{A}}^\times$.

PROOF. Equality (20.15a) is clearly true for $z \in h^\times$. Therefore it is sufficient to prove it when z is a prime element of $h_{\mathfrak{q}}$ for a prime ideal $\mathfrak{q}$ of h, since such prime elements together with $h^\times$ generate a dense subgroup of $h_{\mathbf{A}}^\times$. We can even exclude finitely many prime ideals, so that we may assume that $\mathfrak{q}$ is unramified in k and A has good reduction modulo $\mathfrak{q}$. Let $\mathfrak{p}$ be a prime ideal of k dividing $\mathfrak{q}$, and σ the generator of $\mathrm{Gal}(k/h)$. Since $\sigma \neq$ id. on K^*, we have $\iota(a)^\sigma = \iota(a^\rho)$ for every $a \in K$ as explained in the proof of Proposition 20.4. Thus, by Lemma 20.6 we have $\alpha(x^\sigma) = \alpha(x)^\rho$ and $f(x^\sigma) = f(x)^\rho$, so that

$$\chi(x^\sigma) = \chi(x)^\rho \quad \text{for every} \quad x \in k_{\mathbf{A}}^\times. \tag{20.15b}$$

Assuming $\mathfrak{p}^\sigma \neq \mathfrak{p}$, take a prime element π of $k_\mathfrak{p}$, viewed as an element of $k_\mathbf{A}^\times$; put $z = \pi\pi^\sigma$. Then z is a prime element of $h_\mathfrak{q}$, and $\chi(z) = \alpha(z) = \alpha(\pi)\alpha(\pi^\sigma) = \alpha(\pi)\alpha(\pi)^\rho = |z|^{-1}$ by a formula in Theorem 19.8, which proves (20.15a) for such a z. Next assume $\mathfrak{p}^\sigma = \mathfrak{p}$ and $N(\mathfrak{p}) = N(\mathfrak{q})^2$; take a prime element π of $h_\mathfrak{q}$ and put $\beta = \alpha(\pi)$. Consider reduction modulo $\mathfrak{p}$, and let φ be the $N(\mathfrak{q})$-th power endomorphism of $\tilde{A}$. Then $\varphi^2 = \tilde{\iota}(\beta)$ as shown in the proof of Theorem 19.11. Now $\beta^\rho = \alpha(\pi)^\rho = \alpha(\pi^\sigma) = \alpha(\pi) = \beta$, and hence $\beta^2 = \beta\beta^\rho = \alpha(\pi)\alpha(\pi)^\rho = N(\mathfrak{p}) = N(\mathfrak{q})^2$. Thus $\beta = \pm N(\mathfrak{q})$. Let us now identify K with $\tilde{\iota}(K)$. Since $\varphi \cdot \tilde{\iota}(a) = \widetilde{\iota(a)^\sigma}\varphi = \tilde{\iota}(a^\rho)\varphi$, we see that $\varphi \notin \tilde{\iota}(K)$, and hence $F + F\varphi$ is an F-algebra of degree 2 over F. By Lemma 20.11 $F[\varphi]$ is a totally imaginary field. Therefore $\beta = -N(\mathfrak{q})$, and hence $\chi(\pi) = \alpha(\pi) = -N(\mathfrak{q}) = \psi(\pi)|\pi|^{-1}$, which completes the proof.

20.16. Corollary. *In the setting of Theorems 20.7 and 20.15 we have $L(s, \chi_\tau) = L(s, \chi_{\tau\rho})$ for every $\tau \in J$.*

PROOF. From (20.15b) we obtain $\chi_\tau(x^\sigma) = \chi_{\tau\rho}(x)$ for every $x \in k_\mathbf{A}^\times$, from which we easily obtain our equality.

20.17. Proposition. *If (A, ι) is of type (K, Φ), then every polarization of A satisfying (18.4b) is rational over any field of rationality for (A, ι_F).*

PROOF. Let h be a field of rationality for (A, ι_F), and X a basic polar divisor of a polarization $\mathcal{C}$ satisfying (18.4b). By Proposition 10 of §4.1 we may assume that X is rational over a finite normal extension k of h. Put $G = \mathrm{Gal}(k/h)$; let $\sigma \in G$. Then $\iota(a)^\sigma = \iota(a)$ for every $a \in F$. As seen in §20.14, condition (20.8a) is satisfied with $k_0 = h$, and hence $\iota(a)^\sigma = \iota(a^\pi)$ with $\pi \in \mathrm{Gal}(K/F)$. Define φ_X as in (4) in §1.3. By (5) in §1.3 we have $\varphi_X \cdot \iota(a) = {}^t\iota(a^\rho)\varphi_X$ for every $a \in K$. We easily see that the last equality holds with X^σ in place of X, since $\pi\rho = \rho\pi$. Put $Y = \sum_{\sigma\in G} X^\sigma$. Then $\varphi_Y \cdot \iota(a) = {}^t\iota(a^\rho)\varphi_Y$ for every $a \in K$, and Y is h-rational. Now X and Y correspond to elements ζ and ζ' of K by relation (18.4c). As explained in §6.3, φ_X and φ_Y are represented by $\Phi(\zeta)$ and $\Phi(\zeta')$, and hence $\varphi_X = \varphi_Y \cdot \iota(\zeta/\zeta')$. Since $\zeta/\zeta' \in F$, this shows that φ_X is h-rational. Therefore X^σ is algebraically equivalent to X for every $\sigma \in G$. This shows that $Y \in \mathcal{C}$, which completes the proof.

In general, let A be an abelian variety of dimension n defined over an algebraic number field k such that $\mathrm{End}_\mathbf{Q}(A)$ contains a commutative semisimple algebra of degree $2n$ over $\mathbf{Q}$. One can then pose the problem of determining the zeta function of A over k, with no assumption on the field of rationality for endomorphisms. Theorem 20.15 gives an answer at least when A is simple. In [Y81] Yoshida determined the zeta function in the general case of A which is not necessarily simple.

21. Models over the Field of Moduli and Models with Given Hecke Characters

We start with a proposition that explains a basic principle of finding a model of a polarized abelian variety over a given field.

21.1. Proposition. *Given $\mathcal{P} = (A, \mathcal{C}, \iota; \{t_i\}_{i=1}^r)$ as in §17.1, let k_0 be its field of moduli. If* $\mathrm{Aut}(\mathcal{P})$ *consists of the identity map, then $\mathcal{P}$ has a model rational over k_0, that is, $\mathcal{P}$ is isomorphic to a structure $\mathcal{P}'$ rational over k_0. Moreover, such a $\mathcal{P}'$ is unique up to isomorphisms over k_0.*

PROOF. Let $\mathcal{P}_1$ and $\mathcal{P}_2$ be structures, defined over a field k, isomorphic to $\mathcal{P}$, and let g be an isomorphism of $\mathcal{P}_1$ onto $\mathcal{P}_2$. Since $\mathrm{Aut}(\mathcal{P}_1)$ is trivial, we have $g^\sigma = g$ for every $\sigma \in \mathrm{Aut}(\mathbf{C}/k)$, and hence g is rational over k by Lemma 17.5 (iii). In particular, this proves the uniqueness of $\mathcal{P}'$.

To prove the existence, we first assume that $\mathcal{P}$ is rational over an algebraic extension k of k_0. Clearly we may assume that k is normal over k_0. Put $G = \mathrm{Gal}(k/k_0)$. Then, for each $\sigma \in G$ we can find an isomorphism f_σ of $\mathcal{P}$ onto $\mathcal{P}^\sigma$. Our assumption on $\mathrm{Aut}(\mathcal{P})$ implies that such an f_σ is unique for σ, and rational over k as shown above. Then we can easily verify that $f_{\sigma\tau} = (f_\sigma)^\tau \circ f_\tau$ for every $\sigma, \tau \in G$. Applying a well-known criterion of Weil [55, Theorem 3] (cf. also the argument we employed in the proof of Theorem 3 in §4.4) to the present setting, we find a variety B rational over k_0 and a biregular map f of B to A rational over k such that $f_\sigma = f^\sigma \circ f^{-1}$ for every $\sigma \in G$. Let e denote the zero element of A. Then e^σ is the zero element of A^σ and $f_\sigma(e) = e^\sigma$. Put $y = f^{-1}(e)$. Then y is rational over k, and $y^\sigma = (f^\sigma)^{-1}(e^\sigma) = (f^\sigma)^{-1}\big(f_\sigma(e)\big) = f^{-1}(e) = y$. Thus y is rational over k_0. Taking y to be the zero element of B we can make B an abelian variety rational over k_0. Define a structure $\mathcal{P}' = (B, \mathcal{C}', \iota'; \{t_i'\}_{i=1}^r)$ so that f is an isomorphism of $\mathcal{P}'$ to $\mathcal{P}$. By the same type of argument as for y we easily see that t_i' and $\iota'(a)$ for $a \in L$ are rational over k_0. Finally put $Y = \sum_{\sigma \in G} f^{-1}(X)^\sigma$ with a divisor X in $\mathcal{C}$ rational over k. Then Y is a divisor on B rational over k_0. Moreover, $f(Y) = \sum_{\sigma \in G} f_\sigma^{-1}(X^\sigma) \in \mathcal{C}$, since f_σ^{-1} sends $\mathcal{C}^\sigma$ to $\mathcal{C}$. Thus $Y \in \mathcal{C}'$, and so $\mathcal{P}'$ has the required properties.

In the general case $\mathcal{P}$ is defined over a finitely generated extension k of k_0. Let k_1 be the algebraic closure of k_0 in k. Then k is a regular extension of k_1. Take an isomorphism σ of k onto a field k' over k_1. Then $\mathcal{P}^\sigma$ is defined over k', and there is a unique isomorphism of $\mathcal{P}$ to $\mathcal{P}^\sigma$, which is defined over kk' as observed at the beginning. Therefore we can apply another criterion of [55] to $\mathcal{P}$ to find a structure $\mathcal{P}_1 = (A_1, \mathcal{C}_1, \iota_1, \{s_i\})$ isomorphic to $\mathcal{P}$ such that $(A_1, \iota_1, \{s_i\})$ is rational over k_1 and $\mathcal{C}_1$ is stable under $\mathrm{Aut}(\mathbf{C}/k_1)$. The procedure is similar to that of finding the above $\mathcal{P}'$. By Proposition 10 of §4.1, $\mathcal{C}_1$ is rational over a finite algebraic extension of k_1, and therefore $\mathcal{C}_1$ must be rational over k_1. Thus the problem can be reduced to the case we already treated. This completes the proof.

Since $\mathrm{Aut}\big((A, \mathcal{C})\big)$ is finite by Proposition 17 of §4.4, our proposition is applicable to $\mathcal{P}$ if, for example, the t_i generate the group of all points on A annihilated by an integer $m > 2$. Now, given $\mathcal{Q} = (A, \mathcal{C}, \iota)$ and its field of moduli k^*, take such t_i. Then we can find a model of $\mathcal{Q}$ rational over a finite algebraic extension of k^*, by virtue of the above proposition combined with Proposition 17.2 (iv). This fact can be seen more directly; see the argument of [S65, p. 128], which can be employed in the above proof in place of the second criterion of [55]. In Theorem 27.13 below we shall construct such a model of $(A, \mathcal{C})$ by means of theta functions.

21.2. Theorem. *Given $\mathcal{P} = (A, \mathcal{C}, \iota; \{t_i\}_{i=1}^r)$, let k be a field containing the field of moduli of $\mathcal{P}$ and $\det(\alpha)$ for every $\alpha \in \mathrm{Aut}(\mathcal{P})$, where $\det(\alpha)$ is the determinant of the representation of α on the space $\mathfrak{D}_0(A)$ defined in §2.6. Suppose that the map $\alpha \mapsto \det(\alpha)$ of $\mathrm{Aut}(\mathcal{P})$ into $k^\times$ is injective. Then $\mathcal{P}$ has a model rational over k.*

PROOF. As remarked above, we may assume that $\mathcal{P}$ is defined over a finite Galois extension h of k. Put $G = \mathrm{Gal}(h/k)$. For each $\sigma \in G$ we can find an isomorphism f_σ of $\mathcal{P}$ onto $\mathcal{P}^\sigma$. Then $f_{\sigma\tau} = (f_\sigma)^\tau \circ f_\tau \circ \alpha_{\sigma,\tau}$ with $\alpha_{\sigma,\tau} \in \mathrm{Aut}(\mathcal{P})$. Take a nonzero holomorphic n-form ξ on A rational over h, where n is the dimension of A. Then $\xi^\sigma \circ f_\sigma = c_\sigma \xi$ with $c_\sigma \in h$, and $c_{\sigma\tau} = c_\sigma^\tau c_\tau \det(\alpha_{\sigma,\tau})$. Let s be the order of $\mathrm{Aut}(\mathcal{P})$. Then $c_{\sigma\tau}^s = (c_\sigma^s)^\tau c_\tau^s$, so that there is an element $b \in h$ such that $c_\sigma^s = b/b^\sigma$ for every $\sigma \in G$. Consider an extension $h(d)$ with an element d such that $d^s = b$. Let γ be an isomorphism of $h(d)$ onto its conjugate over k, and let σ be the restriction of γ to h. Then $c_\sigma d^\gamma/d$ is an s-th root of unity, so that our assumption guarantees an element $\alpha \in \mathrm{Aut}(\mathcal{P})$ such that $c_\sigma \det(\alpha) = d/d^\gamma$. This shows that $h(d)$ is normal over k. Put $G' = \mathrm{Gal}\big(h(d)/k\big)$ and denote by $r(\gamma)$ the restriction of $\gamma \in G'$ to h. For each $\gamma \in G'$ we found an element $\alpha_\gamma \in \mathrm{Aut}(\mathcal{P})$ such that $c_{r(\gamma)} \det(\alpha_\gamma) = d/d^\gamma$. Put $g_\gamma = f_{r(\gamma)} \circ \alpha_\gamma$. Then g_γ is an isomorphism of $\mathcal{P}$ onto $\mathcal{P}^\gamma$, and $(d\xi)^\gamma \circ g_\gamma = d^\gamma \xi^\gamma \circ f_{r(\gamma)} \circ \alpha_\gamma = d^\gamma c_{r(\gamma)} \det(\alpha_\gamma)\xi = d\xi$. From this and our injectivity assumption we obtain $g_{\gamma\delta} = g_\gamma^\delta g_\delta$ for every $\gamma, \delta \in G'$. Therefore, by the same argument as in the proof of Theorem 21.1, we obtain the desired model over k.

21.3. Corollary. *Given $\mathcal{P} = (A, \mathcal{C}, \iota; \{t_i\}_{i=1}^r)$, suppose that $\mathrm{Aut}(\mathcal{P}) = \{\pm 1\}$ and A is of odd dimension. Then $\mathcal{P}$ has a model rational over its field of moduli.*

This is an immediate consequence of Theorem 21.2. Thus an odd-dimensional generic polarized abelian variety $(A, \mathcal{C})$ has a model over its field of moduli.

However, there is an even-dimensional generic polarized abelian variety which has no model over its field of moduli (see [S71c]).

Coming back to the case of complex multiplication, we first prove the following result, due to W. Casselman:

21.4. Theorem. *Let (K, Φ) and (K^*, Φ^*) be as in Section 20. Let k be a finite algebraic extension of K^*, and χ a Hecke character of $k_{\mathbf{A}}^{\times}$ satisfying (19.10a, b). Then there exists a structure $(A, \mathcal{C}, \iota)$ of type $(K, \Phi, \mathfrak{a}, \zeta)$ rational over k which determines χ.*

PROOF. First we take a structure $\mathcal{P} = (A, \mathcal{C}, \iota)$ of type $(K, \Phi, \mathfrak{a}, \zeta)$ rational over $\overline{\mathbf{Q}}$. Put $r = \xi \circ q$ with ξ and q as in (18.4a). For every $\sigma \in \mathrm{Gal}\big(\overline{\mathbf{Q}}/k\big)$ take $y \in k_{\mathbf{A}}^{\times}$ so that $\sigma = [y, k]$ on k_{ab}. Define ξ' as in (2) of Theorem 18.6 for $s = N_{k/K^*}(y)$ and put $r' = \xi' \circ q$. In view of (19.10b) we can define a continuous homomorphism β: $k_{\mathbf{A}}^{\times} \to K^{\times}$ by $\beta(x) = \chi(x_{\mathbf{h}})$. By our assumptions (19.10a, b) we have $\beta(y)g(s)^{-1}\mathfrak{a} = \mathfrak{a}$ and $\beta(y)\beta(y)^{\rho} = N(s\mathfrak{r})$. Hence there is an isomorphism λ of A onto A^{σ} such that $\lambda\big(r(w)\big) = r'(\beta(y)^{-1}w)$ for every $w \in K_{\mathbf{R}}/\mathfrak{a}$. Clearly λ is an isomorphism of $\mathcal{P}$ onto $\mathcal{P}^{\sigma}$. Let $c = [\beta(y)f(y)^{-1}]_{\mathbf{h}}$. If $y \in k^{\times}$, then $\beta(y)f(y)^{-1} = \chi(y_{\mathbf{h}})\chi(y_{\mathbf{a}}) = 1$. Therefore we see that $c = 1$ if y belongs to the closure of $k^{\times}k_{\mathbf{a}}^{\times}$. Consequently c is uniquely determined by σ independently of the choice of y. Since $r(w)^{\sigma} = \lambda\big(r(cw)\big)$, λ is also uniquely determined by σ. Put $c = c_{\sigma}$ and $\lambda = \lambda_{\sigma}$. Then we easily find that $c_{\sigma\tau} = c_{\sigma}c_{\tau}$ and $\lambda_{\sigma\tau} = \lambda_{\sigma}^{\tau}\lambda_{\tau}$ for $\sigma, \tau \in \mathrm{Gal}\big(\overline{\mathbf{Q}}/k\big)$. Therefore, for the same reason as in the proof of Proposition 21.1, we can find a k-rational structure $\mathcal{P}_1 = (A, \mathcal{C}, \iota)$ and a $\overline{\mathbf{Q}}$-rational isomorphism μ of $\mathcal{P}$ to $\mathcal{P}_1$ such that $\mu = \mu^{\sigma} \circ \lambda_{\sigma}$ for all $\sigma \in \mathrm{Gal}\big(\overline{\mathbf{Q}}/k\big)$. Put $\xi_1 = \mu \circ \xi$ and $r_1 = \mu \circ r$. Then $r_1 = \xi_1 \circ q$ and for $\sigma = [y, k]$ we have $r_1(w)^{\sigma} = \mu^{\sigma}\big(\lambda_{\sigma}(r(c_{\sigma}w))\big) = r_1\big(\beta(y)f(y)^{-1}w\big)$ for every $w \in K/\mathfrak{a}$, that is, (19.10g) is satisfied. Hence $\mathcal{P}_1$ determines χ.

21.5. Theorem. *Given $\mathcal{P} = (A, \mathcal{C}, \iota)$ of type (K, Φ), define M_F and ι_F as in §20.1; let h be an algebraic number field containing M_F, but not K^*, and let $k = K^*h$. Then $[k : h] = 2$. Further, let χ be a Hecke character of $k_{\mathbf{A}}^{\times}$ satisfying (19.10a, b) and (20.15a). Then there exists a structure $\mathcal{P}' = (A', \mathcal{C}', \iota')$ which is isomorphic to $\mathcal{P}$, rational over k, and determines χ, and such that $(A', \mathcal{C}', \iota'_F)$ is rational over h.*

PROOF. We see that (20.5a) holds in the present case since $M_F \subset h$, and so $[k : h] = 2$. Let σ be the generator of $\mathrm{Gal}(k/h)$. We observe that (20.15b) follows from (19.10a, b) and (20.15a). In fact, for $z \in h_{\mathbf{A}}^{\times}$ (resp. $z \in k_{\mathbf{A}}^{\times}$) let $|z|$ (resp. $|z|_{\mathbf{A}}^{\times}$) denote the idele norm of z in $h_{\mathbf{A}}^{\times}$ (resp. in $k_{\mathbf{A}}^{\times}$). Then, for $x \in k_{\mathbf{A}}^{\times}$ we have $\psi(xx^{\sigma}) = 1$ since $N_{k/h}(k_{\mathbf{A}}^{\times}) \subset \mathrm{Ker}(\psi)$, and hence, by (20.15a), $\chi(x)\chi(x^{\sigma}) = \chi(xx^{\sigma}) = |xx^{\sigma}|^{-1} = |x|_{\mathbf{A}}^{-1}$. On the other hand, by (19.10a,

b) we have $\chi(x)\chi(x)^{\rho} = \chi(x_{\mathbf{h}})\chi(x_{\mathbf{h}})^{\rho}\chi(x_{\mathbf{a}})\chi(x_{\mathbf{a}})^{\rho} = |x_{\mathbf{h}}|_{\mathbf{A}}^{-1}|x_{\mathbf{a}}|_{\mathbf{A}}^{-1} = |x|_{\mathbf{A}}^{-1}$. Thus we obtain $\chi(x^{\sigma}) = \chi(x)^{\rho}$, that is, (20.15b) holds in the present cdase. By Theorem 21.4 we may assume that $\mathcal{P}$ is rational over k and determines χ. Since $M_F \subset h$, there is an isomorphism λ of $(A, \mathcal{C}, \iota_F)$ onto $(A, \mathcal{C}, \iota_F)^{\sigma}$. Then (20.3a) in the present case can be written $\lambda \cdot \iota(a^{\gamma}) = \iota^{\sigma}(a)\lambda$ for every $a \in K$ with $\gamma \in \mathrm{Gal}(K/F)$. We have seen in the proof of Propositions 20.3 and 20.4 that $\sigma \mapsto \gamma$ defines an injection of $\mathrm{Gal}(k/h)$ into $\mathrm{Gal}(K/F)$. Since $\sigma \neq \mathrm{id.}$ on K^*, we have $\gamma = \rho$. Thus if we put $\mathcal{P}^* = (A^{\sigma}, \mathcal{C}^{\sigma}, \iota^*)$ with $\iota^*(a) = \iota(a^{\rho})^{\sigma}$ for $a \in K$, then λ is an isomorphism of $\mathcal{P}$ to $\mathcal{P}^*$. Let χ' be the Hecke character determined by $\mathcal{P}^*$. By Proposition 19.10 and Lemma 20.6, for $x \in k_{\mathbf{A}}^{\times}$ we have $\chi'(x) = \big[\alpha(x^{\sigma})/f(x^{\sigma})\big]^{\rho} = \chi(x^{\sigma})^{\rho}$, which equals $\chi(x)$ by (20.15b). Thus $\mathcal{P}^*$ determines χ. Therefore by Lemma 19.12 λ is rational over k. Then we easily see that λ^{σ} is an isomorphism of $\mathcal{P}^*$ to $\mathcal{P}$, and hence $\lambda^{\sigma}\lambda = \iota(\zeta)$ with a root of unity ζ in K. Then $\iota(\zeta) = \lambda^{-1}\lambda\lambda^{\sigma}\lambda = \lambda^{-1}\iota(\zeta)^{\sigma}\lambda = \iota(\zeta^{\rho})$, and hence $\zeta = \pm 1$. Suppose $\zeta = -1$. Then $\lambda\lambda^{\sigma} = \iota^*(-1)$. Take a prime ideal $\mathfrak{p}$ of k and put $\mathfrak{q} = \mathfrak{p} \cap h$; we can take these so that A has good reduction modulo $\mathfrak{p}$, and $\mathfrak{q}$ remains prime in k. We now consider reduction modulo $\mathfrak{p}$, and indicate by a superscript q the action of the $N(\mathfrak{q})$-th power automorphism of the residue field modulo $\mathfrak{p}$. Let φ (resp. φ') be the $N(\mathfrak{q})$-th power homomorphism of $\tilde{A}$ to $\tilde{A}^q$ (resp. of $\tilde{A}^q$ to $\tilde{A}$). We easily see that $\tilde{A}^q = (A^{\sigma})^{\sim}$ and $\tilde{\lambda}^q = (\lambda^{\sigma})^{\sim}$. Put $\mu = \varphi'\tilde{\lambda}$. Then $\mu \in \mathrm{End}(\tilde{A})$ and $\mu^2 = \varphi'\tilde{\lambda}\varphi'\tilde{\lambda} = \varphi'\tilde{\lambda}\tilde{\lambda}^q\varphi = -\varphi'\varphi$. Now $\varphi'\varphi$ is the $N(\mathfrak{p})$-th power endomorphism of $\tilde{A}$. Let π be a prime element of $h_{\mathfrak{q}}$. Condition (20.15a) implies that $\alpha(\pi) = \chi(\pi) = -N(\mathfrak{q})$. Now recall a formula $r(w)^{[x,k]} = r\big(\alpha(x)f(x)^{-1}w\big)$, valid for $x \in k_{\mathbf{A}}^{\times}$ and $w \in K/\mathfrak{a}$, proved in Theorem 19.8. If ℓ is a rational prime not divisible by $\mathfrak{p}$ and if $u \in (K \otimes \mathbf{Q}_{\ell})/(\mathfrak{a} \otimes \mathbf{Z}_{\ell})$, then $r(u)^{[\pi,k]} = r\big(\alpha(\pi)u\big) = -N(\mathfrak{q})r(u)$. Taking reduction modulo $\mathfrak{p}$, we find that $\varphi'\varphi = \tilde{\iota}\big(-N(\mathfrak{q})\big)$, so that $\mu^2 = \tilde{\iota}\big(N(\mathfrak{q})\big)$. On the other hand we have

$$\mu \cdot \tilde{\iota}(a) = \varphi'\tilde{\lambda} \cdot \tilde{\iota}(a) = \varphi'\tilde{\iota}(a^{\rho})^q\tilde{\lambda} = \tilde{\iota}(a^{\rho})\varphi'\tilde{\lambda} = \tilde{\iota}(a^{\rho})\mu \qquad (a \in K).$$

Therefore $\mu \notin \tilde{\iota}(K)$. By Lemma 20.11 $F[\mu]$ must be a totally imaginary field, which is a contradiction, since $\mu^2 = \tilde{\iota}(N(\mathfrak{q}))$. Therefore $\zeta = 1$, so that $\lambda^{\sigma}\lambda = \iota(1)$. Applying Weil's criterion of descent, we find an abelian variety A' rational over h and an isomorphism η of A onto A' rational over k such that $\eta = \eta^{\sigma} \circ \lambda$. Let $\mathcal{C}'$ be the image of $\mathcal{C}$ under η, and let $\iota'(a) = \eta \cdot \iota(a)\eta^{-1}$ for $a \in K$. Then $(A', \mathcal{C}', \iota')$ has all the required properties.

Theorems 20.7 and 21.5 give a necessary and sufficient condition for χ to be determined by $(A, \mathcal{C}, \iota)$ for which $(A, \mathcal{C}, \iota_F)$ is defined over a field h such that $k = K'h \neq h$. There are two sets of natural questions:

(Q1) *Can we generalize Theorem 21.5 to the case of $(A, \mathcal{C}, \iota_D)$ with a proper subfield D of K other than F?*

(Q2) *Can we take k to be M_K in Theorem 21.4 or h to be M_F in Theorem 21.5? In other words, can we find a Hecke character χ of M_K satisfying (19.10a, b) with $k = M_K$? When M_F does not contain K^*, can we find such a character satisfying (20.15a)? What about the case of D as in (Q1)?*

There are no clear-cut answers to these questions, and in fact there are counter-examples. It should be noted that Theorem 21.2 and Corollary 21.3 are applicable to the odd-dimensional case. Here we content ourselves with mentioning two facts without proof:

(F1) *Given Ω as in (18.4d), suppose that $\xi\mathfrak{a} = \mathfrak{a}$ for every root of unity ξ in K, or that the number of roots of unity in K is squarefree. Then one can construct explicitly a Hecke character of M_K satisfying (19.10a, b), which produces a structure of type Ω defined over M_K, whose points of finite order are all rational over K^*_{ab}.* (Here Theorem 19.13 is relevant.)

(F2) *There is an example of $(A, \mathcal{C})$ of type (C) in §8.4 which has no model over its field of moduli.*

For these and other related results, the reader is referred to [S71a, b], [S82], [Y82].

22. The Case of Elliptic Curves

The above theorems in the one-dimensional case can be stated in simpler forms. Our object of study is now an elliptic curve with complex multiplication. Thus we fix an imaginary quadratic field K embedded in $\mathbf{C}$, and take Φ to be the identity embedding of K into $\mathbf{C}$.

22.1. Theorem. *Let $\mathfrak{a}$ be a $\mathbf{Z}$-lattice in K, j the invariant of an elliptic curve isomorphic to $\mathbf{C}/\mathfrak{a}$, and k a finite algebraic extension of $K(j)$.*

(I) *If E is a k-rational elliptic curve isomorphic to $\mathbf{C}/\mathfrak{a}$, then $Z(s, E/k) = L(s, \chi)L(s, \chi_\rho)$ with a Hecke character χ of k satisfying*

(22.1a) $$\chi(x) = N_{k/K}(x)^{-1} \quad \textit{for every} \quad x \in k_{\mathbf{a}}^{\times},$$

(22.1b) *If x in $k_{\mathbf{h}}^{\times}$, then $\chi(x) \in K^{\times}$ and $\chi(x)\mathfrak{a} = N_{k/K}(x)\mathfrak{a}$.*

(II) *Conversely, let χ be a Hecke character of k satisfying (22.1a, b). Then there exists a k-rational elliptic curve isomorphic to $\mathbf{C}/\mathfrak{a}$ which determines χ.*

22.2. Theorem. *With K, $\mathfrak{a}$, and j as above, let h be a finite algebraic extension of $\mathbf{Q}(j)$ not containing K and let $k = Kh$. Further let ψ be the quadratic Hecke character of h corresponding to the extension k/h.*

(I) *If E is an h-rational elliptic curve isomorphic to $\mathbf{C}/\mathfrak{a}$, then $Z(s, E/h) = L(s, \chi)$ with a Hecke character χ of k determined by E over k. This character satisfies, in addition to (22.1a, b), the following condition:*

$$\text{(22.2a)} \qquad \chi(y) = \psi(y)|y|^{-1} \quad \textit{for every} \quad y \in h_{\mathbf{A}}^{\times},$$

where $|y|$ is the idele norm of y in $h_{\mathbf{A}}^{\times}$.

(II) *Conversely, if χ is a Hecke character of k satisfying (22.1a, b) and (22.2a), then there exists an h-rational elliptic curve isomorphic to $\mathbf{C}/\mathfrak{a}$ which determines χ over k.*

These are special cases of Theorems 19.11, 20.7, 20.15, 21.4, and 21.5. Notice that the equality $\chi(x)\chi(x)^{\rho} = |x|_{\mathbf{A}}^{-1}$ in (19.10b) is automatic in the present case. Part I of Theorem 22.1 and Part I of Theorem 22.2 without the characterization of χ by (22.2a) are essentially due to Deuring [12]. Some more related results can be found in [S71b, pp. 521 and 526].

CHAPTER VI

Families of Abelian Varieties and Modular Functions

23. Symplectic and Unitary Groups

23.1. For a commutative ring A with identity element we put

(23.1a) $$Sp(n, A) = \{ \alpha \in GL_{2n}(A) \mid {}^t\alpha J \alpha = J \}, \quad J = J_n = \begin{bmatrix} 0 & -1_n \\ 1_n & 0 \end{bmatrix},$$

(23.1b) $$Gp(n, A) = \{ \alpha \in GL_{2n}(A) \mid {}^t\alpha J \alpha = \nu(\alpha) J \ \textit{with}\ \nu(\alpha) \in A^\times \}.$$

We are going to treat various objects related to symplectic groups and unitary groups. These are the groups of the above types and those defined by

(23.1c) $$U(r, s) = \{ \alpha \in GL_m(\mathbf{C}) \mid \alpha^* I_{r,s} \alpha = I_{r,s} \}, \quad I_{r,s} = \mathrm{diag}[1_r, -1_s],$$

(23.1d) $$GU(r, s) = \{ \alpha \in GL_m(\mathbf{C}) \mid \alpha^* I_{r,s} \alpha = \nu(\alpha) I_{r,s} \ \textit{with}\ \nu(\alpha) \in \mathbf{R}^\times \}.$$

Here r and s are nonnegative integers and $m = r + s > 0$; we understand that $I_{r,s} = 1_m$ if $rs = 0$; we put $Z^* = {}^t\overline{Z}$ for a complex matrix Z. We easily see that if α belongs to $Gp(n, A)$ or $GU(r, s)$, then ${}^t\alpha$ belongs to the same group and $\nu({}^t\alpha) = \nu(\alpha)$. If $n = 1$, we have $Gp(1, A) = GL_2(A)$ and $\nu(\alpha) = \det(\alpha)$.

Thus our treatment will be made in these two cases, referred to as Case SP and Case U. To make our treatment uniform, we shall often, but not always, employ the same symbols in both cases. We now define domains $\mathfrak{H}_n$ and $\mathfrak{B}_{r,s}$ by

(23.1e) $$\mathfrak{H} = \mathfrak{H}_n = \{ z \in \mathbf{C}_n^n \mid {}^tz = z, \mathrm{Im}(z) > 0 \}.$$

(23.1f) $$\mathfrak{B} = \mathfrak{B}_{r,s} = \mathfrak{B}(r, s) = \{ z \in \mathbf{C}_s^r \mid 1_s - z^* z > 0 \} \qquad (rs > 0).$$

Here for a hermitian matrix ξ (in particular, for a real symmetric matrix) we write $\xi > 0$ (resp. $\xi \geq 0$) to indicate that ξ is positive definite (resp. nonnegative). We then put

(23.1g) $$B(z) = \begin{bmatrix} \overline{z} & z \\ 1_n & 1_n \end{bmatrix}, \quad \xi(z) = \eta(z) = (2i)^{-1}(z - \overline{z})$$
$$({}^tz = z \in \mathbf{C}_n^n, \text{ Case SP}),$$

(23.1h) $$B(z) = \begin{bmatrix} 1_r & z \\ z^* & 1_s \end{bmatrix}, \quad \xi(z) = 1_r - \bar{z} \cdot {}^t z,$$
$$\eta(z) = 1_s - z^* z \quad (z \in \mathbf{C}_s^r, \text{ Case U}).$$

We easily see that

(23.1i) $$B(z)^* J B(z) = \operatorname{diag}[\bar{z} - z, z - \bar{z}] \qquad \text{(Case SP)},$$

(23.1j) $$B(z)^* I_{r,s} B(z) = \operatorname{diag}[{}^t\xi(z), -\eta(z)] \qquad \text{(Case U)},$$

(23.1k) $$\det\left[B(z)\right] = \begin{cases} \det(\bar{z} - z) & \text{(Case SP)}, \\ \det\left[\xi(z)\right] = \det\left[\eta(z)\right] & \text{(Case U)}. \end{cases}$$

The last formula follows from a relation of the type $\left[\begin{smallmatrix} 1 & -z \\ 0 & 1 \end{smallmatrix}\right] B(z) = \left[\begin{smallmatrix} {}^t\xi(z) & 0 \\ z^* & 1 \end{smallmatrix}\right]$.

Now if $\eta(z) > 0$ in Case U, then $\det\left[B(z)\right] \neq 0$, and hence the left-hand side of (23.1j) has signature (r, s), so that $\xi(z) > 0$. Similarly $\eta(z) > 0$ if $\xi(z) > 0$.

23.2. Lemma. Case SP: *Let $\mathfrak{X}$ be the set of all matrices of the form $X = [\bar{u} \quad u]$ with $u \in \mathbf{C}_n^{2n}$ such that $iX^*JX = \operatorname{diag}[v, -w]$ with $0 < v = v^* \in \mathbf{C}_n^n$ and $0 < w = w^* \in \mathbf{C}_n^n$. Then for $X = [\bar{u} \quad u]$ with $u \in \mathbf{C}_n^{2n}$ we have $X \in \mathfrak{X}$ if and only if*

(23.2a) $${}^t u J u = 0 \text{ and } i \cdot {}^t u J \bar{u} > 0.$$

Moreover the map $(z, \mu) \mapsto B(z) \operatorname{diag}[\bar{\mu}, \mu]$ gives a bijection of $\mathfrak{H} \times GL_n(\mathbf{C})$ onto $\mathfrak{X}$.

Case U: *Let $\mathfrak{X}$ be the set of all $X \in \mathbf{C}_m^m$ such that $X^* I_{r,s} X = \operatorname{diag}[v, -w]$ with $0 < v = v^* \in \mathbf{C}_r^r$ and $0 < w = w^* \in \mathbf{C}_s^s$. Then the map $(z, \lambda, \mu) \mapsto B(z) \operatorname{diag}[\bar{\lambda}, \mu]$ gives a bijection of $\mathfrak{B} \times GL_r(\mathbf{C}) \times GL_s(\mathbf{C})$ onto $\mathfrak{X}$.*

PROOF. We prove only Case SP; Case U can be proved in the same manner. The first assertion is clear. That the image of our map is indeed in $\mathfrak{X}$ can be seen from (23.1i, j). Let ${}^t u = [g \quad h]$ with $g, h \in \mathbf{C}_n^n$. Then (23.2a) is equivalent to

(*) $$h \cdot {}^t g = g \cdot {}^t h \quad \text{and} \quad i(h \cdot {}^t\bar{g} - g \cdot {}^t\bar{h}) > 0.$$

If (*) holds, then for $0 \neq x \in \mathbf{C}_n^1$ we have

$$0 < ix\left(h \cdot {}^t\bar{g} - g \cdot {}^t\bar{h}\right){}^t\bar{x} = i\left(xh \cdot {}^t\overline{xg} - xg \cdot {}^t\overline{xh}\right).$$

Therefore $xg \neq 0$ and $xh \neq 0$, and hence both g and h are invertible. Put $z = h^{-1}g$. Then from (*) we obtain ${}^t z = z$ and $i(\bar{z} - z) = ih^{-1}(h \cdot {}^t\bar{g} - g \cdot {}^t\bar{h}){}^t\bar{h}^{-1} > 0$, that is, $z \in \mathfrak{H}$. Thus ${}^t u = h[z \quad 1]$, which proves the surjectivity of the map in our lemma. The injectivity is obvious.

23.3. Let $\alpha \in Gp(n, \mathbf{R})$ (resp. $\alpha \in GU(r, s), rs > 0$) with $\nu(\alpha) > 0$. We easily see that $\alpha\mathfrak{X} \subset \mathfrak{X}$ for $\mathfrak{X}$ of Lemma 23.2. Now, if $z \in \mathfrak{H}$, (resp. $z \in \mathfrak{B}$) then by the above lemma $B(z) \in \mathfrak{X}$, and so $\alpha B(z) \in \mathfrak{X}$. By the same lemma we can put $\alpha B(z) = B(w)\,\mathrm{diag}[\overline{\lambda}, \mu]$ with unique $w \in \mathfrak{H}$ and $\lambda = \mu \in GL_n(\mathbf{C})$ (resp. $w \in \mathfrak{B}$, $\lambda \in GL_r(\mathbf{C})$, and $\mu \in GL_s(\mathbf{C})$). Let us put $w = \alpha z = \alpha(z)$, $\lambda = \lambda(\alpha, z)$, and $\mu = \mu(\alpha, z)$. Then we have

$$\alpha B(z) = B(\alpha z)\,\mathrm{diag}\left[\overline{\lambda(\alpha, z)}, \mu(\alpha, z)\right]. \tag{23.3a}$$

For $\alpha = \left[\begin{smallmatrix} a & b \\ c & d \end{smallmatrix}\right]$ with a of size n (resp. size r) let us put $a = a_\alpha, b = b_\alpha, c = c_\alpha$, and $d = d_\alpha$. Then we immediately see that $a_\alpha z + b_\alpha = \alpha(z)\mu(\alpha, z)$ and $\mu(\alpha, z) = c_\alpha z + d_\alpha$. Since μ is invertible, we obtain

$$\alpha z = (a_\alpha z + b_\alpha)(c_\alpha z + d_\alpha)^{-1}, \tag{23.3b}$$

$$\lambda(\alpha, z) = \mu(\alpha, z) = c_\alpha z + d_\alpha \qquad \text{(Case SP)}, \tag{23.3c}$$

$$\lambda(\alpha, z) = \overline{a}_\alpha + \overline{b}_\alpha \cdot {}^t z, \quad \mu(\alpha, z) = c_\alpha z + d_\alpha \qquad \text{(Case U)}. \tag{23.3d}$$

This means that we can let α act on $\mathfrak{H}$ or $\mathfrak{B}$ by defining αz by (23.3b). Applying another element β with $\nu(\beta) > 0$ to (23.3a), we see that $\beta(\alpha z) = (\beta\alpha)z$ and

$$\lambda(\beta\alpha, z) = \lambda(\beta, \alpha z)\lambda(\alpha, z), \quad \mu(\beta\alpha, z) = \mu(\beta, \alpha z)\mu(\alpha, z). \tag{23.3e}$$

From (23.3a) and (23.1i, j) we obtain

$$\begin{aligned} \lambda(\alpha, z)^*\xi(\alpha z)\lambda(\alpha, z) &= \nu(\alpha)\xi(z), \\ \mu(\alpha, z)^*\eta(\alpha z)\mu(\alpha, z) &= \nu(\alpha)\eta(z). \end{aligned} \tag{23.3f}$$

We note here an easy but essential relation:

$$\det(\alpha)\det\left[\lambda(\alpha, z)\right] = \nu(\alpha)^r \det\left[\mu(\alpha, z)\right] \qquad \text{(case U)}, \tag{23.3g}$$

which follows immediately from (23.1k) and (23.3a, f).

So far we have assumed $rs > 0$ in Case U. We now make the following convention: if $rs = 0$ in Case U, then $\mathfrak{B}(r, s)$ consists of the single element 0, our group acts on it trivially, and $B(0) = 1_m$. Then (23.3a) is valid if we put

$$\lambda(\alpha, z) = \overline{\alpha} \qquad \textit{if} \quad s = 0, \tag{23.3h}$$

$$\mu(\alpha, z) = \alpha \qquad \textit{if} \quad r = 0. \tag{23.3i}$$

We ignore $\mu(\alpha, z)$ if $s = 0$ and $\lambda(\alpha, z)$ if $r = 0$.

23.4. We now take a totally real algebraic number field F of finite degree and denote by $\mathbf{a}$ the set of all archimedean primes of F. In case U we take a CM-type (K, τ) with K containing F as its maximal real subfield, and denote by ρ the generator of $\mathrm{Gal}(K/F)$ as before. Then τ can be written $\tau = \{\tau_v\}_{v\in\mathbf{a}}$ with an embedding τ_v: $K \to \mathbf{C}$ which coincides with v on F. Hereafter we fix τ and for $a \in K$ denote by a_v the image of a under τ_v. Then we identify $\mathbf{a}$ with τ and view $\mathbf{a}$ also as the set of all archimedean primes of K.

Given a set X, we denote by $X^{\mathbf{a}}$ the product of $\mathbf{a}$ copies of X, that is, the set of all indexed elements $(x_v)_{v\in\mathbf{a}}$ with x_v in X. Then all the embeddings of F into $\mathbf{R}$ (resp. K into $\mathbf{C}$) given by the elements of $\mathbf{a}$ determine an isomorphism of $F \otimes_{\mathbf{Q}} \mathbf{R}$ onto $\mathbf{R}^{\mathbf{a}}$ (resp. $K \otimes_{\mathbf{Q}} \mathbf{R}$ onto $\mathbf{C}^{\mathbf{a}}$.). Similarly we obtain embeddings of F_n^m and K_s^r into $(\mathbf{R}_n^m)^{\mathbf{a}}$ and $(\mathbf{C}_s^r)^{\mathbf{a}}$. We view the former sets as subsets of the latter sets. Thus for $\alpha \in F_n^m$ and $v \in \mathbf{a}$ the v-component α_v of α considered in $(\mathbf{R}_n^m)^{\mathbf{a}}$ is the v-th conjugate of α. If $\alpha \in K_s^r$, then α_v is the image of α under τ_v.

We now define algebraic groups G and G_1 by

(23.4a) $$G = Gp(n, F), \qquad G_1 = Sp(n, F) \qquad \text{(Case SP)},$$

(23.4b) $$G = \{\alpha \in GL_m(K) \mid \alpha T \alpha^* = \nu(\alpha) T \ \ with \ \ \nu(\alpha) \in F^\times\} \qquad \text{(Case U)},$$

(23.4c) $$G_1 = \{\alpha \in G \mid \nu(\alpha) = 1\} \qquad \text{(Case U)}.$$

Here T is a fixed element of $GL_m(K)$ such that $T^* = -T$, and Z^* is defined to be ${}^tZ^\rho$.

In Case U, let (r_v, s_v) be the signature of $-iT_v$. We then put

(23.4d) $$F_{\mathbf{a}} = F \otimes_{\mathbf{Q}} \mathbf{R} = \mathbf{R}^{\mathbf{a}}, \qquad K_{\mathbf{a}} = K \otimes_{\mathbf{Q}} \mathbf{R} = \mathbf{C}^{\mathbf{a}},$$

(23.4e) $$\mathfrak{G} = Gp(n, \mathbf{R})^{\mathbf{a}}, \quad \mathfrak{G}_1 = Sp(n, \mathbf{R})^{\mathbf{a}}, \quad \mathcal{H} = \mathfrak{H}_n^{\mathbf{a}} \qquad \text{(Case SP)},$$

(23.4f) $$\mathfrak{G} = \prod_{v\in\mathbf{a}} GU(r_v, s_v), \quad \mathfrak{G}_1 = \prod_{v\in\mathbf{a}} U(r_v, s_v),$$
$$\mathcal{H} = \prod_{v\in\mathbf{a}} \mathfrak{B}(r_v, s_v) \qquad \text{(Case U)},$$

(23.4g) $$\mathfrak{G}_+ = \{\alpha \in \mathfrak{G} \mid \nu(\alpha) \gg 0\}, \quad G_+ = G \cap \mathfrak{G}_+.$$

Here, for $c \in F_{\mathbf{a}} = \mathbf{R}^{\mathbf{a}}$ we write $c \gg 0$ if $c_v > 0$ for every $v \in \mathbf{a}$. We view G as a subgroup of $\mathfrak{G}$, which makes the intersection $G \cap \mathfrak{G}_+$ meaningful. In Case SP there is a natural embedding of G into $\mathfrak{G}$ as explained above. In Case U we take an element $Q_v \in GL_m(\overline{\mathbf{Q}})$ so that

(23.4h) $$-iT_v = Q_v^\rho I_{r_v, s_v} \cdot {}^tQ_v.$$

(The reason why we take $GL_m(\overline{\mathbf{Q}})$ instead of $GL_m(\mathbf{C})$ will be explained later.) Clearly the map

$$\alpha \mapsto \left(Q_v^{-1}\alpha_v^\rho Q_v\right)_{v\in\mathbf{a}} \tag{23.4i}$$

embeds G into $\mathfrak{G}$. We then let G_+ act on $\mathcal{H}$ through the embedding $G_+ \to \mathfrak{G}_+$ in both cases. For $\alpha \in G_+$ and $z = (z_v)_{v\in\mathbf{a}} \in \mathcal{H}$ we put

$$\mu_v(\alpha, z) = \mu(\alpha_v, z_v) \quad \text{(Case SP)}, \tag{23.4j}$$

$$\begin{aligned} \lambda_v(\alpha, z) &= \lambda\left(Q_v^{-1}\alpha_v^\rho Q_v, z_v\right), \\ \mu_v(\alpha, z) &= \mu\left(Q_v^{-1}\alpha_v^\rho Q_v, z_v\right) \quad \text{(Case U)}. \end{aligned} \tag{23.4k}$$

Then, in Case U, from (23.3a) we obtain

$$\alpha_v^\rho Q_v B(z_v) = Q_v B\left((\alpha z)_v\right) \cdot \operatorname{diag}\left[\overline{\lambda_v(\alpha, z)}, \mu_v(\alpha, z)\right]. \tag{23.4l}$$

23.5. Proposition. (1) *$\mathfrak{G}_1$ acts transitively on $\mathcal{H}$.*

(2) *G_1 is dense in $\mathfrak{G}_1$.*

(3) *For every fixed $w \in \mathcal{H}$ the points $\alpha(w)$ for all $\alpha \in G_1$ form a dense subset of $\mathcal{H}$.*

PROOF. To prove (1), it is sufficient to show the transitivity of $Sp(n, \mathbf{R})$ and $U(r, s)$ on $\mathfrak{H}_n$ and $\mathfrak{B}(r, s)$. Let $z = x + iy \in \mathfrak{H}_n$. Since $y > 0$, we can find $a \in GL_n(\mathbf{R})$ so that ${}^t aa = y$. Put $\gamma = \left[\begin{smallmatrix} {}^t a & xa^{-1} \\ 0 & a^{-1} \end{smallmatrix}\right]$. It is then easy to verify that $\gamma \in Sp(n, \mathbf{R})$ and $\gamma(i1_n) = z$, which proves (1) in Case SP. Next, given $z \in \mathfrak{B}(r, s)$, take $\lambda = \lambda^* > 0$ and $\mu = \mu^* > 0$ so that ${}^t\xi(z) = \lambda^2$ and $\eta(z) = \mu^2$. Putting $\gamma = B(z)\operatorname{diag}[\lambda, \mu]^{-1}$, from (23.1j) we see that $\gamma \in U(r, s)$. Clearly $\gamma(0) = z$, which proves (1) in Case U. We prove (2) only in Case U; Case SP can be handled in the same way. Put

$$Z = \left\{\zeta \in K_m^m \mid 1 + \zeta \in GL_m(K), \zeta T = -T\zeta^*\right\},$$

$$S = \{\sigma \in G_1 \mid 1 + \sigma \in GL_m(K)\}.$$

Define $G_{1\mathbf{a}}$ by (23.4b, c) with $K_\mathbf{a}$ in place of K. Define $Z_\mathbf{a}$ and $S_\mathbf{a}$ similarly with $G_{1\mathbf{a}}$ and $K_\mathbf{a}$ in place of G_1 and K. For $\sigma \in S$ (resp. $\sigma \in S_\mathbf{a}$) put $\zeta = (1-\sigma)(1+\sigma)^{-1}$. Then $\zeta \in Z$ (resp. $\zeta \in Z_\mathbf{a}$) and $\sigma = (1-\zeta)(1+\zeta)^{-1}$. Clearly Z is dense in $Z_\mathbf{a}$ and so S is dense in $S_\mathbf{a}$. This proves (2), since $S_\mathbf{a}$ is dense in $G_{1\mathbf{a}}$, and $G_{1\mathbf{a}}$ can be identified with $\mathfrak{G}_1$. Assertion (3) follows imediately from (1) and (2).

24. Families of Polarized Abelian Varieties

24.1. Let us now consider $\mathcal{P} = (A, \mathcal{C}, \iota; \{t_i\}_{i=1}^r)$ of §17.1 by taking W to be F (Case SP) or K (Case U) of §23.4, and classify all such structures. If we take W to be $\mathbf{Q}$, then this amounts to the classification of all polarized abelian varieties. The reader is reminded that $(A, \mathcal{C})$ is a polarized abelian variety in the sense of §§4.1, 4.2, ι is a ring-injection of W into $\mathrm{End}_{\mathbf{Q}}(A)$, and $\{t_i\}_{i=1}^r$ is an ordered set of points of A of finite order. We assume (17.3b). Then we obtain (17.3d, e) in the present case with $W = F$ or $W = K$. By (17.3i) and Lemma 2 of §5.1, ρ must be the identity map or coincides with the present ρ according as $W = F$ or $W = K$, and so (17.3f) is ${}^tT = -T$ if $W = F$ and $T^* = -T$ if $W = K$.

Let us first assume $W = F$. Then changing the coordinate system, we may assume that $T = J_n$ with $n = m/2$. This means that m must be even; then $\dim(A) = n[F : \mathbf{Q}]$. Also any $\mathbf{Q}$-rational representation of F must be equivalent to a multiple of the regular representation of F over $\mathbf{Q}$, and hence (17.3j) shows that Ψ must be equivalent to the direct sum of n copies of the regular representation of F over $\mathbf{Q}$. Namely, we can decompose $\mathbf{C}^d$ into the direct sum $\bigoplus_{v\in\mathbf{a}} V_v$ so that each V_v is isomorphic to $\mathbf{C}^n$ and $\Psi(a)$ acts on V_v as a scalar a_v for each $a \in F$.

Next take $W = K$. Let r_v resp. s_v be the multiplicity of τ_v resp. $\rho\tau_v$ in Ψ. Then (17.3j) shows that $r_v + s_v$ must be the same for all $v \in \mathbf{a}$. Then $r_v + s_v = m$ and $2d = m[K : \mathbf{Q}]$. This time we can decompose $\mathbf{C}^d$ into the direct sum $\bigoplus_{v\in\mathbf{a}} V_v$ so that each V_v is isomorphic to $\mathbf{C}^m$ and $\Psi(a)$ acts on V_v as $\mathrm{diag}[a_v 1_{r_v}, \overline{a}_v 1_{s_v}]$ for each $a \in K$. Making a convention that $m = 2n$ in Case SP and $m = n$ in Case U, we can thus put

(24.1a) $$\mathbf{C}^d = (\mathbf{C}^n)^{\mathbf{a}}, \qquad \Psi(a) = \mathrm{diag}\big[\Psi_v(a)\big]_{v\in\mathbf{a}},$$

(24.1b) $$\Psi_v(a) = \begin{cases} a_v 1_n & (a \in F, \text{ Case SP}), \\ \mathrm{diag}[a_v 1_{r_v}, \overline{a}_v 1_{s_v}] & (a \in K, \text{ Case U}), \end{cases}$$

(24.1c) $$\Omega = \big\{\, W, \Psi, L, T, \{u_i\}_{i=1}^r \,\big\}$$

with a $\mathbf{Z}$-lattice L in W_m^1 and a finite subset $\{u_i\}_{i=1}^r$ of W_m^1. Then (17.3d) can be written

(24.1d) $$\begin{array}{ccccccccc} 0 & \longrightarrow & L & \longrightarrow & (W_{\mathbf{a}})_m^1 & \longrightarrow & (W_{\mathbf{a}})_m^1/L & \longrightarrow & 0 \\ & & \downarrow & & \downarrow{\scriptstyle q} & & \downarrow & & \\ 0 & \longrightarrow & \Lambda & \longrightarrow & (\mathbf{C}^n)^{\mathbf{a}} & \xrightarrow{\ \xi\ } & A & \longrightarrow & 0 \end{array}$$

with a map q such that

(24.1e) $$q(ax) = \Psi(a)q(x) \qquad (a \in W_{\mathbf{a}}, x \in (W_{\mathbf{a}})_m^1),$$

(24.1f) $E_X(q(x), q(y)) = \mathrm{Tr}_{W_\mathbf{a}/\mathbf{R}}(xT \cdot {}^t y^\rho)$ for every $x, y \in (W_\mathbf{a})^1_m$,

(24.1g) $q(u_i) = t_i$ for every i.

Conditions (R1, 3) in §3.1 (with strict positivity) can be written in the forms

(24.1h) $\mathrm{Tr}_{W/\mathbf{Q}}(xT \cdot {}^t y^\rho) \in \mathbf{Z}$ for every $x, y \in L$,

(24.1i) *$E_X(q(x), i \cdot q(y))$ is symmetric in (x, y) and positive definite.*

Condition (24.1h) concerns only L, and so we consider Ω with L satisfying (24.1h). We shall later prove the following compatibility condition on Ψ and T:

(24.1j) *If a structure of type Ω exists in Case U, then the hermitian form $-iT_v$ has signature (r_v, s_v) for every $v \in \mathbf{a}$.*

Once this is established, we take Q_v as in (23.4h), consider G, G_1, and other symbols as in §23.4 with the present T and (r_v, s_v).

24.2. We are going to determine all $\mathcal{P}$ of type Ω of (24.1c), which amounts to the classification of all possible maps q. Let $\{ e_k \}_{k=1}^m$ be the standard basis of W^1_m. Given any map q: $(W_\mathbf{a})^1_m \to (\mathbf{C}^n)^\mathbf{a}$ satisfying (24.1e) (but not necessarily injective), we observe that q is determined by the vectors $q(e_k)$, and define a matrix $X_v(q) \in \mathbf{C}^m_m$ for each $v \in \mathbf{a}$ by

(24.2a) $$X_v(q) = \begin{bmatrix} x_v^1 & \cdots & x_v^m \\ \overline{x}_v^1 & \cdots & \overline{x}_v^m \end{bmatrix}, \quad q(e_k)_v = x_v^k \qquad \text{(Case SP)},$$

(24.2b) $$X_v(q) = \begin{bmatrix} x_v^1 & \cdots & x_v^m \\ \overline{y}_v^1 & \cdots & \overline{y}_v^m \end{bmatrix}, \quad q(e_k)_v = \begin{bmatrix} x_v^k \\ y_v^k \end{bmatrix} \qquad \text{(Case U)}.$$

Here $x_v^k \in \mathbf{C}^n$ in Case SP and $x_v^k \in \mathbf{C}^{r_v}$, $y_v^k \in \mathbf{C}^{s_v}$ in Case U. We easily see that given any $\omega \in (\mathbf{C}^n_{2n})^\mathbf{a}$, (resp. $Y \in (\mathbf{C}^m_m)^\mathbf{a}$) we can find q in Case SP (resp. Case U) satisfying (24.1e) such that $X_v(q) = \left[\begin{smallmatrix}\omega_v \\ \overline{\omega}_v\end{smallmatrix}\right]$ (resp. $X_v(q) = Y_v$) for every $v \in \mathbf{a}$. For the moment we disregard (24.1f, g, i).

24.3. Lemma. (1) *For q as above and $\beta \in (W_\mathbf{a})^m_m$ put $q^\beta(x) = q(x\beta)$ for $x \in (W_\mathbf{a})^1_m$. Then q^β satisfies (24.1e) and $X_v(q^\beta) = X_v(q) \cdot {}^t\beta_v$.*

(2) *The map q is injective (and hence surjective) if and only if* $\det X_v(q) \neq 0$ *for every $v \in \mathbf{a}$.*

PROOF. Assertion (1) can easily be verified. Suppose $q\left(\sum_{i=1}^m b_i e_i\right) = 0$ for some $b = (b_i) \in (W_\mathbf{a})^1_m$. Let β be the element of $(W_\mathbf{a})^m_m$ whose rows are all equal to b. Then $q^\beta(e_k) = q(e_k\beta) = 0$ for every k, and hence $0 = X_v(q^\beta) = X_v(q) \cdot {}^t\beta_v$. If $\det X_v(q) \neq 0$ for every $v \in \mathbf{a}$ then $\beta_v = 0$ for every $v \in \mathbf{a}$, that is, q is injective. To prove the converse, we take q_0 so that $X_v(q_0) = \begin{bmatrix} 1_n & i1_n \\ 1_n & -i1_n \end{bmatrix}$ in Case SP and $X_v(q_0) = 1_m$ in Case U for every $v \in \mathbf{a}$. This is possible by virtue of the observation at the end of §24.2. Then q_0 is injective since $X_v(q_0)$ is invertible. Given an injective q, $q_0^{-1}q$ is an $F_{\mathbf{a}}$-automorphism of $(W_{\mathbf{a}})^1_m$, and so we have $(q_0^{-1}q)(x) = x\beta$ with $\beta \in GL_m(W_{\mathbf{a}})$. Then $q = q_0^{\beta}$, and hence $X_v(q) = X_v(q_0) \cdot {}^t\beta_v$, which is invertible. This completes the proof.

24.4. Given an injective q, we see that the map $x \mapsto q^{-1}(i \cdot q(x))$ is a $W_{\mathbf{a}}$-automorphism of $(W_{\mathbf{a}})^1_m$, and hence we have $i \cdot q(x) = q(xC)$ with $C \in GL_m(W_{\mathbf{a}})$. Then $i \cdot q(e_k) = q^C(e_k)$, and so from (24.2a, b) and (1) of Lemma 24.3 we obtain

$$\text{(24.4a)} \qquad i \cdot I_v X_v(q) = X_v(q) \cdot {}^tC_v \quad \textit{with} \quad I_v = \mathrm{diag}[1_{r_v}, -1_{s_v}],$$

where $r_v = s_v = n$ in Case SP. We now take (24.1f, i) into account. We have

$$\begin{aligned} E_X\big(q(x), i \cdot q(y)\big) &= E_X\big(q(x), q(yC)\big) = \mathrm{Tr}_{W_{\mathbf{a}}/\mathbf{R}}\big(xT \cdot {}^t(yC)^{\rho}\big) \\ &= [W : F] \sum_{v \in \mathbf{a}} \mathrm{Re}\,\big(x_v T_v C_v^* \, y_v^*\big). \end{aligned}$$

This must be symmetric in (x, y) and positive definite, which is so if and only if $T_v C_v^*$ is hermitian and positive definite for every $v \in \mathbf{a}$. Fixing our attention to one v, dropping the subscript v, and putting simply $X = X_v(q)$, from (24.4a) we obtain $TC^* = -iT\overline{X}^{-1}I\overline{X}$. This is hermitian if and only if $T\overline{X}^{-1}I\overline{X} = {}^tXI \cdot {}^tX^{-1}T$. Taking the inverse, we see that this is so if and only if $\overline{X}T^{-1} \cdot {}^tX$ is of the form $\mathrm{diag}[\beta, \gamma]$ with $\beta \in GL_r(\mathbf{C})$ and $\gamma \in GL_s(\mathbf{C})$. Then

$$\overline{X}(TC^*)^{-1} \cdot {}^tX = iI\overline{X}T^{-1} \cdot {}^tX = \mathrm{diag}[i\beta, -i\gamma],$$

so that both $i\beta$ and $-i\gamma$ are hermitian and positive definite. We note here that in Case SP if we put $X = \begin{bmatrix} \omega \\ \overline{\omega} \end{bmatrix}$ with $\omega \in \mathbf{C}^n_{2n}$, then $\omega J \cdot {}^t\omega = 0$ and

$$\text{(24.4b)} \quad X(J \cdot {}^tC)^{-1} \cdot {}^t\overline{X} = iIXJ \cdot {}^t\overline{X} = \begin{bmatrix} i\omega J \cdot {}^t\overline{\omega} & 0 \\ 0 & -i\overline{\omega}J \cdot {}^t\omega \end{bmatrix}.$$

Now in both cases $i\overline{X}T^{-1} \cdot {}^tX = \mathrm{diag}[i\beta, i\gamma]$. This proves (24.1j). Applying Lemma 23.2 to X^* (resp. $\overline{Q}^{-1} \cdot {}^tX$), and reinstating the subscript v, we thus obtain

$$\text{(24.4c)} \qquad X_v(q) = \begin{cases} \mathrm{diag}[\xi_v, \overline{\xi}_v]B(z_v)^* & \text{(Case SP)}, \\ \mathrm{diag}[\eta_v, \zeta_v]B(z_v)Q_v^* & \text{(Case U)}, \end{cases}$$

with $\xi_v \in GL_n(\mathbf{C})$, $\eta_v \in GL_{r_v}(\mathbf{C})$, $\zeta_v \in GL_{s_v}(\mathbf{C})$, and $(z_v) \in \mathcal{H}$. Conversely, given $X_v(q)$ in this fashion, we see that q is injective, and reversing our reasoning, we find that $T_v C_v^*$ is hermitian and positive definite, and hence (24.1i) holds.

24.5. Given $z = (z_v) \in \mathcal{H}$, define p: $(W_{\mathbf{a}})^1_m \times \mathcal{H} \to (\mathbf{C}^n)^{\mathbf{a}}$ and p_z: $(W_{\mathbf{a}})^1_m \to (\mathbf{C}^n)^{\mathbf{a}}$ by specifying $X_v(p_z)$ as follows:

(24.5a)
$$X_v(p_z) = \begin{cases} B(z_v)^* & \text{(Case SP)}, \\ B(z_v)Q_v^* & \text{(Case U)}, \end{cases}$$

(24.5b)
$$p(x, z) = p_z(x) \qquad \bigl(x \in (W_{\mathbf{a}})^1_m,\ z \in \mathcal{H}\bigr).$$

It can easily be seen that $p_z(x)$ is holomorphic in z. In particular, in Case SP we have

(24.5c)
$$p(x, z) = p_z(x) = \bigl([z_v \quad 1_n] \cdot {}^t x_v\bigr)_{v \in \mathbf{a}} \qquad \bigl(x \in (F_{\mathbf{a}})^1_{2n},\ z \in \mathfrak{H}^{\mathbf{a}}\bigr).$$

As observed at the end of §24.4, from p_z we obtain a structure $\mathcal{P}_z$ of type Ω for which (24.1d) holds with p_z as q. To be more precise, by Lemma 24.3 (2), $p_z(L)$ is a lattice in $(\mathbf{C}^n)^{\mathbf{a}}$ so that $(\mathbf{C}^n)^{\mathbf{a}}/p_z(L)$ is a complex torus. Define E_z by

(24.5d)
$$E_z\bigl(p_z(x), p_z(y)\bigr) = \mathrm{Tr}_{W_{\mathbf{a}}/\mathbf{R}}(xTy^*).$$

Since (24.1h, i) are satisfied, E_z is a Riemann form, so that $(\mathbf{C}^n)^{\mathbf{a}}/p_z(L)$ has a structure of abelian variety; call it A_z and denote by $\mathcal{C}_z$ the polarization of A_z given by E_z. For $a \in W$ denote by $\iota_z(a)$ the element of $\mathrm{End}_{\mathbf{Q}}(A_z)$ represented by $\Psi(a)$ of (24.1a, b) and by $t_i(z)$ the point of A_z represented by $p_z(u_i)$. Thus we obtain a nonempty family of polarized abelian varieties

(24.5e)
$$\mathcal{F}(\Omega) = \{\,\mathcal{P}_z \mid z \in \mathcal{H}\,\}, \qquad \mathcal{P}_z = (A_z, \mathcal{C}_z, \iota_z; \{t_i(z)\}_{i=1}^r)$$

under the condition, which we hereafter assume, that

(24.5f) *In Case U the hermitian form $-iT_v$ has signature (r_v, s_v) for every $v \in \mathbf{a}$.*

24.6. Theorem. (1) *$\mathcal{P}_z$ is of type (24.1c) for every $z \in \mathcal{H}$.*

(2) *A structure of type (24.1c) is isomorphic to $\mathcal{P}_z$ for some $z \in \mathcal{H}$.*

(3) *$\mathcal{P}_z$ and $\mathcal{P}_w$ are isomorphic if and only if $w = \gamma z$ for some $\gamma \in \Gamma$, where*

(24.6a)
$$\Gamma = \{\,\alpha \in G_1 \mid L\alpha = L \text{ and } u_i\alpha - u_i \in L \text{ for every } i\,\}.$$

PROOF. Assertion (1) is obvious. Given $\mathcal{P}$ of type Ω, take q as in (24.1d); then we obtain ξ_v, η_v, ζ_v, and $z \in \mathcal{H}$ as in (24.4c). Define $S \in GL_n(\mathbf{C})^{\mathbf{a}}$ by $S = \mathrm{diag}[\kappa_v]_{v \in \mathbf{a}}$ with $\kappa_v = \mathrm{diag}[\xi_v]$ in Case SP and $\kappa_v = \mathrm{diag}[\eta_v, \overline{\zeta}_v]$ in Case U. Clearly $S\Psi(a) = \Psi(a)S$ for every $a \in W$. Now (24.4c) and (24.5a) show that $q(e_k) = Sp_z(e_k)$ for every k, that is, $q = S \circ p_z$. We easily see that S gives an isomorphism of $\mathcal{P}_z$ onto $\mathcal{P}$, which proves (2). Before proving (3) we make some preliminary observations.

If $\sum_{v\in\mathbf{a}} r_v s_v = 0$ in Case U, then $\mathcal{H}$ consists of a single point. Thus, under (24.5f) there is exactly one isomorphism-class of structures $\mathcal{P}$ of type Ω. This $\mathcal{P}$ is isogenous to the product of m copies of an abelian variety belonging to a CM-type as will be shown in Theorem 24.15 below.

24.7. For $\alpha \in G_+$ and $z \in \mathcal{H}$ define $M(\alpha, z) \in GL_n(\mathbf{C})^{\mathbf{a}}$ by

$$(24.7a) \qquad M(\alpha, z) = \begin{cases} \operatorname{diag}\left[\mu_v(\alpha, z)\right]_{v\in\mathbf{a}} & \text{(Case SP)}, \\ \operatorname{diag}\left[\lambda_v(\alpha, z), \mu_v(\alpha, z)\right]_{v\in\mathbf{a}} & \text{(Case U)}. \end{cases}$$

The transpose of (23.3a) or (23.4l) combined with (24.5a) and Lemma 24.3 (1) gives

$$(24.7b) \quad X_v(p_z^\alpha) = X_v(p_z)\cdot {}^t\alpha_v = \operatorname{diag}\left[{}^t\lambda_v(\alpha, z), \mu_v(\alpha, z)^*\right]X_v(p_{\alpha z}).$$

This means that $p_z^\alpha(e_k) = {}^tM(\alpha, z)p_{\alpha z}(e_k)$ for every k, and hence $p_z(x\alpha) = p_z^\alpha(x) = {}^tM(\alpha, z)p_{\alpha z}(x)$, that is,

$$(24.7c) \quad p(x\alpha, z) = {}^tM(\alpha, z)p(x, \alpha z) \qquad \left(x \in (W_{\mathbf{a}})_m^1, \alpha \in G_+, z \in \mathcal{H}\right).$$

If α belongs to Γ of (24.6a), then we easily see that ${}^tM(\alpha, z)$ defines an isomorphism of $\mathcal{P}_{\alpha z}$ onto $\mathcal{P}_z$, which proves the "if" part of Theorem 24.6 (3). Conversely, suppose that there is an isomorphism of $\mathcal{P}_w$ to $\mathcal{P}_z$; represent it by $S \in GL_n(\mathbf{C})^{\mathbf{a}}$. Then $p_z^{-1} \circ S \circ p_w$ defines an element α of $GL_m(W_{\mathbf{a}})$, that is, $Sp_w(x) = p_z(x\alpha)$. Since S sends $\left(p_w(L), \mathcal{C}_w, p_w(u_i)\right)$ to $\left(p_z(L), \mathcal{C}_z, p_z(u_i)\right)$, we see that $\alpha \in \Gamma$ by virtue of (17.6a). This completes the proof of Theorem 24.6.

24.8. Let $\mathfrak{g}$ denote the maximal order of W. Given an integral ideal $\mathfrak{c}$ of W, we put

$$(24.8a) \qquad \Gamma(\mathfrak{c}) = \left\{\alpha \in G_1 \cap SL_m(\mathfrak{g}) \mid \alpha - 1_{2n} \in (\mathfrak{c})_m^m\right\},$$

and call a subgroup Γ of G_+ an *arithmetic subgroup* of G if $W^\times\Gamma$ contains $W^\times\Gamma(\mathfrak{c})$ as a subgroup of finite index for some $\mathfrak{c}$. Clearly the group of (24.6a) is arithmetic.

It is well known that for an arithmetic subgroup Γ of G the quotient space $\Gamma\backslash\mathcal{H}$ has a compactification, called *the Satake compactification*, which has a structure of complex analytic space, and as such, is isomorphic to a projective variety V^*; moreover, $\Gamma\backslash\mathcal{H}$ is mapped onto a Zariski open subset V of V^* (see [BB]). Let φ denote the Γ-invariant map $\mathcal{H} \to V$ that gives this isomorphism. We then call (V, φ) a *model* of $\Gamma\backslash\mathcal{H}$.

24.9. Theorem. *The notation being as in Theorem 24.6, there exist an algebraic number field k_Ω and a model (V, φ) of $\Gamma\backslash\mathcal{H}$ with the following properties:*

(1) *If $\mathcal{P}$ is of type Ω, then for $\sigma \in \mathrm{Aut}(\mathbf{C})$, $\mathcal{P}^\sigma$ is of type Ω if and only if $\sigma = id.$ on k_Ω.*
(2) *V is rational over k_Ω.*
(3) *For every $z \in \mathcal{H}$ the field $k_\Omega(\varphi(z))$ is the field of moduli of $\mathcal{P}_z$.*
(4) *In Case SP, k_Ω is a finite abelian extension of $\mathbf{Q}$; in Case U, k_Ω is a finite abelian extension of K', where K' is the field generated over $\mathbf{Q}$ by $\mathrm{tr}\left[\Psi(a)\right]$ for all $a \in K$.*

In fact, we can consider a family of polarized abelian varieties by taking any division algebra with a positive involution as W, and can prove the generalizations of Theorems 24.6 and 24.9 for such a family. For these the reader is referred to [S63] and [S66]. See also Theorem 26.3 below. In Case U we have $K' = \mathbf{Q}$ if and only if $r_v = s_v$ for every $v \in \mathbf{a}$; otherwise K' is a CM-field.

24.10. We now need the notion of *CM-points* on $\mathfrak{H}^{\mathbf{a}}$. We take a CM-algebra $Y = K_1 \oplus \cdots \oplus K_t$ with CM-fields K_i as in §18.7, assuming that $W \subset K_i$ for every i and $m = [Y : W]$. We denote by ρ the automorphism of Y that coincides with the Galois involution of K_i/F_i for every i. Let us now consider a W-linear ring-injection h: $Y \to W_m^m$ satisfying

$$h(a^\rho) = Th(a)^*T^{-1} \qquad (a \in Y). \tag{24.10a}$$

Put $Y^u = \{\, a \in Y \mid aa^\rho = 1 \,\}$. Then clearly $h(Y^u) \subset G_1$. Since Y^u is contained in a compact subgroup of $(Y \otimes_{\mathbf{Q}} \mathbf{R})^\times$, we easily see that $h(Y^u)$ is contained in a compact subgroup of $\mathfrak{G}_+$, and hence $h(Y^u)$ has a common fixed point in $\mathcal{H}$. Moreover, $h(Y^u)$ has only one common fixed point, as will be shown in §24.14 below. We call a point on $\mathcal{H}$ which is obtained as such a fixed point a *CM-point on $\mathcal{H}$ with respect to G.* If w is a CM-point and $\gamma \in G_+$, then $\gamma(w)$ is also a CM-point, since h' defined by $h'(a) = \gamma h(a)\gamma^{-1}$ satisfies (24.10a) and $\gamma(w)$ is fixed by $h'(Y^u)$.

Let us now show that in Case SP an injection h of type (24.10a) always exists for any given Y. Take an element ζ of $Y^\times$ such that $\zeta^\rho = -\zeta$. Then $(x, y) \mapsto \mathrm{Tr}_{Y/F}(\zeta x y^\rho)$ is an F-valued nondegenerate alternating form on $Y \times Y$. Therefore we can find an F-linear bijection r: $Y \to F_{2n}^1$ such that $\mathrm{Tr}_{Y/F}(\zeta x y^\rho) = r(x)J \cdot {}^t r(y)$. Since multiplication by $a \in Y$ is an F-linear endomorphism of Y, we can define an F-linear map h: $Y \to F_{2n}^{2n}$ by the relation $r(ax) = r(x)h(a)$ for every $a, x \in Y$. Then we can easily verify (24.10a).

In Case U the matter is not so simple. Given (Y, Ψ) as in §18.7, we can at least find a member of our family which is of type (Y, Ψ) *provided T is suitably chosen.* We shall need this fact in the proof of the main theorem on period relations in §33.7, where the procedure will be explained in detail. We also note that given Ω satisfying (24.5f), we can find CM-points on $\mathcal{H}$ associated with infinitely many different CM-fields Y, as shown in [S64, Proposition 4.10]. See also Lemma 24.16 below for a different type of result.

Now let w denote the CM-point obtained as the common fixed point of $h(Y^u)$ in either case. Putting $X_v = X_v(p_w)$, from (24.7b, c) we obtain

$$\text{(24.10b)} \quad h(\alpha)_v^\rho X_v^* = X_v^* \operatorname{diag}\left[\overline{\lambda_v\big(h(\alpha), w\big)}, \mu_v\big(h(\alpha), w\big)\right] \qquad (v \in \mathbf{a}),$$

$$\text{(24.10c)} \quad p_w\big(xh(\alpha)\big) = {}^tM\big(h(\alpha), w\big)p_w(x) \qquad (x \in (W_\mathbf{a})_m^1, \alpha \in Y^u).$$

Since Y^u spans Y over $\mathbf{Q}$, we can extend these equalities $\mathbf{Q}$-linearly to Y and define $\psi_v\colon Y \to \mathbf{C}_{r_v}^{r_v}$, $\varphi_v\colon Y \to \mathbf{C}_{s_v}^{s_v}$, and $\Phi\colon Y \to \operatorname{End}\big((\mathbf{C}^n)^\mathbf{a}\big)$ so that

$$\text{(24.10d)} \quad h(\alpha)_v^\rho X_v^* = X_v^* \operatorname{diag}\left[\overline{\psi_v(\alpha)}, \varphi_v(\alpha)\right] \qquad (\alpha \in Y, v \in \mathbf{a}),$$

$$\text{(24.10e)} \quad p_w\big(xh(\alpha)\big) = {}^t\Phi(\alpha)p_w(x) \qquad (x \in (W_\mathbf{a})_m^1, \alpha \in Y),$$

$$\text{(24.10f)} \quad \Phi(\alpha) = \operatorname{diag}[\Phi_v(\alpha)]_{v\in\mathbf{a}},$$

$$\Phi_v(\alpha) = \begin{cases} \varphi_v(\alpha) & \text{(Case SP)}, \\ \operatorname{diag}\left[\psi_v(\alpha), \varphi_v(\alpha)\right] & \text{(Case U)}. \end{cases}$$

(In Case SP we have $\psi_v = \varphi_v$ and $r_v = s_v = n$.) Notice that the last equality follows from (24.7a). From (24.10d, f) we see that $\Phi_v(a) = \Psi_v(a)$ for every $a \in K$. Therefore the restriction of Φ to K is Ψ. Clearly ${}^t\Phi(\alpha)$ for each $\alpha \in Y$ defines an element of $\operatorname{End}_\mathbf{Q}(A_w)$. Moreover, from (24.5d) and (24.10a, e) we see that $E_w\big({}^t\Phi(\alpha)u, v\big) = E_w\big(u, {}^t\Phi(\alpha^\rho)v\big)$. Thus we find that $\mathcal{P}_w$ is a structure considered in §18.7 with ${}^t\Phi$ as Ψ there, and obtain a CM-type (K_i, Φ_i) for each i such that Φ is equivalent to the direct sum of $\Phi_1, \ldots, \Phi_t$ in the sense of (18.7a). Denote by K^* the field generated over $\mathbf{Q}$ by $\operatorname{tr}\big(\Phi(\alpha)\big)$ for all $\alpha \in Y$. Then we can define a map $g\colon (K^*)_\mathbf{A}^\times \to Y_\mathbf{A}^\times$ by (18.7c), which will be needed in Section 26.

24.11. Theorem. *Let the notation be as in Theorem 24.9 and let w be a CM-point on $\mathcal{H}$. Define K^* as above. Then $k_\Omega\big(\varphi(w)\big) \subset K_{ab}^*$.*

This follows immediately from Theorems 18.9 and 24.9.

24.12. If $n = 1$ in Case SP, we have $G = GL_2(F)$, $G_1 = SL_2(F)$, and the space is $\mathfrak{H}_1^\mathbf{a}$, and the CM-points can be obtained in a simpler manner. We naturally consider a CM-field K containing F such that $[K : F] = 2$. In this case (24.10a) is satisfied for every F-linear injection $h : K \to F_2^2$. This is because ${}^t\xi J_1 \xi = \det(\xi) J_1$ for every $\xi \in F_2^2$ and $h(\alpha\alpha^\rho) = h\big(N_{K/F}(\alpha)\big) = \det\big(h(\alpha)\big)$. We see that φ_v of (24.10f) satisfies $\varphi_v(a) = a_v$ for every $a \in F$, and hence it is an injection $K \to \mathbf{C}$ which extends $v\colon F \to \mathbf{R}$. We thus obtain a CM-type (K, Φ) with $\Phi = \{\varphi_v\}_{v\in\mathbf{a}}$.

24.13. Proposition. *Let $(K, \{\tau_v\}_{v\in\mathbf{a}})$ be a CM-type with K containing F as the maximal real subfield and an injection τ_v: $K \to \mathbf{C}$ which extends v: $F \to \mathbf{R}$. Take an element w_0 of K so that* $\operatorname{Im}(\tau_v(w_0)) > 0$ *for every $v \in \mathbf{a}$, and put $w = (w_v)_{v\in\mathbf{a}}$ with $w_v = \tau_v(w_0)$. Then w is a CM-point on $\mathfrak{H}_1^{\mathbf{a}}$ with respect to $GL_2(F)$, and (A_w, ι_w) is of type $(K, \{\tau_v\}_{v\in\mathbf{a}})$. Conversely every CM-point on $\mathfrak{H}_1^{\mathbf{a}}$ can be obtained in such a manner.*

PROOF. Let such a CM-type and w_0 be given. Since $\{w_0, 1\}$ is a basis of K over F, we can define an F-linear injection h: $K \to F_2^2$ by $h(\alpha)\left[\begin{smallmatrix} w_0 \\ 1 \end{smallmatrix}\right] = \left[\begin{smallmatrix} w_0 \\ 1 \end{smallmatrix}\right]\alpha$ for $\alpha \in K$. Taking the image under τ_v, we obtain (24.10d) for every $\alpha \in K$ with τ_v in place of φ_v, which shows that w is a fixed point of $h(K^\times)$ and (A_w, ι_w) is of type $(K, \{\tau_v\}_{v\in\mathbf{a}})$. Thus w is a CM-point. Conversely, consider a CM-point w obtained from an F-linear ring-injection h: $K \to F_2^2$. Then we obtain maps φ_v: $K \to \mathbf{C}$ satisfying (24.10d, f). Take $\alpha \in K$ so that $h(\alpha) = \left[\begin{smallmatrix} a & b \\ c & d \end{smallmatrix}\right]$ with $c \neq 0$. Then (24.10d) shows that $c_v w_v + d_v = \varphi_v(\alpha)$. Put $w_0 = c^{-1}(\alpha - d)$. Then $w_v = \varphi_v(w_0)$. This proves the converse part, and our proof is complete.

24.14. Let us now prove that $h(Y^u)$ of §24.10 has only one common fixed point on $\mathcal{H}$. We first consider Case SP. Put $\xi_0 = \left[\begin{smallmatrix} 1_n & -i1_n \\ 1_n & i1_n \end{smallmatrix}\right]$ and $\Xi = \xi_0 \cdot Sp(n, \mathbf{R})$. For every $\xi \in \Xi$ and $z \in \mathfrak{H}$ put $\xi(z) = (a_\xi z + b_\xi)(c_\xi z + d_\xi)^{-1}$. It is well known that $c_\xi z + d_\xi$ is indeed invertible and $z \mapsto \xi(z)$ gives a bijection of $\mathfrak{H}$ onto

$$\mathfrak{D} = \left\{ z \in \mathbf{C}_n^n \mid {}^t z = z,\, 1 - {}^t\bar{z}z > 0 \right\}. \tag{24.14a}$$

Moreover, put $\mathfrak{S}(z) = \{\gamma \in Sp(n, \mathbf{R}) \mid \gamma z = z\}$ for $z \in \mathfrak{H}$. If $\xi = \xi_0\beta$ with $\beta \in Sp(n, \mathbf{R})$ such that $\beta z = i1_n$, then $\xi(z) = 0$ and it can easily be seen that

$$\xi\mathfrak{S}(z)\xi^{-1} = \{\operatorname{diag}[u, \bar{u}] \mid u \in U(n)\},\, U(n) = \left\{ u \in GL_n(\mathbf{C}) \mid {}^t\bar{u}u = 1 \right\}.$$

Now let w be a fixed point of $h(Y^u)$ as in §24.10. Then for each $v \in \mathbf{a}$ we can find $\xi_v \in \Xi$ such that $\xi_v(w_v) = 0$ and

$$\xi_v h(\alpha)_v \xi_v^{-1} = \operatorname{diag}[\sigma_v(\alpha), \bar{\sigma}_v(\alpha)] \qquad (\alpha \in Y^u) \tag{24.14b}$$

with a map σ_v: $Y^u \to U(n)$. Suppose w' is another fixed point of $h(Y^u)_v$ on $\mathfrak{H}$; put $z' = \xi_v(w')$. Then $z' = \sigma_v(\alpha)z' \cdot {}^t\sigma_v(\alpha)$ for every $\alpha \in Y^u$. Let $c_{v1}, \dots, c_{vn}$ be the characteristic roots of $\sigma_v(\alpha)$. Diagonalizing $\sigma_v(\alpha)$, we easily see that z' must be 0 if $c_{vj}c_{vk} \neq 1$ for every (j, k). From (24.14b) we see that $c_{v1}, \dots, c_{vn}, \bar{c}_{v1}, \dots, \bar{c}_{vn}$ are the characteristic roots of $h(\alpha)_v$. Since $[Y : F] = 2n$, it is an easy exercise to find an element α of Y^u such that $c_{vj}c_{vk} \neq 1$ for every (j, k) and every $v \in \mathbf{a}$. Then $w' = w_v$, which proves the desired fact.

In Case U we take an element β of $\mathfrak{G}_1$ so that $\beta(w) = 0$. Then taking $\mathfrak{B}(r_v, s_v)$ itself in place of $\mathfrak{D}$, we can prove the uniqueness of the fixed point in the same manner.

24.15. Theorem. *Let (A, ι) be of type (K, Ψ) with a CM-field K and Ψ as in (24.1a, b). (Thus $\dim(A) = d = m[F : \mathbf{Q}]$ and $r_v + s_v = m$.) If $\sum_{v \in \mathbf{a}} r_v s_v = 0$, then A is isogenous to the product of m copies of an abelian variety belonging to a CM-type (K, φ) with φ such that Ψ is equivalent to the sum of m copies of φ.*

PROOF. This may be viewed as an improved version of Theorem 3 in §6.2. Let $\tau = \{ \tau_v \}$ be as in §24.1. Changing τ_v for $\rho\tau_v$ if necessary, we may assume that $s_v = 0$ for all v. Then Ψ is equivalent to the sum of m copies of τ. Take q and L as in (24.1d) with $W = K$ for the present A. Let $\{ e_k \}_{k=1}^m$ be the standard basis of K_m^1. Take a fractional ideal $\mathfrak{a}$ in K, and put $M_k = \Psi(\mathfrak{a})q(e_k)$ and $U_k = \Psi(K_\mathbf{a})q(e_k)$. Then $\mathbf{C}^d = \bigoplus_{k=1}^m U_k$, and the map $a \mapsto \Psi(a)q(e_k)$ gives an isomorphism of $K_\mathbf{a}$ onto U_k, and hence M_k is a lattice in U_k. Take $c \in K_\mathbf{a}$ so that $c^{\tau_v} = i$ for every v. Then $\Psi(c) = i1_d$. Therefore we see that U_k is a complex vector subspace of $\mathbf{C}^d$, and so U_k/M_k is a complex torus on which the maximal order of K acts through Ψ. Let φ_k be the representation of K on U_k. Then, as seen in Theorem 3 in §6.2, U_k/M_k is an abelian variety of type (K, φ_k). Clearly A is isogenous to $\prod_{k=1}^m (U_k/M_k)$, and Ψ must be equivalent to the sum of the φ_k. Then each φ_k must be equivalent to τ. This completes the proof.

24.16. Lemma. *Let (K, Φ_ν) for $1 \le \nu \le m$ be m CM-types with K as above, and let (A_ν, ι_ν) be of type (K, Φ_ν). If Ψ of (24.1c) in Case U is equivalent to the direct sum of $\Phi_1, \dots, \Phi_m$, then there exists a member $\mathcal{P}_z$ of $\mathcal{F}(\Omega)$ with a CM-point z such that A_z is isogenous to $\prod_{\nu=1}^m A_\nu$.*

PROOF. Let us find $\zeta_1, \dots, \zeta_m$ in K such that T is equivalent to $\mathrm{diag}[\zeta_1, \dots, \zeta_m]$ and $\mathrm{Im}(\zeta_\nu^\sigma) > 0$ for every $\sigma \in \Phi_\nu$. If $m = 1$, then $\Psi = \Phi_1$ and in view of (24.5f) we can simply take $T = \zeta_1$. We now prove the desired fact by induction on m. Assuming $m > 1$ and changing τ_v for $\rho\tau_v$ if necessary, we may assume that Φ_1 coincides with τ of §23.4. Then $r_v > 0$ for every $v \in \mathbf{a}$. In this case we can find $x \in K_m^1$ such that $-i(xTx^*)_v > 0$ for every v. Put $\zeta_1 = xTx^*$. Then T is equivalent to $\mathrm{diag}[\zeta_1, T']$ with some T' of size $m-1$ such that $-iT'_v$ has signature $(r_v - 1, s_v)$ for every v. Then by induction we find $\zeta_2, \dots, \zeta_m$. Now take Y to be the direct sum of m copies of K. Changing the coordinate system, we may assume that $T = \mathrm{diag}[\zeta_1, \dots, \zeta_m]$. Then the diagonal embedding h of Y into K_m^m satisfies (24.10a), and we obtain a CM-point z on $\mathcal{H}$. With $\{ e_k \}$ and $\mathfrak{a}$ as in the proof of Theorem 24.15, put $U_k = p_z(K_\mathbf{a} e_k)$ and $M_k = p_z(\mathfrak{a} e_k)$. Then $\mathbf{C}^d = \bigoplus_{k=1}^m U_k$. Take $c_k \in K_\mathbf{a}$ so that $c_k^\sigma = i$ for every $\sigma \in \Phi_k$, and put $c = (c_1, \dots, c_m)$. Then $\Psi(c) = i1_d$. Since $U_k = \Psi(K_\mathbf{a})p_z(e_k)$, we see that U_k is a complex vector subspace of $\mathbf{C}^d$, and A_z is isogenous to $\prod_{k=1}^m U_k/M_k$. Let φ_k be the representation of $K_\mathbf{a}$ on U_k.

Then U_k/M_k is an abelian variety of type (K, φ_k), and Ψ is the sum of the φ_k, so that $c_k^\sigma = i$ for every $\sigma \in \varphi_k$. This shows that $\Phi_k = \varphi_k$, which completes the proof.

25. Modular Forms and Functions

25.1. To define modular forms on $\mathcal{H}$, we first put

(25.1a) $$\mathbf{e}(c) = \exp(2\pi i c) \qquad (c \in \mathbf{C}),$$

(25.1b) $$\mathbf{e_a}(x) = \exp\left(2\pi i \sum_{v \in \mathbf{a}} x_v\right) \qquad (x \in \mathbf{C}^{\mathbf{a}}).$$

(25.1c) $$j_\alpha(z) = j(\alpha, z) = \big(j_v(\alpha, z)\big)_{v \in \mathbf{a}},$$

$$j_v(\alpha, z) = \det\big(\mu_v(\alpha, z)\big) \qquad (\alpha \in G_+, z \in \mathcal{H}),$$

where μ is defined by (23.3c) and (23.4d). To simplify our notation, for $a, b \in \mathbf{C}^{\mathbf{a}}$ we put

(25.1d) $$a^b = \prod_{v \in \mathbf{a}} a_v^{b_v},$$

whenever the factors $a_v^{b_v}$ are well defined. If $0 < a_v \in \mathbf{R}$, we always put $a_v^{b_v} = \exp\big(b_v \log a_v\big)$ with real $\log a_v$.

We naturally assume that $\dim(\mathcal{H}) > 0$. For $k \in \mathbf{Z}^{\mathbf{a}}$, $\alpha \in G_+$, and a function $f \colon \mathcal{H} \to \mathbf{C}$ we define $f\|_k\alpha \colon \mathcal{H} \to \mathbf{C}$ by

(25.1e) $$(f\|_k\alpha)(z) = \nu(\alpha)^{nk/2} j_\alpha(z)^{-k} f(\alpha z) \qquad (z \in \mathfrak{H}^{\mathbf{a}}),$$

where $\nu(\alpha)^{nk/2}$ and $j_\alpha(z)^{-k}$ should be understood in the sense of (25.1d). We easily see that $f\|_k(\alpha\beta) = (f\|_k\alpha)\|_k\beta$.

25.2. Given an arithmetic subgroup Γ of G and $k \in \mathbf{Z}^{\mathbf{a}}$, we denote by $\mathcal{M}_k(\Gamma)$ the set of all functions $f \colon \mathcal{H} \to \mathbf{C}$ satisfying the following conditions:

(25.2a) *f is holomorphic;*
(25.2b) *$f\|_k\gamma = f$ for every $\gamma \in \Gamma$;*
(25.2c) *f is holomorphic at every cusp.*

Condition (25.2c) is necessary or meaningful only in the following exceptional cases: (Case SP) $F = \mathbf{Q}$ *and* $n = 1$; (Case U) $F = \mathbf{Q}$, $m = 2$, *and* $x^*Tx = 0$ *for some* $x \in K^2, \neq 0$. In these cases Γ is commensurable with a "conjugate" of $SL_2(\mathbf{Z})$. For the precise meaning of (25.2c) the reader is referred to [S71a, §2.1].

An element of $\mathcal{M}_k(\Gamma)$ is called a *(holomorphic) modular form of weight k with respect to Γ*. For our later purposes it is necessary to generalize this notion by considering vector-valued modular forms. Thus we take a rational representation

$$\text{(25.2d)} \qquad \rho : GL_n(\mathbf{C})^{\mathbf{a}} \to GL(V) \qquad \text{(Case SP)},$$

$$\text{(25.2e)} \qquad \rho : \prod_{v\in\mathbf{a}} \left[GL_{r_v}(\mathbf{C}) \times GL_{s_v}(\mathbf{C})\right] \to GL(V) \qquad \text{(Case U)}$$

with a finite-dimensional complex vector space V. Given a map $f\colon \mathcal{H} \to V$ and $\alpha \in G_+$, we define $f\|_\rho\alpha\colon \mathcal{H} \to V$ by

$$\text{(25.2f)} \qquad (f\|_\rho\alpha)(z) = \rho\left(\nu(\alpha)^{1/2}M(\alpha, z)^{-1}\right) f(\alpha z) \qquad (z \in \mathcal{H})$$

with $M(\alpha, z)$ of (24.7a). If $\rho(x) = \det(x)^k$ (resp. $\rho(x, y) = \det(y)^k$) in the sense of (25.1e), then $f\|_\rho\alpha = f\|_k\alpha$. We then define the set $\mathcal{M}_\rho(\Gamma)$ to be the set of all holomorphic maps $\mathcal{H} \to V$ satisfying $f\|_\rho\gamma = f$ for every $\gamma \in \Gamma$ and also (25.2c).

25.3. Since it is often convenient not to specify Γ, we denote by $\mathcal{M}_\rho$ (resp. $\mathcal{M}_k$) the union of $\mathcal{M}_\rho(\Gamma)$ (resp. $\mathcal{M}_k(\Gamma)$) for all arithmetic subgroups Γ, and put

$$\text{(25.3a)} \qquad \mathcal{A}_\rho = \bigcup_e \left\{ g^{-1}f \mid f \in \mathcal{M}_{\tau_e}, 0 \neq g \in \mathcal{M}_e \right\},$$

$$\text{(25.3b)} \qquad \mathcal{A}_\rho(\Gamma) = \left\{ h \in \mathcal{A}_\rho \mid h\|_\rho\gamma = h \ \textit{for every}\ \gamma \in \Gamma \right\},$$

where e runs over $\mathbf{Z}^{\mathbf{a}}$, and τ_e denotes the representation defined by $\tau_e(x) = \det(x)^e\rho(x)$ in Case Sp and $\tau_e(y, x) = \det(x)^e\rho(y, x)$ in Case U. If $\rho(x) = \det(x)^k$ (resp. $\rho(x, y) = \det(y)^k$), we denote these by $\mathcal{A}_k$ and $\mathcal{A}_k(\Gamma)$. If $F = \mathbf{Q}$ in Case SP, that is, if $G_1 = Sp(n, \mathbf{Q})$, the elements of $\mathcal{M}_k$ are called *Siegel modular forms;* if $n = 1$ in Case SP, that is, if $G = GL_2(F)$, they are called *Hilbert modular forms.*

25.4. An element of $\mathcal{A}_0(\Gamma)$ is a Γ-invariant meromorphic function on $\mathcal{H}$, and it is known that conversely every Γ-invariant meromorphic function on $\mathcal{H}$ belongs to $\mathcal{A}_0(\Gamma)$ if we exclude the exceptional cases mentioned in (25.2c). An element of $\mathcal{A}_0(\Gamma)$ is called a *modular function with respect to Γ* or a *Γ-automorphic function* on $\mathcal{H}$. In Case SP it is called a *Siegel modular function* if $F = \mathbf{Q}$ and a *Hilbert modular function* if $n = 1$.

Now, let (V, φ) be a model of $\Gamma\backslash\mathcal{H}$ in the sense of §24.8, and let $\mathfrak{F}$ be the field of all functions on V in the sense of algebraic geometry. Then $\mathcal{A}_0(\Gamma)$ consists of the functions $g \circ \varphi$ for all $g \in \mathfrak{F}$. In this sense $\mathcal{A}_0(\Gamma)$ can be identified with $\mathfrak{F}$ if we identify $\Gamma\backslash\mathfrak{H}^{\mathbf{a}}$ with V. We shall refine this fact in Theorem 26.4 below.

25.5. Hereafter until the end of this chapter we confine ourselves to Case SP; we shall return to Case U toward the end of the book.

Now every element f of $\mathcal{M}_\rho$ (in Case SP) has an expansion of the form

$$f(z) = \sum_{0 \le h \in L} c_h \mathbf{e_a}\big(\mathrm{tr}(hz)\big), \tag{25.5a}$$

where $c_h \in V$, L is a $\mathbf{Z}$-lattice in the vector space

$$S = \{ s \in F_n^n \mid {}^t s = s \}, \tag{25.5b}$$

and we write $h \ge 0$ if $h_v \ge 0$ for every $v \in \mathbf{a}$. Usually we call the right-hand side of (25.5a) *the Fourier expansion of* f, and call the c_h *the Fourier coefficients of* f.

To consider the arithmeticity of modular forms, let us hereafter assume that (V, ρ) has a $\mathbf{Q}$-structure in the sense that $V = V_0 \otimes_{\mathbf{Q}} \mathbf{C}$ with a fixed vector space V_0 over $\mathbf{Q}$, and ρ is the natural extension of a rational representation ρ_0: $GL_n(\mathbf{Q})^{\mathbf{a}} \to GL(V_0)$. (Often $V = \mathbf{C}^m$, $V_0 = \mathbf{Q}^m$, and ρ_0 is a representation $GL_n(\mathbf{Q})^{\mathbf{a}} \to GL_m(\mathbf{Q})$.) Then, given a subfield D of $\mathbf{C}$, we say that f of (25.5a) is *D-rational* if $c_h \in V_0 \otimes_{\mathbf{Q}} D$ for all $h \in L$, and denote by $\mathcal{M}_\rho(D)$ the set of all D-rational elements of $\mathcal{M}_\rho$. Then we put

$$\mathcal{A}_\rho(D) = \bigcup_e \{ g^{-1} f \mid f \in \mathcal{M}_{\tau_e}(D), 0 \ne g \in \mathcal{M}_e(D) \}, \tag{25.5c}$$

$$\mathcal{M}_\rho(\Gamma, D) = \mathcal{M}_\rho(\Gamma) \cap \mathcal{M}_\rho(D), \quad \mathcal{A}_\rho(\Gamma, D) = \mathcal{A}_\rho(\Gamma) \cap \mathcal{A}_\rho(D), \tag{25.5d}$$

where $\tau_e(x) = \det(x)^e \rho(x)$. We write $\mathcal{M}_k$ and $\mathcal{A}_k$ for $\mathcal{M}_\rho$ and $\mathcal{A}_\rho$ in (25.5c, d) if $\rho(x) = \det(x)^k$.

25.6. Proposition. *$\mathcal{M}_k \ne 0$ if and only if either $k_v \ge n/2$ for every $v \in \mathbf{a}$, or $k_v = \kappa$ for every $v \in \mathbf{a}$ with an integer κ such that $0 \le \kappa < n/2$. In particular $\mathcal{M}_0 = \mathbf{C}$.*

For the proof see [S94, Corollary 5.7].

25.7. Let $\mathfrak{T}(L)$ denote the set of all formal series of form (25.5a) with $c_h \in \mathbf{C}$. This has a natural ring-structure. Put $[x] = \mathrm{Tr}_{F/\mathbf{Q}}\big(\mathrm{tr}(x)\big)$ for $x \in F_n^n$ and $m = [F: \mathbf{Q}]n(n+1)/2$; then S of (25.5b) has dimension m over $\mathbf{Q}$. We can find a $\mathbf{Q}$-basis $\{ s_1, \ldots, s_m \}$ of S such that $[s_i L] \subset \mathbf{Z}$ and $s_{iv} > 0$ for every i and every $v \in \mathbf{a}$. Clearly $[s_i h] \ge 0$ if $0 \le h \in L$. Taking m independent indeterminates $\xi_1, \ldots, \xi_m$, for $f \in \mathfrak{T}(L)$ as in (25.5a) we put

$$\omega(f) = \sum_{0 \le h \in L} c_h \prod_{i=1}^m \xi_i^{[s_i h]}. \tag{25.7a}$$

We easily see that ω defines a ring-injection of $\mathfrak{T}(L)$ into the ring $\mathbf{C}[[\xi_1, \dots, \xi_m]]$ of all formal power series in $\xi_1, \dots, \xi_m$ with coefficients in $\mathbf{C}$. Therefore $\mathfrak{T}(L)$ is an integral domain. Given $\sigma \in \mathrm{Aut}(\mathbf{C})$, we obtain automorphisms of $\mathbf{C}[[\xi_1, \dots, \xi_m]]$ and $\mathfrak{T}(L)$ by applying σ to the coefficients; denote by f^σ the image of f under these automorphisms. This means that for $f \in \mathfrak{T}(L)$ as in (25.5a) we have

$$f^\sigma = \sum_{0 \le h \in L} (c_h)^\sigma \mathbf{e_a}\big(\mathrm{tr}(hz)\big), \tag{25.7b}$$

and clearly $\omega(f)^\sigma = \omega(f^\sigma)$. We can in fact define f^σ formally in the same manner even when $c_h \in V$, since σ acts naturally on V. Let us now prove

$$\mathcal{A}_\rho(D) \cap \mathcal{M}_\rho = \mathcal{M}_\rho(D). \tag{25.7c}$$

If $f \in \mathcal{A}_\rho(D) \cap \mathcal{M}_\rho$, then $f = p^{-1}q$ with $0 \neq p \in \mathcal{M}_e(D)$ and $q \in \mathcal{M}_{\tau_e}(D)$. Taking L suitably, we may assume that p and the vector components of q belong to $\mathfrak{T}(L)$. Then for $\sigma \in \mathrm{Aut}(\mathbf{C}/D)$ we have $pf^\sigma = (pf)^\sigma = q^\sigma = q = pf$. Since p is not a zero-divisor, we obtain $f^\sigma = f$, that is, $f \in \mathcal{M}_\rho(D)$.

25.8. Proposition. *Given $f \in \mathcal{M}_\rho$ with expansion (25.5a) and $\sigma \in \mathrm{Aut}(\mathbf{C})$, there exist a representation τ and an element f^σ of $\mathcal{M}_\tau$ whose Fourier expansion is exactly that of (25.7b). The representation τ is determined by ρ and σ; if $\rho(x) = \otimes_{v \in \mathbf{a}} \rho_v(x_v)$ for $x = (x_v)_{v \in \mathbf{a}} \in GL_n(\mathbf{C})^{\mathbf{a}}$ with representations ρ_v of $GL_n(\mathbf{C})$, then $\tau(x) = \otimes_{v \in \mathbf{a}} \tau_v(x_v)$ with $\tau_{v\sigma} = \rho_v$, where $v\sigma \in \mathbf{a}$ is defined by $a_{v\sigma} = (a_v)^\sigma$ for $a \in F$.*

In the setting of this proposition we put $\tau = \rho^\sigma$. We easily see that σ sends $\mathcal{A}_\rho$ onto $\mathcal{A}_\tau$. In particular, if $k \in \mathbf{Z}^{\mathbf{a}}$, the map $f \mapsto f^\sigma$ sends $\mathcal{M}_k$ onto $\mathcal{M}_\ell$ with ℓ determined by $\ell_{v\sigma} = k_v$. This proposition and the following theorem are proved in [S75a, b] and [S78a] in the case $n = 1$ or $F = \mathbf{Q}$; the general case can be treated in the same manner. In these papers the $\mathbf{Q}$-rationality of the Fourier coefficients of certain Eisenstein series is needed, but instead we can employ a fact concerning theta functions, which will be stated in Theorem 27.13 below. The details of the idea are explained in [S83, pp. 455–456].

25.9. Theorem. *Let F' denote the Galois closure of F in $\mathbf{R}$, and E_ρ the subfield of F' such that $\mathrm{Gal}(F'/E_\rho) = \{\sigma \in \mathrm{Gal}(F'/\mathbf{Q}) \mid \rho^\sigma = \rho\}$. Then the following assertions hold:*

(1) $\mathcal{M}_\rho = \mathcal{M}_\rho(E_\rho) \otimes_{E_\rho} \mathbf{C}$.

(2) *Let $f \in \mathcal{M}_\rho(D)$ (resp. $f \in \mathcal{A}_\rho(D)$) with a subfield D of $\mathbf{C}$ containing $\mathbf{Q}_{\mathrm{ab}}$ and E_ρ. Define $f|_\rho\alpha$ for $\alpha \in G_+$ by $(f|_\rho\alpha)(z) = \rho(\mu(\alpha, z)^{-1})f(\alpha z)$. Then $f|_\rho\alpha \in \mathcal{M}_\rho(D)$ (resp. $f|_\rho\alpha \in \mathcal{A}_\rho(D)$.)*

For example, if $\rho(x) = \prod_{v\in\mathbf{a}} \det(x_v)^\kappa$ with $\kappa \in \mathbf{Z}$, then clearly $E_\rho = \mathbf{Q}$. Though we state the facts here, these are closely connected with, and in fact consequences of, the results of the next section.

26. Canonical Models

26.1. We now adelize various objects using the symbols introduced in §18.3. In particular, we remind the reader that given an algebraic number field M of finite degree and an element c of the idele group $M_\mathbf{A}^\times$, we denote by $[c, M]$ the element of Gal (M_{ab}/M) determined by c.

For simplicity, we restrict our exposition to Case SP, though Case U can be treated in the same manner. Thus our basic field is F as before, $G = Gp(n, F)$, and $G_1 = Sp(n, F)$. We denote by $\mathfrak{g}$ the maximal order of F and by $\mathbf{h}$ the set of all nonarchimedean primes of F. For each $v \in \mathbf{a} \cup \mathbf{h}$ we denote by F_v the v-completion of F, and put

$$(26.1\text{a}) \qquad G_\mathbf{A} = Gp(n, F_\mathbf{A}), \qquad G_v = Gp(n, F_v)$$

with Gp as in (23.1b). Then ν of (23.1b) defines homomorphisms $G_\mathbf{A} \to F_\mathbf{A}^\times$ and $G_v \to F_v^\times$. It can easily be seen that $G_\mathbf{A}$ consists of the elements (α_v) of $\prod_{v\in\mathbf{a}\cup\mathbf{h}} Gp(n, F_v)$ such that $\alpha_v \in Gp(n, \mathfrak{g}_v)$ for almost all $v \in \mathbf{h}$, where $\mathfrak{g}_v$ is the v-closure of $\mathfrak{g}$ in F_v. We denote by $G_\mathbf{a}$ (resp. $G_\mathbf{h}$) the subgroup of $G_\mathbf{A}$ consisting of the elements (α_v) such that $\alpha_v = 1$ for every $v \in \mathbf{h}$ (resp. $v \in \mathbf{a}$). Then $G_\mathbf{A} = G_\mathbf{a} G_\mathbf{h}$, and $G_\mathbf{a}$ can be identified with $\prod_{v\in\mathbf{a}} G_v$. For $x \in G_\mathbf{A}$ we denote by $x_\mathbf{a}$ and $x_\mathbf{h}$ its projections to $G_\mathbf{a}$ and $G_\mathbf{h}$. We view G as a subgroup of $G_\mathbf{A}$ by identifying an element α of G with the element of $G_\mathbf{A}$ whose v-component is α for every $v \in \mathbf{a} \cup \mathbf{h}$. (Thus we denote by G the group which is often written $G_\mathbf{Q}$.)

We now put

$$(26.1\text{b}) \quad D = \{x \in G \mid \nu(x) \in \mathbf{Q}^\times\}, \qquad D_\mathbf{A} = \{x \in G_\mathbf{A} \mid \nu(x) \in \mathbf{Q}_\mathbf{A}^\times\}.$$

Then $D_\mathbf{A} = D_\mathbf{a} D_\mathbf{h}$ with $D_\mathbf{a} = D_\mathbf{A} \cap G_\mathbf{a}$ and $D_\mathbf{h} = D_\mathbf{A} \cap G_\mathbf{h}$.

26.2. For $t \in F_\mathbf{A}^\times$ we put $|t|_\mathbf{A} = \prod_{v\in\mathbf{a}\cup\mathbf{h}} |t_v|$ and write $t \gg 0$ if $t_v > 0$ for every $v \in \mathbf{a}$. Now the map ν: $G_\mathbf{A} \to F_\mathbf{A}^\times$ sends $D_\mathbf{A}$ into $\mathbf{Q}_\mathbf{A}^\times$. We then put

$$(26.2\text{a}) \quad F_{\mathbf{a}+}^\times = \{a \in F_\mathbf{a}^\times \mid a \gg 0\}, \quad G_{\mathbf{a}+} = \{x \in G_\mathbf{a} \mid \nu(x) \gg 0\},$$

$$(26.2\text{b}) \qquad G_{\mathbf{A}+} = \{x \in G_\mathbf{A} \mid x_\mathbf{a} \in G_{\mathbf{a}+}\}, \quad G_+ = G \cap G_{\mathbf{A}+},$$

$$(26.2\text{c}) \qquad D_{\mathbf{A}+} = D_\mathbf{A} \cap G_{\mathbf{A}+}, \quad D_+ = D \cap D_{\mathbf{A}+}.$$

We can identify $G_{\mathbf{a}}$ and $G_{\mathbf{a}+}$ with $\mathfrak{G}$ and $\mathfrak{G}_+$ of (23.4e, g). Let S be a subgroup of $G_{\mathbf{A}+}$ such that $S = S_{\mathbf{h}} G_{\mathbf{a}+}$ with a compact subgroup $S_{\mathbf{h}}$ of $G_{\mathbf{h}}$ containing an open subgroup of $D_{\mathbf{h}}$. Since $\mathbf{Q}_{\mathbf{A}}^{\times} \cap F^{\times}\nu(S)$ is an open subgroup of $\mathbf{Q}_{\mathbf{A}}^{\times}$ containing $\mathbf{Q}^{\times}\mathbf{Q}_{\mathbf{a}+}^{\times}$, it corresponds to a subfield of $\mathbf{Q}_{\mathrm{ab}}$, which we denote by k_S. Put also $\Gamma_S = G \cap S$. Clearly Γ_S is an arithmetic subgroup of G. For instance if

$$\text{(26.2d)} \quad S_{\mathbf{h}} = \left\{ y \in D_{\mathbf{h}} \mid y_v \in GL_n(\mathfrak{g}_v),\ y_v - 1 \in (N\mathfrak{g}_v)_n^n \ \textit{for every}\ v \in \mathbf{h} \right\}$$

with a positive integer N, then $\Gamma_S = \Gamma(N\mathfrak{g})$ with the notation of (24.8a), and $k_S = \mathbf{Q}\big(\mathbf{e}(1/N)\big)$.

26.3. Theorem. *To every S of §26.2 (not necessarily of type (26.2d)) we can canonically assign a model (V_S, φ_S) of $\Gamma_S\backslash\mathfrak{H}^{\mathbf{a}}$ with the following properties:*

(1) *V_S is rational over k_S.*

(2) *For Ω as in (24.1c), let*

$$\text{(26.3a)} \quad S = G_{\mathbf{a}+}\{\, y \in D_{\mathbf{h}} \mid Ly = L,\ u_i y \equiv u_i \pmod{L}\ \textit{for every}\ i \,\},$$

where the action of y on the lattices and F_{2n}^1/L are defined in the same manner as in §18.3. Then k_S coincides with k_Ω of Theorem 24.8 and $k_S\big(\varphi_S(z)\big)$ is the field of moduli of $\mathcal{P}_z$ for every $z \in \mathfrak{H}^{\mathbf{a}}$, that is, (V_S, φ_S) can be taken as (V, φ) of that theorem.

In fact, these (V_S, φ_S) can be characterized by certain properties. We call (V_S, φ_S) a *canonical model* of $\Gamma_S\backslash\mathfrak{H}^{\mathbf{a}}$. For details the reader is referred to [S70]. In particular, property (2) is explained in [S70, §5.5, (5.1.3)]. We can describe the function field of V_S in terms of modular functions as follows:

26.4. Theorem. *Given S as in §26.2, put*

$$\text{(26.4a)} \qquad \Delta_S = \left\{ \operatorname{diag}[1_n, t1_n] \,\middle|\, t \in \prod_p \mathbf{Z}_p^{\times},\ [t, \mathbf{Q}] = \mathrm{id.}\ \textit{on}\ k_S \right\}.$$

Suppose that $\Delta_S \subset S$. Then the following assertions hold:

(1) *$\mathcal{A}_0(\Gamma_S, k_S)$ coincides with the set of all functions of the form $g \circ \varphi_S$ with a k_S-rational function g on V_S in the sense of algebraic geometry.*

(2) *$\mathcal{M}_{\kappa u}(\Gamma_S) = \mathcal{M}_{\kappa u}(\Gamma_S, k_S) \otimes_{k_S} \mathbf{C}$ for every $\kappa \in 2\mathbf{Z}, > 0$, where κu denotes the element of $\mathbf{Z}^{\mathbf{a}}$ whose components are all equal to κ.*

(3) *$\mathcal{M}_{\kappa u}(\Gamma) = \mathcal{M}_{\kappa u}(\Gamma, k_S) \otimes_{k_S} \mathbf{C}$ for every odd positive integer κ, if Γ is an arithmetic subgroup such that $F^{\times}\Gamma = \Gamma_S$ and $\mathcal{A}_{\kappa u}(\Gamma, k_S) \neq \{0\}$.*

Clearly (26.4a) is satisfied with S of (26.2d). This theorem is proved in [S75a, b] and [S78a] in the case $n = 1$ or $F = \mathbf{Q}$; the general case can be proved in the same manner. The same comment applies to Theorem 26.10 below.

Since $\mathcal{A}_0(\mathbf{Q}_{ab})$ is the union of $\mathcal{A}_0(\Gamma_S, \mathbf{Q}_{ab})$ for all such S, from Theorem 26.4(1) and Theorem 24.11 we obtain

26.5. Theorem. *Let w be a CM-point on $\mathfrak{H}^{\mathbf{a}}$ and K^* the field defined at w as in §24.10. Then $f(w) \in K^*_{ab}$ for every $f \in \mathcal{A}_0(\mathbf{Q}_{ab})$ finite at w.*

If $n = 1$, Proposition 24.13 makes the statement more clear-cut:

26.6. Theorem. *Suppose $n = 1$, that is, $G = GL_2(F)$; let $\big(K, \{\tau_v\}_{v\in\mathbf{a}}\big)$ be a CM-type with K containing F as the maximal real subfield, and K^* the field generated over $\mathbf{Q}$ by $\sum_{v\in\mathbf{a}} a^{\tau_v}$ for all $a \in K$. Take $w_0 \in K$ so that $\mathrm{Im}(w_0^{\tau_v}) > 0$ for every $v \in \mathbf{a}$ and let $w = (w_0^{\tau_v})_{v\in\mathbf{a}}$. Then $f(w) \in K^*_{ab}$ for every $f \in \mathcal{A}_0(\mathbf{Q}_{ab})$ finite at w.*

We can actually tell how $f(w)$ behaves under $\mathrm{Gal}(K^*_{ab}/K^*)$, as will be shown in Theorem 26.8 and (26.9a) below.

26.7. We put

(26.7a) $$\mathcal{G} = D_{\mathbf{A}}GG_{\mathbf{a}+}, \quad \mathcal{G}_+ = \mathcal{G} \cap G_{\mathbf{A}+}.$$

We can easily show that

(26.7b) $$\mathcal{G}_+ = D_{\mathbf{A}+}G_+G_{\mathbf{a}+} = \big\{x \in G_{\mathbf{A}+} \mid \nu(x) \in \mathbf{Q}_{\mathbf{A}}^\times F^\times F_{\mathbf{a}+}^\times\big\}.$$

Given $x \in \mathcal{G}_+$, take $c \in \mathbf{Q}_{\mathbf{A}}^\times$ so that $\nu(x) \in cF^\times F_{\mathbf{a}+}^\times$. We then define $\sigma(x)$ to be $[c^{-1}, \mathbf{Q}]\big(\in \mathrm{Aut}(\mathbf{Q}_{ab})\big)$. Then we easily see that $\sigma(x)$ is well defined independently of the choice of c, and thus we obtain a homomorphism

(26.7c) $$\sigma\colon \mathcal{G}_+ \longrightarrow \mathrm{Gal}(\mathbf{Q}_{ab}/\mathbf{Q}).$$

26.8. Theorem. *Put $\mathfrak{K} = \mathcal{A}_0(\mathbf{Q}_{ab})$. Then there exists a homomorphism $\tau\colon \mathcal{G}_+ \longrightarrow \mathrm{Aut}(\mathfrak{K})$ with the following properties:*

(1) $\tau(x) = \sigma(x)$ *on* $\mathbf{Q}_{ab}$.

(2) $f^{\tau(\alpha)} = f \circ \alpha$ *for every* $f \in \mathfrak{K}$ *and every* $\alpha \in G_+$.

(3) *For S as in §26.2, let $\mathfrak{K}_S$ denote the subfield of $\mathfrak{K}$ consisting of $p \circ \varphi_S$ with all k_S-rational functions p on V_S. Then $F^\times(S \cap \mathcal{G}) = \{x \in \mathcal{G}_+ \mid \tau(x) =$ id. on $\mathfrak{K}_S\}$.*

(4) *Let Y, h, w, and K^* be as in 24.10. Define $g\colon (K^*)_{\mathbf{A}}^\times \to Y_{\mathbf{A}}^\times$ by (18.7c) as mentioned at the end of §24.10. Then for every $f \in \mathfrak{K}$ defined at w, the value $f(w)$ belongs to K^*_{ab}. Moreover, if $b \in (K^*)_{\mathbf{A}}^\times$, then $f^{\tau(x)}$ with $x = h\big(g(b)^{-1}\big)$ is finite at w and $f(w)^{[b,K^*]} = f^{\tau(x)}(w)$.*

This is a special case of what is given in [S70, §§2.7, 2.8]. Notice that if $b \in (K^*)^\times_{\mathbf{A}}$, then from (18.5c) and (24.10a) we see that $\nu\big(h(g(b))\big) = N_{K^*/\mathbf{Q}}(b)$, and so $h\big(g(b)^{-1}\big) \in D_{\mathbf{A}+}$. It should also be noted that (4), or rather its special case (26.9a) below, is a generalization of the classical theorem concerning the behavior of the singular values of the classical modular function J under automorphisms. The results of the same type as this theorem can be obtained for groups more general than the present G; for details, see [S67] and [S70].

In [S70, II] we considered the closure $\overline{\mathcal{G}}_+$ of $\mathcal{G}_+$ in $G_{\mathbf{A}}$ and defined $\tau\colon \overline{\mathcal{G}}_+ \to \mathrm{Aut}(\mathfrak{K})$ and formulated the results in terms of this map. This is a natural thing to do, but it makes our treatment somewhat cumbersome, and so we adopted an easier formulation. The reader is referred to this paper and Miyake [M] for related problems.

26.9. To illustrate (4) in the simplest case, denote by S_1 the group S such that $S_{\mathbf{h}}$ is given by (26.2d) with $N = 1$. By strong approximation we have $D_{\mathbf{A}+} = D_+(S_1 \cap D_{\mathbf{A}})$. Therefore, given $b \in (K^*)^\times_{\mathbf{A}}$, there exists an element $\alpha \in D_+$ such that $h\big(g(b)\big) \in \alpha(S_1 \cap D_{\mathbf{A}})$. With such an α we obtain from (2, 3, 4) above,

$$\text{(26.9a)}\quad f(w)^{[b,K^*]} = f\big(\alpha^{-1}(w)\big) \quad \textit{for every}\ f \in \mathcal{A}_0(\Gamma(1), \mathbf{Q})\ \textit{and}\ b \in (K^*)^\times_{\mathbf{A}}.$$

We now define a subgroup $\mathfrak{G}$ of $\mathcal{G}_+ \times \mathrm{Gal}(\overline{\mathbf{Q}}/\mathbf{Q})$ by

$$\text{(26.9b)}\qquad \mathfrak{G} = \Big\{ (x, \sigma) \in \mathcal{G}_+ \times \mathrm{Gal}(\overline{\mathbf{Q}}/\mathbf{Q}) \mid \sigma(x) = \sigma \quad on \quad \mathbf{Q}_{\mathrm{ab}} \Big\}.$$

26.10. Theorem. *Each element (x, σ) of $\mathfrak{G}$ gives a ring-automorphism of the graded algebra $\sum_k \mathcal{A}_k\left(\overline{\mathbf{Q}}\right)$, written $f \mapsto f^{(x,\sigma)}$, with the following properties:*

(1) $\big(f^{(x,\sigma)}\big)^{(y,\tau)} = f^{(xy,\sigma\tau)}$.
(2) $f^{(\alpha,1)} = f\|_k\alpha$ *if* $\alpha \in G_+$ *and* $f \in \mathcal{A}_k$, *where* $f\|_k\alpha$ *is defined as in Theorem 25.9(2) with* $\rho(x) = \det(x)^k$.
(3) $f^{(x,\sigma)}$ *coincides with* f^σ *of Proposition 25.8 if* $x = \mathrm{diag}[1_n, t1_n]$ *with* $t \in \prod_p \mathbf{Z}_p^\times$.
(4) $\mathcal{A}_k\left(\overline{\mathbf{Q}}\right)^{(x,\sigma)} = \mathcal{A}_l\left(\overline{\mathbf{Q}}\right)$ *and* $\mathcal{M}_k\left(\overline{\mathbf{Q}}\right)^{(x,\sigma)} = \mathcal{M}_l\left(\overline{\mathbf{Q}}\right)$ *with* $l = k^\sigma$ *of Proposition 25.8.*
(5) *If* $f \in \mathcal{A}_0\left(\mathbf{Q}_{\mathrm{ab}}\right)$, *then* $f^{(x,\sigma)} = f^{\tau(x)}$ *with* τ *of Theorem 26.8.*

For this the reader is referred to [S75a, b] and [S78a].

CHAPTER VII

Theta Functions and Periods on Abelian Varieties

27. Theta Functions

27.1. We first reformulate and refine what was explained in §3.1. Given a complex vector space V of dimension n and a lattice Λ in V, we mean by a *hermitian Riemann form on* V/Λ a positive definite hermitian form $H(u, v)$ on $V \times V$, $\mathbf{C}$-linear in v, such that $\mathrm{Im}\big(H(u, v)\big) \in \mathbf{Z}$ for every $u, v \in \Lambda$. For such an H put

$$E(u, v) = \mathrm{Im}\big(H(u, v)\big). \tag{27.1a}$$

Then E is a nondegenerate Riemann form on V/Λ in the sense of §3.1. It can easily be seen that, conversely, a nondegenerate Riemann form E corresponds to a unique hermitian Riemann form H by (27.1a). Since the present H is $2i$ times H of (2) in §3.1, we have

$$H(u, v) = E(u, iv) + iE(u, v). \tag{27.1b}$$

For $\alpha = \left[\begin{smallmatrix} a & b \\ c & d \end{smallmatrix}\right] \in \mathbf{R}_{2n}^{2n}$ with a, b, c, and d in $\mathbf{R}_n^n$ we put $a_\alpha = a, b_\alpha = b, c_\alpha = c$, and $d_\alpha = d$. Dropping the subscript for the moment, we see that

$$\text{(27.1c)}\quad \alpha \in Sp(n, \mathbf{R}) \iff {}^t ad - {}^t cb = 1,\ \ {}^t ac = {}^t ca,\ \ {}^t bd = {}^t db,$$
$$\iff a \cdot {}^t d - b \cdot {}^t c = 1,\ \ a \cdot {}^t b = b \cdot {}^t a,\ \ c \cdot {}^t d = d \cdot {}^t c.$$

To specialize the results of §§24.4–5 to the case $F = \mathbf{Q}$, we need:

27.2. Lemma. *Let M be a free $\mathbf{Z}$-module of rank $2n$, $Y = M \otimes_{\mathbf{Z}} \mathbf{Q}$, and let $E(x, y)$ be a nondegenerate $\mathbf{Q}$-valued $\mathbf{Q}$-bilinear alternating form on Y such that $E(M, M) \subset \mathbf{Z}$. Then there exists an isomorphism U of $\mathbf{Q}^{2n}$ to Y and positive integers $\varepsilon_1, \ldots, \varepsilon_n$ such that $U\big(\mathbf{Z}^{2n}\big) = M$, $\varepsilon_i / \varepsilon_{i+1}$ for every $i < n$, and*

$$\text{(27.2a)}\quad E(U(x), U(y)) = {}^t x B y \text{ with } B = \begin{bmatrix} 0 & -\delta \\ \delta & 0 \end{bmatrix}, \quad \delta = \mathrm{diag}[\varepsilon_1, \ldots, \varepsilon_n].$$

Moreover, $\varepsilon_1, \ldots, \varepsilon_n$ are uniquely determined by E and M.

This is well known. We call δ *the matrix of elementary divisors of E relative to M.*

27.3. Taking $F = \mathbf{Q}$ in the setting of §24.1, we consider $\Omega = \{\mathbf{Q}, L, J_n\}$ with a $\mathbf{Z}$-lattice L in $\mathbf{Q}^{2n}$. (Here we take $\mathbf{Q}^{2n}$ instead of $\mathbf{Q}^1_{2n}$ employed in §24.) By the above lemma there is an element $U \in GL_{2n}(\mathbf{Q})$ such that $U(\mathbf{Z}^{2n}) = L$ and ${}^tU(x)J_nU(y) = {}^txBy$ with B as in that lemma. Put

(27.3a) $$T = T_\delta = \mathrm{diag}[1_n, \delta].$$

(27.3b) $$\Lambda(z, \delta) = \omega_z T_\delta \mathbf{Z}^{2n} = z\mathbf{Z}^n + \delta\mathbf{Z}^n, \quad \omega_z = [z \quad 1_n] \qquad (z \in \mathfrak{H}_n).$$

Then we see that $T_\delta U^{-1}$ gives an "isomorphism" of Ω to $\{\mathbf{Q}, T_\delta(\mathbf{Z}^{2n}), J_n\}$ in an obvious sense. Therefore we may assume that $\Omega = \{\mathbf{Q}, T_\delta(\mathbf{Z}^{2n}), J_n\}$ without losing generality. Then for each $z \in \mathfrak{H}_n$ we obtain $\mathcal{P}_z = (A_z, \mathcal{C}_z)$ as in §24.5 defined with p_z of (24.5a). Clearly $p_z(x) = \omega_z x$ for $x \in \mathbf{R}^{2n}$, $p_z(L) = \Lambda(z, \delta)$, and so A_z is isomorphic to $\mathbf{C}^n/\Lambda(z, \delta)$, and $\mathcal{C}_z$ is determined by the Riemann form

(27.3c) $$E_z(\omega_z x, \omega_z y) = {}^tx J_n y.$$

Put $\Gamma'_\delta = Sp(n, \mathbf{Q}) \cap T_\delta GL_{2n}(\mathbf{Z})T_\delta^{-1}$. Then, as a special case of Theorem 24.6(3) we obtain

(27.3d) *$\mathcal{P}_z$ and $\mathcal{P}_w$ are isomorphic if and only if $w = \gamma z$ with $\gamma \in \Gamma'_\delta$.*

Now the hermitian Riemann form associated with E_z by (27.1b) is given by

(27.3e) $$H(u, v) = 2i \cdot {}^t\overline{u}\big(z - \overline{z}\big)^{-1}v \qquad (u, v \in \mathbf{C}^n).$$

To prove this, let C be an element of $GL_{2n}(\mathbf{R})$ such that $i \cdot \omega_z = \omega_z C$. This corresponds to ${}^tC$ of §24.4. Write simply ω for ω_z, and put $X = \begin{bmatrix}\omega \\ \overline{\omega}\end{bmatrix}$. Then $JC = iJX^{-1}IX$ as seen in §24.4. For $u = \omega x$ with $x \in \mathbf{R}^{2n}$ we have

$$\begin{aligned}(*)\quad H(u, u) &= E_z(u, iu) = {}^txJCx = {}^tx \cdot {}^t(JC)\overline{x} \\ &= -i \cdot {}^tx \cdot {}^tXI \cdot {}^tX^{-1}J\overline{x} = -i\,[{}^tu \quad {}^t\overline{u}]\,I \cdot {}^tX^{-1}J\overline{X}^{-1}\begin{bmatrix}\overline{u} \\ u\end{bmatrix}.\end{aligned}$$

Taking the inverse of the last equality of (24.4b) we obtain

$$-{}^tX^{-1}J\overline{X}^{-1} = \begin{bmatrix}(\overline{\omega}J \cdot {}^t\omega)^{-1} & 0 \\ 0 & (\omega J \cdot {}^t\overline{\omega})^{-1}\end{bmatrix} = \begin{bmatrix}(z - \overline{z})^{-1} & 0 \\ 0 & (\overline{z} - z)^{-1}\end{bmatrix}.$$

Substituting this into the last expression of $(*)$, we obtain (27.3e) for $u = v$, which proves (27.3e), since its both sides are hermitian.

27.4. Given V/Λ and H as in §27.1, take a map $\psi\colon \Lambda \to \mathbf{T}$, where

(27.4a) $$\mathbf{T} = \{\zeta \in \mathbf{C} \mid |\zeta| = 1\}.$$

We then denote by $T(H, \psi, \Lambda)$ the set of all holomorphic functions f on V satisfying

(27.4b) $$f(u+\ell) = f(u)\psi(\ell)\mathbf{e}\big((2i)^{-1}H(\ell, u+(\ell/2))\big) \qquad (u \in V, \ell \in \Lambda),$$

where $\mathbf{e}$ is as in (25.1a). For $T(H, \psi, \Lambda)$ to be nonzero, ψ cannot be arbitrary; it must satisfy the relation

(27.4c) $$\psi(\ell + m) = \psi(\ell)\psi(m)\mathbf{e}\big(E(\ell, m)/2\big) \qquad (\ell, m \in \Lambda).$$

This is essentially (1) in §3.1, and such an f is a normalized theta function defined there. Let us now take $\Lambda = \Lambda(z, \delta)$ and find an explicit basis of the vector space $T(H, \psi, \Lambda)$ with H of (27.3e). We define the classical theta function θ and its modification φ by

(27.4d) $$\theta(u, z; r, s) = \sum_{g-r\in\mathbf{Z}^n} \mathbf{e}\big(2^{-1}\cdot {}^t gzg + {}^t g(u+s)\big),$$

(27.4e) $$\varphi(u, z; r, s) = \mathbf{e}\big(2^{-1}\cdot {}^t u(z-\bar{z})^{-1}u\big)\theta(u, z; r, s).$$

Here $u \in \mathbf{C}^n$, $z \in \mathfrak{H}_n$, and $r, s \in \mathbf{R}^n$. It is not difficult to show that the right-hand side of (27.4d) is locally uniformly convergent on $\mathbf{C}^n \times \mathfrak{H}_n$, and so defines a holomophic function in (u, z). By purely formal calculations we can easily verify

(27.4f) $$\theta(u + za + b, z; r, s) = \mathbf{e}\big(-2^{-1}\cdot {}^t aza - {}^t a(u+b+s)\big)\theta(u, z; r+a, s+b) \quad (a, b \in \mathbf{R}^n),$$

(27.4g) $$\theta(u + za + b, z; r, s) = \mathbf{e}\big(-2^{-1}\cdot {}^t aza - {}^t au + {}^t rb - {}^t sa\big)\theta(u, z; r, s) \quad (a, b \in \mathbf{Z}^n),$$

(27.4h) $$\theta(u, z; r + a, s + b) = \mathbf{e}({}^t rb)\theta(u, z; r, s) = \theta(u + b, z; r, s) \quad (a, b \in \mathbf{Z}^n),$$

(27.4i) $$\theta(-u, z; r, s) = \theta(u, z; -r, -s).$$

27.5. Proposition. *Put $f(u) = \varphi(u, z; r, s)$ with fixed z, r, s. Then f belongs to $T\big(H, \psi, \Lambda(z, \delta)\big)$ with ψ defined by*

(27.5a) $$\psi(za + \delta b) = \mathbf{e}\big(2^{-1}\cdot {}^t a\delta b - {}^t sa + {}^t r\delta b\big) \quad (a, b \in \mathbf{Z}^n).$$

Moreover, let R be a complete set of representatives for $\delta^{-1}\mathbf{Z}^n/\mathbf{Z}^n$, and let $g_j(u) = \varphi(u, z; r + j, s)$ with fixed z, r, s and $j \in R$. Then $\{g_j\}_{j\in R}$ is a basis of $T\big(H, \psi, \Lambda(z, \delta)\big)$ over $\mathbf{C}$.

PROOF. The first fact can easily be verified by means of (27.4g). As for the second assertion, it is easy to see that the g_j are linearly independent over $\mathbf{C}$, so that they span a vector space whose dimension is $\det(\delta)$. That $\det(\delta)$ is exactly the dimension of $T(H, \psi, \Lambda(z, \delta))$ is classical; see [57, pp. 120–121].

Let us note here that this proposition covers all possible (H, ψ, Λ) in the following sense: First we can put $\Lambda = \Lambda(z, \delta)$ with H as in (27.3e). Take any ψ satisfying (27.4c) with this Λ, and put $\varphi(\ell) = \mathbf{e}(-2^{-1} \cdot {}^t a\delta b)\psi(\ell)$ for $\ell = za + \delta b \in \Lambda$ with $a, b \in \mathbf{Z}^n$. From (27.4c) we can easily derive that $\varphi(\ell + m) = \varphi(\ell)\varphi(m)$. This means that there exist $p, q \in \mathbf{R}^n$ such that $\varphi(za + \delta b) = \mathbf{e}({}^t pa + {}^t qb)$, and hence ψ must be of form (27.5a).

27.6. For a positive integer N denote by $\Gamma(N)$ the subgroup $\Gamma(N\mathbf{Z})$ of $Sp(n, \mathbf{Z})$ defined by (24.8a), and put

$$\text{(27.6a)} \quad \Gamma_\theta = \left\{ \gamma \in \Gamma(1) \mid \{{}^t a_\gamma c_\gamma\} \equiv \{{}^t b_\gamma d_\gamma\} \equiv 0 \pmod{2\mathbf{Z}^n} \right\},$$

where $\{S\}$ is the column vector consisting of the diagonal elements of S. Clearly $\Gamma(2) \subset \Gamma_\theta$. It can easily be seen that Γ_θ consists of the elements γ of $\Gamma(1)$ such that $F((x \ \ y)\gamma) - F(x \ \ y) \in \mathbf{Z}$ for every $(x \ \ y) \in \mathbf{Z}^1_{2n}$, where $F(x \ \ y) = x \cdot {}^t y/2$. Therefore Γ_θ is a subgroup of $\Gamma(1)$. This group is also defined by the condition

$$\text{(27.6b)} \qquad \{a_\gamma \cdot {}^t b_\gamma\} \equiv \{c_\gamma \cdot {}^t d_\gamma\} \equiv 0 \pmod{2\mathbf{Z}^n}.$$

27.7. Theorem. (1) *For every $\gamma \in \Gamma(1)$ we have*

$$\text{(27.7a)} \quad \theta({}^t(c_\gamma z + d_\gamma)^{-1}u, \gamma z; r, s) = \zeta \cdot \det(c_\gamma z + d_\gamma)^{1/2}\mathbf{e}(2^{-1} \cdot {}^t u(c_\gamma z + d_\gamma)^{-1}c_\gamma u)\theta(u, z; r'', s'')$$

with a constant $\zeta \in \mathbf{T}$ depending on r, s, γ, and a suitable choice of branch of $\det(c_\gamma z + d_\gamma)^{1/2}$, *and*

$$\text{(27.7b)} \qquad \begin{bmatrix} r'' \\ s'' \end{bmatrix} = {}^t\gamma \begin{bmatrix} r \\ s \end{bmatrix} + \frac{1}{2}\begin{bmatrix} \{{}^t ac\} \\ \{{}^t bd\} \end{bmatrix}.$$

(2) *For every $\gamma \in \Gamma_\theta$ there is a holomorphic function $h_\gamma(z)$ in $z \in \mathfrak{H}$, written also $h(\gamma, z)$, such that $h_\gamma(z)^2 = \zeta \cdot \det(c_\gamma z + d_\gamma)$ with a constant $\zeta \in \mathbf{T}$ and*

$$\text{(27.7c)} \quad \theta({}^t(c_\gamma z + d_\gamma)^{-1}u, \gamma z; r, s) = \mathbf{e}(2^{-1}({}^t rs - {}^t r's'))h_\gamma(z)\mathbf{e}(2^{-1} \cdot {}^t u(c_\gamma z + d_\gamma)^{-1}c_\gamma u)\theta(u, z; r', s')$$

with $\left[\begin{smallmatrix} r' \\ s' \end{smallmatrix}\right] = {}^t\gamma\left[\begin{smallmatrix} r \\ s \end{smallmatrix}\right]$.

(3) $h_\gamma(z)^4 = \det(c_\gamma z + d_\gamma)^2$ *if* $\gamma \in \Gamma(2)$.

These formulas are classical (see [KP], for example). We shall give a shorter proof in Section 28. See also [S93] for the transformation formula for a more general type of theta function.

27.8. We note that $(c_\gamma z + d_\gamma)^{-1} c_\gamma$, for every $\gamma \in Sp(n, \mathbf{R})$, is symmetric and

$$\big(\gamma(z) - \gamma(\bar{z})\big)^{-1} = (c_\gamma z + d_\gamma)(z - \bar{z})^{-1} \cdot {}^t(c_\gamma \bar{z} + d_\gamma),$$

as already observed in (23.3f). Putting $g(u, z) = \mathbf{e}\big(2^{-1} \cdot {}^t u(z - \bar{z})^{-1} u\big)$ and employing these facts, we can easily verify that

$$g\big({}^t(c_\gamma z + d_\gamma)^{-1} u, \gamma z\big) = g(u, z)\mathbf{e}\big(-2^{-1} \cdot {}^t u(c_\gamma z + d_\gamma)^{-1} c_\gamma u\big). \tag{27.8a}$$

Then it is easy to see that (27.7a) and (27.7c) are equivalent to

$$\varphi\big({}^t(c_\gamma z + d_\gamma)^{-1} u, \gamma z; r, s\big) = \zeta \cdot \det\big(c_\gamma z + d_\gamma\big)^{1/2} \varphi(u, z; r'', s''), \tag{27.8b}$$

$$\varphi\big({}^t(c_\gamma z + d_\gamma)^{-1} u, \gamma z; r, s\big) = \mathbf{e}\big(2^{-1}({}^t r s - {}^t r' s')\big) h_\gamma(z) \varphi(u, z; r', s'). \tag{27.8c}$$

Now we put

$$\theta(z) = \sum_{g \in \mathbf{Z}^n} \mathbf{e}\big(2^{-1} \cdot {}^t g z g\big) \quad (z \in \mathfrak{H}). \tag{27.8d}$$

Then $\theta(z) = \theta(0, z; 0, 0) = \varphi(0, z; 0, 0)$, and hence from (27.7c) we obtain

$$\theta(\gamma z) = h_\gamma(z)\theta(z) \quad (\gamma \in \Gamma_\theta), \tag{27.8e}$$

$$h(\beta\gamma, z) = h_\beta(\gamma z) h_\gamma(z) \quad (\beta, \gamma \in \Gamma_\theta). \tag{27.8f}$$

In the following proposition we shall see that h_γ^2 coincides with j_γ of (25.1c) if $\gamma \in \Gamma(4)$. For every congruence subgroup Γ of $Sp(n, \mathbf{Q})$ contained in Γ_θ we denote by $\mathcal{M}_{1/2}(\Gamma)$ the set of all holomorphic functions f on $\mathfrak{H}$ such that $f^2 \in \mathcal{M}_1$ and

$$f(\gamma z) = h_\gamma(z) f(z) \quad \textit{for every } \gamma \in \Gamma. \tag{27.8g}$$

In fact, the condition $f^2 \in \mathcal{M}_1$ follows from (27.8g) if $n > 1$. If $f \in \mathcal{M}_{1/2}(\Gamma)$, then f has expansion (25.5a). Then for a subfield D of $\mathbf{C}$ we define $\mathcal{M}_{1/2}(\Gamma, D)$ to be the set of all $f \in \mathcal{M}_{1/2}(\Gamma)$ such that c_h of (25.5a) belongs to D for all h. Notice that $f/g \in \mathcal{A}_0(\Gamma, D)$ if $f, g \in \mathcal{M}_{1/2}(\Gamma, D)$ and $g \neq 0$, since $f/g = fg/g^2$ and both fg and g^2 belong to $\mathcal{M}_1(D)$. We denote by $\mathcal{M}_{1/2}(D)$ (resp. $\mathcal{M}_{1/2}$) the union of $\mathcal{M}_{1/2}(\Gamma, D)$ (resp. $\mathcal{M}_{1/2}(\Gamma)$) for all such Γ.

27.9. Proposition. (1) *If* $\gamma \in \Gamma_\theta$ *and* $\det(d_\gamma) \neq 0$, *then*

$$\lim_{z\to 0} h_\gamma(z) = \sum_{x\in A} \mathbf{e}\big(-{}^t x d_\gamma^{-1} c_\gamma x/2\big) = \sum_{x\in B} \mathbf{e}\big({}^t x b_\gamma d_\gamma^{-1} x/2\big), \tag{27.9a}$$

where $A = \mathbf{Z}^n/{}^t d_\gamma \mathbf{Z}^n$ *and* $B = \mathbf{Z}^n/d_\gamma \mathbf{Z}^n$.

(2) *If* $\gamma \in \Gamma_\theta$ *and* $\det(d_\gamma)$ *is odd, then* $h_\gamma(z)^2 = \chi\big(\det(d_\gamma)\big) j_\gamma(z)$, *where* χ *is the Dirichlet character of conductor 4.*

(3) $h(J, z) = \det(-iz)^{1/2}$ *with the choice of branch such that* $h(J, z) > 0$ *for* $\mathrm{Re}(z) = 0$.

(4) *Let* $\alpha \in G_+$ *and* $r(z) = \det(c_\alpha z + d_\alpha)^{1/2}$ *with any choice of branch. Then there is a congruence subgroup* Δ *of* Γ_θ *such that*

$$h(\alpha\gamma\alpha^{-1}, \alpha z) = r(\gamma z) h(\gamma, z) r(z)^{-1} \tag{27.9b}$$

for every $\gamma \in \Delta$.

We shall prove these assertions in Section 28. Clearly (2) implies Theorem 27.7(3). Assertion (1) and an analogue of (2) are proved in [S93, Lemma 2.2 and Proposition 2.4] in a more general setting. (Strictly speaking, we need the fact that the right-hand side of (27.9a) has absolute value $|\det(d_\gamma)|^{1/2}$, which is not stated in [S93, Lemma 2.2]. However, this follows immediately from [S93, p. 1026, lines 6–11].)

27.10. Proposition. *Let* α *and* r *be as in (4) of Proposition 27.9. Then, for every* $f \in \mathcal{M}_{1/2}$ *(resp.* $f \in \mathcal{M}_{1/2}(\mathbf{Q}_{\mathrm{ab}})$*) the function* $r(z)^{-1} f(\alpha z)$ *belongs to* $\mathcal{M}_{1/2}$ *(resp.* $\mathcal{M}_{1/2}(\mathbf{Q}_{\mathrm{ab}})$*).*

PROOF. Let $g(z) = r(z)^{-1} f(\alpha z)$ with $f \in \mathcal{M}_{1/2}$. From (27.9b) we easily see that $g \in \mathcal{M}_{1/2}$. Suppose $f \in \mathcal{M}_{1/2}(\mathbf{Q}_{\mathrm{ab}})$. Let $P = \{\xi \in G \mid c_\xi = 0\}$. As will be shown in Lemma 28.4 below, G is generated by P and J. Therefore it is sufficient to show that $g \in \mathcal{M}_{1/2}(\mathbf{Q}_{\mathrm{ab}})$ for $\alpha \in P$ and $\alpha = J$. This is obvious for $\alpha \in P$. If $\alpha = J$, then $r(z) = \zeta h(J, z)$ with an eighth root of unity ζ. Put $p = f/\theta$. Then $p = f\theta/\theta^2 \in \mathcal{A}_0(\mathbf{Q}_{\mathrm{ab}})$ and $g(z) = r(z)^{-1}\theta(Jz)p(Jz) = \zeta^{-1}h(J, z)^{-1}\theta(Jz)p(Jz) = \zeta^{-1}\theta(z)p(Jz)$. Since $p \circ J \in \mathcal{A}_0(\mathbf{Q}_{\mathrm{ab}})$, we obtain the desired result.

27.11. Proposition. *With* $p, q, r, s \in \mathbf{Q}^n$ *and* $0 < \tau \in \mathbf{Q}$ *put*

$$f(z) = \mathbf{e}\big((\tau/2)\cdot {}^t pzp\big)\theta\big(\tau(zp + q), \tau z; r, s\big) \quad (z \in \mathfrak{H}). \tag{27.11a}$$

Then $f \in \mathcal{M}_{1/2}(\mathbf{Q}_{\mathrm{ab}})$. *In particular* $\theta(0, z; r, s) \in \mathcal{M}_{1/2}(\mathbf{Q}_{\mathrm{ab}})$.

PROOF. Clearly $\theta(0, z; r, s)$ has its Fourier coefficients in $\mathbf{Q}_{\mathrm{ab}}$, and hence the last assertion can be seen from (27.4h) and (27.7c), since $\theta(0, z; r, s) =$

$\theta(0, z; r', s')$ and ${}^t rs - {}^t r's'$ in (27.7c) is an even integer if γ belongs to a sufficiently small congruence subgroup. Now, by (27.4f) we have

$$f(z) = \mathbf{e}\big(-{}^t p(\tau q + s)\big)\theta\big(0, \tau z; r + p, s + \tau q\big), \tag{27.11b}$$

which reduces the problem to the case we have proved.

27.12. Let us now recall a well-known fact, due to Lefschetz [27], that a basis of $T\big(H, \psi, \Lambda(z, \delta)\big)$ defines an embedding of $\mathbf{C}^n/\Lambda(z, \delta)$ into a projective space if

(27.12a) *the entries of δ are divisible by an integer greater than 2*

(see [57, p. 130]). Thus, assuming (27.12a) and using the basis given in Proposition 27.5, we can define a projective embedding of $\mathbf{C}^n/\Lambda(z, \delta)$ by

$$u \mapsto \Theta(u) = \Theta_z(u) = \big(\theta(u, z; j, 0)\big)_{j \in R} \quad (u \in \mathbf{C}^n). \tag{27.12b}$$

Let $A(z, \delta)$ be the image abelian variety. The polarization determined by (27.3c, e) considered on $A(z, \delta)$ is the polarization by its hyperplane sections. Hereafter we denote by the same symbol $A(z, \delta)$ the polarized abelian variety given in this manner.

Define a subgroup Γ_δ of Γ_θ by

$$\Gamma_\delta = \big\{\alpha \in \Gamma_\theta \cap \beta^{-1}\Gamma_\theta\beta \mid a_\alpha - 1 \in \delta\mathbf{Z}_n^n\big\}, \quad \beta = \mathrm{diag}[\delta^{-1}, \delta]. \tag{27.12c}$$

Put $\mathfrak{K} = \mathcal{A}_0(\mathbf{Q}_{\mathrm{ab}})$ and $\mathfrak{K}_\delta = \mathcal{A}_0(\Gamma_\delta, \mathbf{Q})$. For $w \in \mathfrak{H}_n$ we let $\mathfrak{K}_\delta[w]$ resp. $\mathfrak{K}[w]$ denote the field generated over $\mathbf{Q}$ by the values $f(w)$ for all f in $\mathfrak{K}_\delta$ resp. $\mathfrak{K}$ finite at w. From (27.8c) we easily obtain

$$\varphi\big({}^t(c_\gamma z + d_\gamma)^{-1}u, \gamma z; r, 0\big) = h_\gamma(z)\varphi(u, z; r, 0) \quad \text{if } \gamma \in \Gamma_\delta \text{ and } r \in \delta^{-1}\mathbf{Z}^n. \tag{27.12d}$$

This implies in particular that if $\gamma \in \Gamma_\delta$, then $A(z, \delta) = A(\gamma z, \delta)$ and

$$\begin{array}{ccc} \mathbf{C}^n/\Lambda(\gamma z, \delta) & \xrightarrow{\Theta_{\gamma z}} & A(\gamma z, \delta) \\ {\scriptstyle {}^t(c_\gamma z + d_\gamma)}\downarrow & & \downarrow{\scriptstyle \mathrm{id}} \\ \mathbf{C}^n/\Lambda(z, \delta) & \xrightarrow{\Theta_z} & A(z, \delta) \end{array} \tag{27.12e}$$

is a commutative diagram.

27.13. Theorem. *Suppose that (27.12a) is satisfied; let $A(z, \delta)$ be the polarized abelian variety as above. Then for every $w \in \mathfrak{H}$, $A(w, \delta)$ is defined over $\mathfrak{K}_\delta[w]$. Moreover, $\mathfrak{K}[w]$ is the smallest field of rationality for all the points of finite order on $A(w, \delta)$ and is algebraic over the field of moduli of $A(w, \delta)$.*

PROOF. From Proposition 27.11 we see that

$$\theta(zp + q, z; j, 0)/\theta(zp + q, z; j', 0) \in \mathfrak{K} \tag{27.13a}$$

for $j, j' \in R$ if $p, q \in \mathbf{Q}^n$. Let $\mathfrak{F}$ be the field generated over $\mathbf{Q}$ by all the quotients of (27.13a), and $\mathfrak{F}[w]$ the field generated over $\mathbf{Q}$ by the values $f(w)$ for all $f \in \mathfrak{F}$ finite at w. Then $\mathfrak{F} \subset \mathfrak{K}$. Since $\mathfrak{K} = \bigcup_{m=1}^{\infty} \mathcal{A}_0(\Gamma(m), \mathbf{Q}(\mathbf{e}(1/m)))$ and $\mathcal{A}_0(\Gamma(m), \mathbf{Q}(\mathbf{e}(1/m)))$ is the function field of a variety over $\mathbf{Q}(\mathbf{e}(1/m))$, we easily see that $\mathfrak{K}$ is a countable set. Therefore, by Lemma 27.14 below we can find a point $z_0 \in \mathfrak{H}$ such that $f \mapsto f(z_0)$ for $f \in \mathfrak{K}$ defines an isomorphism of $\mathfrak{K}$ onto $\mathfrak{K}[z_0]$. Now every point of finite order on $A(w, \delta)$ is of the form $\big(\theta(wp + q, z; j, 0)\big)_{j \in R}$, and hence such a point is rational over $\mathfrak{F}[w]$. Since all such points are dense in $A(w, \delta)$, we have $A(w, \delta)^\sigma = A(w, \delta)$ for every $\sigma \in \mathrm{Aut}\big(\mathbf{C}/\mathfrak{F}[w]\big)$, and hence $A(w, \delta)$ is rational over $\mathfrak{F}[w]$. We consider a structure $\mathcal{P} = \big(A(w, \delta), \{t_i\}\big)$ with any finite set $\{t_i\}$ of points of finite order on $A(w, \delta)$. Then $\mathcal{P}$ is rational over $\mathfrak{F}[w]$, and hence the field of moduli of $\mathcal{P}$ is contained in $\mathfrak{F}[w]$. This means that $k_S\big(\varphi_S(w)\big) \subset \mathfrak{F}[w]$ for any S of type (26.3a) with $L = T_\delta \mathbf{Z}^{2n}$. Now $f \mapsto f \circ \varphi_S$ gives an isomorphism of the function field of V_S over $\mathbf{Q}_{\mathrm{ab}}$ onto $\mathcal{A}_0(\Gamma_S, \mathbf{Q}_{\mathrm{ab}})$. Since $\mathfrak{K}$ is the union of $\mathcal{A}_0(\Gamma_S, \mathbf{Q}_{\mathrm{ab}})$ for all such S, we obtain $\mathfrak{K}[w] = \mathfrak{F}[w]$, which proves the last assertion of our proposition. Taking w to be z_0, we find that $\mathfrak{K} = \mathfrak{F}$. To prove the first assertion, we have to examine the behavior of $A(w, \delta)$ under $\mathrm{Aut}\big(\mathfrak{K}[w]/\mathfrak{K}_\delta[w]\big)$, employing Theorem 26.3. For this we refer the reader to [S76, Proposition 1.7] and [S77a, pp. 368–369]. In this proof we needed the following fact:

27.14. Lemma. *Let $\{ f_\nu \mid \nu \in N \}$ be a set of meromorphic functions in a connected open subset D of $\mathbf{C}^d$, indexed by an at most countable set N. Let k be a subfield of $\mathbf{C}$ with only countably many elements. Then there exists a point z_0 of D such that the specialization $\{ f_\nu \}_{\nu \in N} \mapsto \{ f_\nu(z_0) \}_{\nu \in N}$ defines an isomorphism of the field $k(f_\nu \mid \nu \in N)$ onto $k\big(f_\nu(z_0) \mid \nu \in N\big)$ over k.*

PROOF. We may assume that $N = \{1, 2, 3, \dots\}$ (finite or not). By induction we can find a subset $M = \{\nu_1, \nu_2, \dots\}$ of N such that: (i) $\nu_1 < \nu_2 < \cdots$; (ii) $f_{\nu_1}, f_{\nu_2}, \dots$ are algebraically independent over k; and (iii) $f_1, \dots, f_n$ are algebraic over $k(f_\nu \mid \nu \in M, \nu \le n)$. Let S_m be the set of all polynomials $P(X_1, \dots, X_m) \neq 0$ in m indeterminates with coefficients in k, and W_ν the set of the points of D where f_ν is not holomorphic. For each $P \in S_m$ put

$$E_P = \left\{ z \in D - \bigcup_{i=1}^{m} W_{\nu_i} \,\middle|\, P\big(f_{\nu_1}(z), \ldots, f_{\nu_m}(z)\big) = 0 \right\}.$$

The closure of E_P in D has no interior point of D. Now observe that S_m has only countably many elements. Recall the well-known fact that if D is covered by countably many closed subsets, then at least one of them has an interior point. Therefore we find a point z_0 of D not belonging to the countable union $\left[\bigcup_{\nu\in N} W_\nu\right] \cup \left[\bigcup_{m=1}^{\infty} \bigcup_{P\in S_m} E_P\right]$. Then our construction shows that $k(f_1, \ldots, f_n)$ has the same transcendence degree as $k\big(f_1(z_0), \ldots, f_n(z_0)\big)$ over k for every n. Therefore the specialization $f_\nu \mapsto f_\nu(z_0)$ defines an isomorphism of these fields as expected.

27.15. Proposition. *The field $\mathfrak{K}$ is generated by all the quotients of the form*

(27.14a) $\theta(0, z;\ j+p, q)/\theta(0, z;\ j'+p, q) \quad (j, j' \in \delta^{-1}\mathbf{Z}^n/\mathbf{Z}^n;\ p, q \in \mathbf{Q}^n)$

provided δ satisfies (27.12a).

This is contained in the proof of Theorem 27.13, since the quotient of (27.14a) equals that of (27.13a) in view of (27.11b).

We note here that we can prove Theorem 25.9 and other related results of the same nature (at least when $F = \mathbf{Q}$) by employing theta functions with the same ideas as above, without relying on the general theory of canonical models. For example, if $\alpha \in G_+$, we see from (24.7c) that ${}^tM(\alpha, w)^{-1}$ times a suitable positive integer gives an isogeny of $A(w, \delta)$ to $A(\alpha w, \delta)$, and hence $\mathfrak{K}[\alpha w] \subset \mathfrak{K}[w]$. Taking w to be z_0, we find that $\mathfrak{K} = \mathcal{A}_0(\mathbf{Q}_{ab})$ is stable under $f \mapsto f \circ \alpha$ for every $\alpha \in G_+$.

That $A(w, \delta)$ is rational over $\mathfrak{K}_\delta[w]$ is not so essential in our later discussion, and in fact, will not be required in the proof of the period relations in §32.

28. Proof of Theorem 27.7 and Proposition 27.9

28.1. We first verify formula (27.7a) for the elements γ of $\Gamma(1) = Sp(n, \mathbf{Z})$ of the forms

(28.1a) $$\begin{bmatrix} a & 0 \\ 0 & d \end{bmatrix}, \quad \begin{bmatrix} 1 & b \\ 0 & 1 \end{bmatrix}, \quad \begin{bmatrix} 0 & -1 \\ 1 & 0 \end{bmatrix}.$$

If $\gamma = \mathrm{diag}[a, d] \in \Gamma(1)$, then $d = {}^ta^{-1} \in GL_n(\mathbf{Z})$ and $\gamma(z) = az \cdot {}^ta$, and hence we easily see that

(28.1b) $$\theta(au, az \cdot {}^ta;\ r, s) = \theta(u, z;\ {}^tar, a^{-1}s).$$

Next, if $\gamma = \begin{bmatrix} 1 & b \\ 0 & 1 \end{bmatrix} \in \Gamma(1)$, then ${}^t b = b$ and $\gamma(z) = z + b$. Observing that ${}^t xbx/2 \equiv {}^t x\{b\}/2 \pmod{\mathbf{Z}}$ if $x \in \mathbf{Z}^n$ (with { } defined as in §27.6), we obtain

$$\text{(28.1c)} \quad \theta(u, z+b; r, s) = \mathbf{e}\big(-2^{-1}({}^t rbr + {}^t r\{b\})\big)\theta\big(u, z; r, s+br+2^{-1}\{b\}\big).$$

If $\gamma = \begin{bmatrix} 0 & -1 \\ 1 & 0 \end{bmatrix}$, then $\gamma(z) = -z^{-1}$. To discuss this case, we first recall the formula

$$\text{(28.1d)} \quad \int_{\mathbf{R}^n} \mathbf{e}(2^{-1}z[x])\mathbf{e}(-{}^t vx)\, dx = \det(-iz)^{-1/2}\mathbf{e}\big(-2^{-1}z^{-1}[v]\big) \quad (v \in \mathbf{R}^n, z \in \mathfrak{H}_n),$$

where $z[x] = {}^t xzx$ and dx is the standard volume element of $\mathbf{R}^n$. If $z = i \cdot {}^t aa$ with $a \in GL_n(\mathbf{R})$, this follows immediately from the self-reciprocity of $\exp\big(-\pi x^2\big)$. Now the left-hand side of (28.1d) is convergent and defines a holomorphic function in z; the right-hand side is clearly holomorphic in z. Since they coincide on "the imaginary axis" of $\mathfrak{H}$, we obtain (28.1d) on the whole $\mathfrak{H}$.

Put $v = y - u - s$ in (28.1d) with real vectors y, u, s, and

$$f^*(y) = \int_{\mathbf{R}^n} f(x)\mathbf{e}(-{}^t xy)\, dx, \qquad f(x) = \mathbf{e}\big(2^{-1}z[x] + {}^t x(u+s)\big).$$

Then (28.1d) shows that

$$f^*(y) = \det(-iz)^{-1/2}\mathbf{e}\big(-2^{-1}z^{-1}[y-u-s]\big).$$

We apply the Poisson summation formula

$$\sum_{g \in \mathbf{Z}^n} f(r+g) = \sum_{h \in \mathbf{Z}^n} \mathbf{e}({}^t hr) f^*(h) \quad (r \in \mathbf{R}^n)$$

to the present situation to find that

$$\text{(28.1e)} \quad \det(-iz)^{1/2}\mathbf{e}\big(2^{-1}z^{-1}[u]\big)\theta(u, z; r, s) = \mathbf{e}({}^t rs)\theta(z^{-1}u, -z^{-1}; -s, r)$$

for real u, and hence for all $u \in \mathbf{C}^n$, since both sides are holomorphic in u. Thus we know that (27.7a) holds for γ of the forms of (28.1a). Taking $r = s = 0$ in (28.1e), we obtain the formula of Proposition 27.9(3).

28.2. Lemma. (1) *$\Gamma(1)$ is generated by the elements of the forms of (28.1a).*

(2) *Let $\Gamma' = \big\{\gamma \in \Gamma(1) \mid b_\gamma \equiv c_\gamma \equiv 0 \pmod{2\mathbf{Z}_n^n}\big\}$. Then Γ' is generated by the elements of the forms*

$$\text{(28.2a)} \quad \begin{bmatrix} a & 0 \\ 0 & d \end{bmatrix}, \quad \begin{bmatrix} 1 & b \\ 0 & 1 \end{bmatrix}, \quad \begin{bmatrix} 1 & 0 \\ c & 1 \end{bmatrix}, \quad b \equiv c \equiv 0 \pmod{2\mathbf{Z}_n^n}.$$

(3) *Let Γ^* be the subgroup of $\Gamma(1)$ generated by the elements of the forms*

$$(28.2b)\qquad \begin{bmatrix} a & 0 \\ 0 & d \end{bmatrix}, \quad \begin{bmatrix} 0 & -1 \\ 1 & 0 \end{bmatrix}, \quad \begin{bmatrix} 1 & b \\ 0 & 1 \end{bmatrix}, \quad b \equiv 0 \pmod{2\mathbf{Z}_n^n}.$$

Then $\Gamma' \subset \Gamma^*$.

PROOF. We first prove (2). For $x \in \mathbf{Z}^n$, let $[x]$ denote the greatest common divisor of its components. We put $[0] = 0$. Also, for $a \in \mathbf{Z}_n^n$ let a^j denote its j-th column, and a_i^j its (i, j)-entry. Now let $\gamma = \left[\begin{smallmatrix} a & b \\ c & d \end{smallmatrix}\right] \in \Gamma'$. Our idea is to reduce $[a^1]$ and $[c^1]$ by multiplying by elements of the forms listed in (2). Clearly $[c^1]$ is even, and hence $[a^1]$ is odd, since $\gamma \in SL_{2n}(\mathbf{Z})$. First suppose $[a^1] < [c^1]$. Considering $\left[\begin{smallmatrix} u & 0 \\ 0 & v \end{smallmatrix}\right]\left[\begin{smallmatrix} a & b \\ c & d \end{smallmatrix}\right]$ instead of $\left[\begin{smallmatrix} a & b \\ c & d \end{smallmatrix}\right]$ with a suitable $u = {}^t v^{-1} \in GL_n(\mathbf{Z})$, we may assume that $a_1^1 > 0, a_2^1 = \cdots = a_n^1 = 0$. Then a_1^1 is odd and $< [c^1]$. For each k we can find an integer s_k^1 such that $|c_k^1 + 2s_k^1 a_1^1| \le a_1^1$. Take any $s = {}^t s \in \mathbf{Z}_n^n$ with such s_k^1 and put $\left[\begin{smallmatrix} 1 & 0 \\ 2s & 1 \end{smallmatrix}\right]\left[\begin{smallmatrix} a & b \\ c & d \end{smallmatrix}\right] = \left[\begin{smallmatrix} a & b \\ p & q \end{smallmatrix}\right]$. Then we find $[p^1] < [a^1]$. Next assume that $0 < [c^1] < [a^1]$. Then, first considering vc with a suitable $v \in GL_n(\mathbf{Z})$, and then $\left[\begin{smallmatrix} 1 & 2s \\ 0 & 1 \end{smallmatrix}\right]\gamma$ with a suitable s, we can reduce this to the case $[a^1] < [c^1]$. Repeating these procedures, we obtain an element, written again $\gamma = \left[\begin{smallmatrix} a & b \\ c & d \end{smallmatrix}\right]$, with $c^1 = 0$. Then $[a^1] = 1$. For the same reason as above, we may assume that $a_1^1 = 1, a_2^1 = \cdots = a_n^1 = 0$. Since $a \cdot {}^t d - b \cdot {}^t c = 1$, we see that $d_1^1 = 1$. Then left multiplication by $\mathrm{diag}[u, v]$ with a suitable $v \in GL_n(\mathbf{Z})$ produces $d^1 = {}^t(1\ 0\ \cdots\ 0)$ without changing a^1 and c^1. Further, left multiplication by $\left[\begin{smallmatrix} 1 & 2s \\ 0 & 1 \end{smallmatrix}\right]$ with a suitable s produces $b^1 = 0$. We obtain in this way an element of the form

$$(28.2c)\qquad \begin{bmatrix} 1 & 0 & 0 & 0 \\ 0 & a' & 0 & b' \\ 0 & 0 & 1 & 0 \\ 0 & c' & 0 & d' \end{bmatrix}$$

with $\left[\begin{smallmatrix} a' & b' \\ c' & d' \end{smallmatrix}\right] \in Sp(n-1, \mathbf{Z}), b' \equiv c' \equiv 0 \pmod{2\mathbf{Z}_{n-1}^{n-1}}$. The proof of (2) is therefore completed by induction on n, since if $\left[\begin{smallmatrix} a' & b' \\ c' & d' \end{smallmatrix}\right]$ is of a type belonging to (28.2a), then so is the matrix of (28.2c). To prove (1) and (3), we note

$$(28.2d)\qquad \begin{bmatrix} 0 & -1 \\ 1 & 0 \end{bmatrix} = \begin{bmatrix} 1 & -1 \\ 0 & 1 \end{bmatrix}\begin{bmatrix} 1 & 0 \\ 1 & 1 \end{bmatrix}\begin{bmatrix} 1 & -1 \\ 0 & 1 \end{bmatrix},$$

$$(28.2e)\qquad \begin{bmatrix} 1 & 0 \\ -b & 1 \end{bmatrix} = \begin{bmatrix} 0 & -1 \\ 1 & 0 \end{bmatrix}\begin{bmatrix} 1 & b \\ 0 & 1 \end{bmatrix}\begin{bmatrix} 0 & -1 \\ 1 & 0 \end{bmatrix}^{-1}.$$

In view of (28.2e) we obtain (3) immediately from (2). As for (1), we employ the same type of argument as in the proof of (2). Since we have J in (28.1a), we can use it in addition to the matrices of (28.2a) without congruence conditions, again in view of (28.2e). Now left multiplication by J changes (a, c) into $(-c, a)$. We first assume that $0 < [a^1] \leq [c^1]$ and repeat the above argument with the following modification: take $s_k^1 \in \mathbf{Z}$ so that $0 \leq c_k^1 + s_k^1 a_1^1 < a_1^1$, and use s instead of $2s$. Then we can reduce the problem to the case $[c^1] \leq [a^1]$, and further to the case $c_1 = 0$, and eventually to (28.2c). If $\left[\begin{smallmatrix} a' & b' \\ c' & d' \end{smallmatrix}\right] = J_{n-1}$, then (28.2c) does not belong to the three types of (28.1a), but applying (28.2d) to J_{n-1} and employing (28.2e), we can justify our induction.

28.3. As noted in §27.8, formulas (27.7a, c) are equivalent to (27.8b, c). Therefore our result of §28.1 shows that (27.8b) holds for γ of (28.1a). Thus, by Lemma 28.2(1) we have

$$(28.3a) \qquad \varphi\big({}^t(c_\gamma z + d_\gamma)^{-1}u, \gamma z; r, s\big) = \zeta \cdot \det\big(c_\gamma z + d_\gamma\big)^{1/2}\varphi(u, z; r'', s'')$$

for every $\gamma \in \Gamma(1)$ with some r'', s'', and $\zeta \in \mathbf{T}$, since it is easy to see that (28.3a) is "associative" with respect to successive applications of elements of $\Gamma(1)$. Thus our task is to determine r'' and s''. For this purpose we first note that if $w \in \mathfrak{H}$, $f(u) = \varphi(u, w; r, s)$, and $\ell = wp + q$ with p and q in $\mathbf{Z}^n$, then

$$(28.3b) \quad f(u+\ell) = f(u)\mathbf{e}\big(2^{-1}\cdot {}^t pq - {}^t sp + {}^t rq + {}^t\overline{\ell}(w - \overline{w})^{-1}(u + 2^{-1}\ell)\big).$$

This follows from (27.4g). Observe that r and s are determined modulo $\mathbf{Z}^n$ by this formula. Suppressing the subscript γ, put $w = \gamma(z)$ and

$$g(u) = f\big({}^t(cz + d)^{-1}u\big) = \varphi\big({}^t(cz + d)^{-1}u, \gamma z; r, s\big).$$

Observe that ${}^t(cz + d)^{-1}(zp + q) = (\gamma z)p' + q'$ if $\left[\begin{smallmatrix} p \\ q \end{smallmatrix}\right] = {}^t\gamma\left[\begin{smallmatrix} p' \\ q' \end{smallmatrix}\right]$. Therefore if $m = zp + q$ with p and q in $\mathbf{Z}^n$, then from (28.3b) we obtain

$$(28.3c) \quad g(u+m) = g(u)\mathbf{e}\big(2^{-1}\cdot {}^t pq - {}^t s_1 p + {}^t r_1 q + {}^t\overline{m}(z - \overline{z})^{-1}(u + 2^{-1}m)\big).$$

with

$$\begin{bmatrix} r_1 \\ s_1 \end{bmatrix} = {}^t\gamma\begin{bmatrix} r \\ s \end{bmatrix} + \frac{1}{2}\begin{bmatrix} -\{{}^t ac\} \\ \{{}^t bd\} \end{bmatrix},$$

after a somewhat lengthy but straightforward calculation. On the other hand, $g(u) = \zeta \cdot \det(cz + d)^{1/2}\varphi(u, z; r'', s'')$, so that we have (28.3c) with r'' and s'' instead of r_1 and s_1. Therefore $r'' \equiv r_1$ and $s'' \equiv s_1 \pmod{\mathbf{Z}^n}$. This combined with (27.4h) proves (1) of Theorem 27.7.

Next assume that both $\{{}^t ac\}$ and $\{{}^t bd\}$ belong to $2\mathbf{Z}^n$. Then we can put

$$(28.3d) \quad \varphi\big({}^t(cz + d)^{-1}u, \gamma z; 0, 0\big) = \lambda_\gamma \det(cz + d)^{1/2}\varphi(u, z; 0, 0)$$

with a constant $\lambda_\gamma \in \mathbf{T}$. From (27.4e, f) we obtain

$$\text{(28.3e)}\quad \varphi(u, z; r, s) = \mathbf{e}({}^t rs/2)\mathbf{e}\big(-2^{-1}\cdot{}^t(\bar{z}r+s)(z-\bar{z})^{-1}(2u+zr+s)\big) \cdot\varphi(u+zr+s, z; 0, 0).$$

Combining this with (28.3d), we obtain (2) of Theorem 27.7, or rather (27.8c), after a few lines of straightforward calculation.

Finally, to prove (3) of Theorem 27.7, we observe that $\lambda_\gamma^4 = 1$ for the first two types of elements of (28.2a). As for the third type, making substitutions $z \mapsto z - 2c$ and $z \mapsto -z^{-1}$ in (28.1e), we find that

$$\theta\big(z(2cz+1)^{-1}\big) = \pm\det(2cz+1)^{1/2}\theta(z) \quad \textit{if}\ \ {}^t c = c \in \mathbf{Z}_n^n$$

with θ of (27.8d). Thus $\lambda_\gamma = \pm 1$ for the third type. (We shall prove a stronger result, Proposition 27.9(2), in §28.7.) The proof of Theorem 27.7 is now complete.

28.4. Lemma. *Let $P = \{\alpha \in G \mid c_\alpha = 0\}$ and $P_1 = G_1 \cap P$. Then $G = P\Gamma(1)$, $G_1 = P_1\Gamma(1)$, and G (resp. G_1) is generated by P (resp. P_1) and J.*

PROOF. Since $G = PG_1$ and the first two matrices of (28.1a) belong to P_1, we only have to show that $G_1 = P_1\Gamma(1)$. Put $Y = \{x \in \mathbf{Q}_{2n}^n \mid xJ\cdot{}^t x = 0,\ \mathrm{rank}(x) = n\}$. Then $(0\ \ 1_n) \in Y$ and $GL_n(\mathbf{Q})YG \subset Y$, and hence $(0\ \ 1_n)\xi \in Y$ for every $\xi \in G$. Observe also that $(0\ \ 1_n)\xi$ is the lower half of ξ. Now, given $\alpha \in G_1$, by elementary divisor theory we find $u \in GL_n(\mathbf{Z})$ and $\beta \in SL_{2n}(\mathbf{Z})$ such that $u(c_\alpha\ \ d_\alpha)\beta^{-1} = (0\ \ p)$ with $p \in GL_n(\mathbf{Q})$. Then $(0\ \ 1_n)\beta = p^{-1}u(c_\alpha\ \ d_\alpha) \in Y$. Therefore we see that $\beta J\cdot{}^t\beta = \left[\begin{smallmatrix} f & -e \\ {}^t e & 0\end{smallmatrix}\right]$ with $e, f \in \mathbf{Z}_n^n$. Since $\beta \in SL_n(\mathbf{Z})$, we have $e \in GL_n(\mathbf{Z})$. Put $\gamma = \mathrm{diag}[e^{-1}, 1]$. Then $\gamma\beta J\cdot{}^t(\gamma\beta) = \left[\begin{smallmatrix} g & -1 \\ 1 & 0\end{smallmatrix}\right]$ with ${}^t g = -g \in \mathbf{Z}_n^n$. We can find $h \in \mathbf{Z}_n^n$ so that $g = {}^t h - h$. Putting $\delta = \left[\begin{smallmatrix} 1 & h \\ 0 & 1\end{smallmatrix}\right]$, we obtain $\delta\gamma\beta J\cdot{}^t(\delta\gamma\beta) = J$, so that $\delta\gamma\beta \in \Gamma(1)$. Now $(0\ \ 1_n)\delta\gamma\beta\alpha^{-1} = (0\ \ 1_n)\beta\alpha^{-1} = (0\ \ p^{-1}u)$. This shows that $\delta\gamma\beta\alpha^{-1} \in P_1$, which proves that $\alpha \in P_1\Gamma(1)$ as expected.

28.5. Let us now prove (1) and (2) of Proposition 27.9. To simplify the notation, suppress the subscript γ, and put $L = \mathbf{Z}^n$, $A = L/{}^t dL$, $B = L/dL$, and $s[x] = {}^t xsx$ for $s \in \mathbf{Q}_n^n$ and $x \in \mathbf{Q}^n$. Since $c\cdot{}^t d = d\cdot{}^t c$, we see that $c\cdot{}^t dL \subset dL$. Therefore $x \mapsto cx$ sends A into B. Since $\gamma \in SL_{2n}(\mathbf{Z})$, we have $cL + dL = L$, and hence the map is surjective. Comparing the orders of the groups, we find that the map gives an isomorphism of A onto B. If $y = cx$, we have, by (27.1c),

$$\begin{aligned} bd^{-1}[y] &= {}^t cbd^{-1}c[x] = ({}^t ad - 1)d^{-1}c[x] \\ &= {}^t ac[x] - d^{-1}c[x] \equiv -d^{-1}c[x] \pmod 2, \end{aligned}$$

since the diagonal elements of ${}^t ac$ are even. This shows that the two sums of (27.9a) are the same.

From (27.8e) and (28.1b) we obtain $h_\alpha(z) = 1$ for $\alpha = \mathrm{diag}[a, d] \in \Gamma(1)$, and hence $h_{\alpha\gamma} = h_\gamma$ for every $\gamma \in \Gamma_\theta$ and such an α. Therefore, to prove (2) and the first equality of (1), we may assume that $\det(d) > 0$. Under this assumption, put $w = {}^t d^{-1} z(cz+d)^{-1}$ and $p = bd^{-1}$. Then, by (27.1c),

$$\begin{aligned}(\gamma z - p)(cz+d) &= az + b - bd^{-1}(cz+d) = az - b \cdot {}^t c \cdot {}^t d^{-1} z \\ &= (a \cdot {}^t d - b \cdot {}^t c) \cdot {}^t d^{-1} z = {}^t d^{-1} z,\end{aligned}$$

and so $\gamma(z) = w + p$. With θ of (27.8d) we thus have

$$\text{(28.5a)} \quad h_\gamma(z)\theta(z) = \theta(\gamma z) = \theta(w+p) = \sum_{x \in L} \mathbf{e}\big((1/2)(w+p)[x]\big).$$

Putting $x = v + dg$ with $v \in B$ and $g \in L$, we find

$$\begin{aligned}\theta(w+p) &= \sum_v \sum_g \mathbf{e}\big((p/2)[v+dg] + (w/2)[v+dg]\big) \\ &= \sum_v \mathbf{e}\big((p/2)[v]\big)\theta\big(0, z(cz+d)^{-1}d; d^{-1}v, 0\big),\end{aligned}$$

since $p[v+dg] \equiv p[v] \pmod 2$. Now (28.1e) shows that

$$\det(-iz)^{1/2}\theta(0, z;\ r, s) = \mathbf{e}({}^t rs)\theta(0, -z^{-1};\ -s, r).$$

Put $z = i\tau 1_n$ with $0 < \tau \in \mathbf{R}$ and observe that

$$\lim_{\tau \to 0} \tau^{n/2}\theta(0, i\tau 1_n; r, s) = \mathbf{e}({}^t rs)\delta(s),$$

where $\delta(s) = 1$ or 0 according as $s \in L$ or $s \notin L$. Taking the limit of $\tau^{n/2}$ times (28.5a) as τ tends to 0, we obtain (27.9a). Before proceeding further, we recall a classical result:

28.6. Lemma. *Let a be an odd positive integer and b an integer prime to a. Let $\varepsilon(a)$ be 1 or i according as $a \equiv 1 \pmod 4$ or $a \equiv -1 \pmod 4$. Then*

$$\sum_{x=1}^{a} \mathbf{e}(bx^2/a) = \left(\frac{b}{a}\right)\varepsilon(a)\sqrt{a}.$$

For the proof, see [E, p. 59], for example.

28.7. Given $\gamma \in \Gamma_\theta$, assume that $0 < \det(d) \equiv 1 \pmod 2$, and put $f = \det(d)$, $g = fd^{-1}$, and $s = -fd^{-1}c$. Then both g and s are integral, f is odd, and $ds \cdot {}^t d = -fc \cdot {}^t d$, and hence the diagonal elements of s are even. Denote by λ the first sum of (27.9a) and put $\sigma = h_\gamma(z)^2/j_\gamma(z)$. Then $\sigma = \lambda^2/\det(d)$. Now

$$\lambda = \sum_{x \in A} \mathbf{e}\big(s[x]/(2f)\big) = [{}^t dL : fL]^{-1} \sum_{x \in L/fL} \mathbf{e}\big(s[x]/(2f)\big).$$

It is an elementary fact that one can find $u \in \mathbf{Z}_n^n$ such that $\det(u)$ is a positive integer prime to f and ${}^t usu \equiv \mathrm{diag}[r_1, \ldots, r_n] \pmod f$. Then each r_i must be even, and

$$\lambda = f^{1-n} \prod_{\nu=1}^{n} \sum_{x=1}^{f} \mathbf{e}\big(r_\nu x^2/(2f)\big).$$

Put $r_\nu/f = 2b_\nu/a_\nu$ with relatively prime integers a_ν and b_ν; take $a_\nu > 0$. Then

$$\sum_{x=1}^{f} \mathbf{e}\big(r_\nu x^2/(2f)\big) = (f/a_\nu) \sum_{x=1}^{a_\nu} \mathbf{e}\big(x^2 b_\nu/a_\nu\big) = f\left(\frac{b_\nu}{a_\nu}\right) \varepsilon(a_\nu) a_\nu^{-1/2}$$

by Lemma 28.6. Thus we obtain

$$\lambda = f \prod_{\nu=1}^{n} \left(\frac{b_\nu}{a_\nu}\right) \varepsilon(a_\nu) a_\nu^{-1/2}.$$

Since $cL + dL = L$, we have $sL + fL = gL$. For a rational prime p put $L_p = \mathbf{Z}_p^n$. Then $gL_p = L_p$ if $p \nmid f$. If p/f, then ${}^t uL_p = uL_p = L_p$, and hence ${}^t ugL_p = {}^t usuL_p + fL_p$. From this we easily see that the elementary divisors of g are $\{(f, r_\nu)\}_{\nu=1}^n$. Since $a_\nu = |f/(f, r_\nu)|$ and $d = fg^{-1}$, we thus know that the a_ν are exactly the elementary divisors of d and

$$\lambda = \det(d)^{1/2} \prod_{\nu=1}^{n} \left(\frac{b_\nu}{a_\nu}\right) \varepsilon(a_\nu), \tag{28.7a}$$

so that $\sigma = \lambda^2/\det(d) = \chi\big(\det(d)\big)$, which proves (2) of Proposition 27.9.

28.8. Assertion (4) of Proposition 27.9 was proved in [S93, Proposition 1.4] for $\alpha \in G_1$. Unfortunately the proof requires many preliminary observations, and so we do not include it here. Now every element of G_+ is of the form $\varepsilon\beta$ with $\varepsilon = \mathrm{diag}[1_n, r1_n]$, $0 < r \in \mathbf{Q}$ and $\beta \in G_1$. Thus our task is to show that $h(\varepsilon\gamma\varepsilon^{-1}, r^{-1}z) = h(\gamma, z)$ for γ in a sufficiently small congruence subgroup. Clearly it is sufficient to prove the case where $r \in \mathbf{Z}$. We may again assume that $\det(d_\gamma) > 0$. We may also assume that $\det(d_\gamma)$ is prime to r. Now replace

c_γ by rc_γ in (27.9a). Then the value of the sum is multiplied by $\left(\frac{r}{\det(d_\gamma)}\right)$. This fact can be seen from (28.7a). Therefore, if we take $\gamma \in \Gamma(4r)$, then the sum in (27.9a) is the same for γ and $\varepsilon\gamma\varepsilon^{-1}$. This completes the proof. One can also give a proof by invoking the congruence subgroup property of $Sp(n, \mathbf{Z})$ (cf. [S76, pp. 680–681]).

29. Theta Functions with Complex Multiplication

29.1. We now consider V/Λ and $T(H, \psi, \Lambda)$ of §27.4 with ψ under the following condition:

(29.1a) $\psi(\ell)$ *is a root of unity for every* $\ell \in \Lambda$.

This means that r and s of (27.5a) belong to $\mathbf{Q}^n$. For $f \in T(H, \psi, \Lambda)$ we define a (nonholomorphic) function f_* on V by

(29.1b) $$f_*(u) = \mathbf{e}\big((i/4)H(u, u)\big)f(u) \quad (u \in V).$$

Then from (27.4b) we easily see that

(29.1c) $$f_*(u + \ell) = \psi(\ell)\mathbf{e}\big(E(\ell, u)/2\big)f_*(u) \quad (\ell \in \Lambda).$$

We now consider a CM-algebra Y and $\mathcal{P} = (A, \mathcal{C}, \iota)$ as in §18.7. We take a complex torus V/Λ isomorphic to A and H on V corresponding to a divisor in $\mathcal{C}$. Defining K^* as in §18.7, we call an element f of $T(H, \psi, \Lambda)$ *arithmetic* if $f_*(u) \in K^*_{\mathrm{ab}}$ for every $u \in \mathbf{Q}L$, and denote by $T_a(H, \psi, \Lambda)$ the set of all such f. To obtain explicit generators of this space, we take V/L in the form $\mathbf{C}^n/\Lambda(z_0, \delta)$ with H given by (27.3e) and a point z_0 in $\mathfrak{H}_n$. (We can start from Y and h as in §24.10 in Case SP, and take z_0 to be the CM-point obtained there.)

29.2. Proposition. (1) *Let* $f(u) = \varphi(u, z_0; r, s)$ *with* φ *of (27.4e) and* $r, s \in \mathbf{Q}^n$, *and let* g *be an element of* $\mathcal{M}_{1/2}(\mathbf{Q}_{\mathrm{ab}})$ *such that* $g(z_0) \neq 0$. *Then* $g(z_0)^{-1}f \in T_a(H, \psi, \Lambda)$ *with* ψ *of (27.5a).*

(2) $T_a(H, \psi, \Lambda)$ *with* ψ *of (27.5a) can be spanned by* $g(z_0)^{-1}\varphi(u, z_0; r+j, s)$ *over* K^*_{ab} *for all* $j \in \delta^{-1}\mathbf{Z}^n/\mathbf{Z}^n$.

Proof. By Proposition 27.5 we have $f \in T(H, \psi, \Lambda)$. If $u = z_0a + \delta b$ with $a, b \in \mathbf{Q}^n$, then

$$f_*(u) = \mathbf{e}\big(2^{-1} \cdot {}^t a(z_0a + \delta b)\big)\theta(z_0a + \delta b, z_0; r, s).$$

By Proposition 27.11 this is the value at z_0 of a function belonging to $\mathcal{M}_{1/2}(\mathbf{Q}_{\mathrm{ab}})$. Therefore (1) follows from Theorem 26.5. Then (2) is immediate from Proposition 27.5.

29.3. If $f \in T(H, \psi, \Lambda)$ and M is a $\mathbf{Z}$-lattice in $\mathbf{Q}\Lambda$, then (29.1b) shows that f_* defines a function on M/M' for a sufficiently small $\mathbf{Z}$-lattice M' in $\mathbf{Q}L$. Denote by $(\mathbf{Q}\Lambda)_{\mathbf{h}}$ the nonarchimedean part of $(\mathbf{Q}\Lambda)_{\mathbf{A}}$. This is a subset of $\prod_p(\mathbf{Q}\Lambda \otimes_{\mathbf{Q}} \mathbf{Q}_p)$, where p runs over all the rational primes. For every $x = (x_p) \in (\mathbf{Q}\Lambda)_{\mathbf{h}}$ we can find a $\mathbf{Z}$-lattice M in $\mathbf{Q}\Lambda$ such that $x_p \in M_p$ for all p, where $M_p = M \otimes_{\mathbf{Z}} \mathbf{Z}_p$. Take M' as above. Then there is an element $y \in M$ such that $y - x_p \in M'_p$ for all p. We then define $f_*(x)$ to be the same as $f_*(y)$. Clearly this is well defined, and thus f_* as a function on $(\mathbf{Q}\Lambda)_{\mathbf{h}}$ is meaningful. We view $\mathbf{Q}\Lambda$ as a subset of $(\mathbf{Q}\Lambda)_{\mathbf{h}}$.

We can now let every element s of $Y_{\mathbf{A}}^\times$ act on $(\mathbf{Q}\Lambda)_{\mathbf{h}}$ in an obvious fashion by ignoring $s_{\mathbf{a}}$. Then for $u \in \mathbf{Q}\Lambda$ the value $f_*(su)$ is meaningful. As explained in §§18.3 and 18.7, we can define $s\Lambda$ to be a lattice in V, and so if ψ is given as above, then ψ' can be defined on $s\Lambda$ by $\psi'(s\ell) = \psi(\ell)$.

To simplify our notation we write a^x for a^σ when $a \in K^*_{\mathrm{ab}}$, $x \in (K^*)_{\mathbf{A}}^\times$, and $\sigma = [x, K^*]$.

29.4. Theorem. *Define $g\colon (K^*)_{\mathbf{A}}^\times \to Y_{\mathbf{A}}^\times$ by (18.7c). Then every element x of $(K^*)_{\mathbf{A}}^\times$ defines a $\mathbf{Q}$-linear map $f \mapsto f^x$ of $T_a(H, \psi, \Lambda)$ onto $T_a\big(N(x\mathfrak{r})H, \psi', g(x)^{-1}\Lambda\big)$ satisfying the following conditions (1–6), where $\mathfrak{r}$ is the maximal order of K^* and $\psi'(\ell) = \psi\big(g(x)\ell\big)^x$:*

(1) $(af + bh)^x = a^x f^x + b^x h^x$ *if* $a, b \in K^*_{\mathrm{ab}}$ *and* $f, h \in T_a(H, \psi, \Lambda)$.
(2) $(f^x)^y = f^{xy}$ *for every* $x, y \in (K^*)_{\mathbf{A}}^\times$.
(3) $f_*(u)^x = (f^x)_*\big(g(x)^{-1}u\big)$ *for every* $u \in \mathbf{Q}\Lambda$.
(4) $f^x(u) = f\big(g(x)u\big)$ *if* $x \in K^{*\times}$.
(5) $f^x = f$ *if* $x \in (K^*)_{\mathbf{a}}^\times$.
(6) $\{x \in (K^*)_{\mathbf{A}}^\times \mid f^x = f\}$ *is an open subgroup of* $(K^*)_{\mathbf{A}}^\times$ *for every fixed* f.

This is essentially a reformulation, or a refinement, of Theorems 18.8 and 26.8(4). In fact, we can express the commutative diagram of Theorem 18.6(2) in terms of the projective embedding Θ of (27.12b) and its image $A(z, \delta)$. For the proof and such an expression, the reader is referred to [S76, pp. 686–688]. (See also [S77a, Remark 2.6], [S78b, Proposition 3.13], and also the paragraph after that proposition.) We note here also that employing the above theorem, we can give another proof of Theorem 21.4 as follows:

Take $Y = K$, $\psi = 1$, and take δ so that the entries of δ are divisible by an even integer > 2. The notation being as in Theorem 21.4, let T_b be the k_{ab}-linear span of $T_a(H, \psi, \Lambda)$. Given $y \in k_{\mathbf{A}}^\times$ and $f = \sum_h c_h h \in T_b$ with $c_h \in k_{\mathrm{ab}}$ and $h \in T_a(H, \psi, \Lambda)$, put $f^y(u) = \sum_h c_h^y h^x\big(\chi(y_{\mathbf{h}})^{-1}u\big)$ with $x = N_{k/K^*}(y)$. Then it can be seen that $f^y \in T_b$ and f^y depends only on f and $[y, k]$. Put $U = \{f \in T_b \mid f^y = f$ *for every* $y \in k_{\mathbf{A}}^\times\}$. Then $T_b = U \otimes_k k_{\mathrm{ab}}$. A basis of U gives a projective embedding of V/Λ, whose image is the desired model. This method works even when Y is not necessarily a field.

One can also describe the Hecke character determined by the model $A(w, \delta)$ when it belongs to a CM-type. For this, see Rumely [Ru83a, b].

30. The Periods of Differential Forms on Abelian Varieties

30.1. We are going to consider the periods of integrals of invariant 1-forms over 1-cycles on an abelian variety. Let $\mathcal{P} = (A, \mathcal{C})$ be an n-dimensional polarized abelian variety defined over a subfield k of $\mathbf{C}$. We denote by $H_1(A, \mathbf{Z})$ the first homology group of A with coefficients in $\mathbf{Z}$. We define $H_1(A, \mathbf{Q})$ and $H_1(A, \mathbf{R})$ in a similar fashion.

As remarked in the paragraph after the proof of Proposition 21.1 (or by Theorem 27.13), we can find a model of $\mathcal{P}$ rational over a finite algebraic extension of its field of moduli. Replacing $\mathcal{P}$ by this model, we may assume that k is such an extension. We denote by $\bar{k}$ the algebraic closure of k in $\mathbf{C}$. By Proposition 10 of §4.1, $\bar{k}$ depends only on the isogeny class of A.

Now let $\{ \eta_1, \dots, \eta_n \}$ be a k-rational basis of invariant 1-forms on A (see §§2.6 and 2.7), and $\{ c_1, \dots, c_{2n} \}$ a $\mathbf{Z}$-basis of $H_1(A, \mathbf{Z})$. Then we consider the following $n \times 2n$-matrix

$$\text{(30.1a)} \qquad \Pi = \begin{bmatrix} \int_{c_1} \eta_1 & \cdots & \int_{c_{2n}} \eta_1 \\ \cdots & \cdots & \cdots \\ \int_{c_1} \eta_n & \cdots & \int_{c_{2n}} \eta_n \end{bmatrix}.$$

We easily see that the coset in $GL_n(\bar{k})\backslash\mathbf{C}^n_{2n}/GL_{2n}(\mathbf{Z})$ represented by Π is completely determined by the isomorphism class of $\mathcal{P}$ independently of the choice of a model, $\{ \eta_i \}$, and $\{c_j\}$. (We can replace $GL_{2n}(\mathbf{Z})$ by a smaller group by assuming $\{c_j\}$ to be "normalized" with respect to the polarization. Also, if we take $GL_{2n}(\mathbf{Q})$ instead of $GL_{2n}(\mathbf{Z})$, then the coset is determined by the isogeny class of A.) We call this coset *the period coset of* $\mathcal{P}$. Then we ask:

(Q) *Can one determine the period coset of* $\mathcal{P}$ *in an explicit way?*

The purpose of this section is to answer this question by exhibiting a matrix belonging to the period coset. We need modular forms in the sense of (25.3b, f) with respect to the congruence subgroup Γ_δ of $Sp(n, \mathbf{Q})$ defined in (27.12c).

30.2. Lemma. *Define a representation* $\omega : GL_n(\mathbf{C}) \to GL(\mathbf{C}^n_n)$ *by*

$$\text{(30.2a)} \qquad \omega(x)y = xy \quad \textit{for} \quad x \in GL_n(\mathbf{C}),\ y \in \mathbf{C}^n_n.$$

Then the following assertions hold:

(1) *Given a point* $w \in \mathfrak{H}$ *and* δ *satisfying (27.12a), there exists an element* T *of* $\mathcal{A}_\omega(\Gamma_\delta, \mathbf{Q})$ *holomorphic at* w *such that* $\det[T(w)] \neq 0$.

(2) *If* $U \in \mathcal{A}_\omega(\mathbf{Q}_{ab})$ *and* $\alpha \in G_+$, *then* $U\|_\omega \alpha \in \mathcal{A}_\omega(\mathbf{Q}_{ab})$.

PROOF. With φ as in (27.4e) put

$$\psi(u, z; r, s) = \frac{1}{2\pi i}\begin{bmatrix} \partial\varphi/\partial u_1 \\ \vdots \\ \partial\varphi/\partial u_n \end{bmatrix}. \tag{30.2b}$$

From (27.8c) and (27.12d) we obtain

$$\begin{aligned} \text{(30.2c)} \quad \psi(u, \gamma z; r, s) &= \mathbf{e}\big(2^{-1}({}^t rs - {}^t r's')\big)h_\gamma(z)(c_\gamma z + d_\gamma) \\ &\quad \cdot \psi\big({}^t(c_\gamma z + d_\gamma)u, z; r', s'\big) \\ &\quad \textit{with } [{}^t r' \ \ {}^t s'] = [{}^t r \ \ {}^t s]\gamma \ \textit{ if } \ \gamma \in \Gamma_\theta, \\ \text{(30.2d)} \quad \psi(u, \gamma z; r, 0) &= h_\gamma(z)(c_\gamma z + d_\gamma)\psi\big({}^t(c_\gamma z + d_\gamma)u, z; r, 0\big) \\ &\quad \textit{if } \gamma \in \Gamma_\delta \ \textit{ and } \ r \in \delta^{-1}\mathbf{Z}^n. \end{aligned}$$

Assuming (27.12a), denote the elements of $R = \delta^{-1}\mathbf{Z}^n/\mathbf{Z}^n$ by $r_1, \ldots, r_m$, and put

$$\varphi_k(u, z) = \varphi(u, z; r_k, 0), \quad \psi_k(u, z) = \psi(u, z; r_k, 0) \quad (1 \le k \le m).$$

Since $u \mapsto \big(\varphi_k(u, z)\big)_{k=1}^m$ is the embedding of (27.12b) which is biregular, we have

$$\operatorname{rank}\begin{bmatrix} \varphi_1(u, z) & \cdots & \varphi_m(u, z) \\ \psi_1(u, z) & \cdots & \psi_m(u, z) \end{bmatrix} = n + 1$$

for every $(u, z) \in \mathbf{C}^n \times \mathfrak{H}$. Therefore, changing the order of r_k, we may assume that $\det\big[\psi_1(0, w) \ \cdots \ \psi_n(0, w)\big] \neq 0$. Also we can find an index j such that $\varphi_j(0, w) \neq 0$. Define T by

$$T(z) = \varphi_j(0, z)^{-1}\big[\psi_1(0, z) \ \cdots \ \psi_n(0, z)\big]. \tag{30.2e}$$

Then (30.2d) together with (27.12d) shows that $T \in \mathcal{A}_\omega(\Gamma_\delta)$. We easily see that $\psi_k(0, z)$ has Fourier coefficients in $\mathbf{Q}$. This proves (1).

Since $\varphi = \theta$ and $\partial\varphi/\partial u_i = \partial\theta/\partial u_i$ at $u = 0$, we have

$$(30.2\text{f}) \quad T(z) = \frac{1}{2\pi i \cdot \theta_j(0, z)} \begin{bmatrix} \partial\theta_1(u, z)/\partial u_1 & \cdots & \partial\theta_n(u, z)/\partial u_1 \\ \cdots & \cdots & \cdots \\ \partial\theta_1(u, z)/\partial u_n & \cdots & \partial\theta_n(u, z)/\partial u_n \end{bmatrix}_{u=0},$$

where $\theta_k(u, z) = \theta(u, z; r_k, 0)$.

Now (2) is clear if $\alpha \in P$. Therefore, by Lemma 28.4 it is sufficient to prove (2) for $\alpha = J$. Define T by (30.2e) or (30.2f) and put $S = T^{-1}U$. Then the entries of S belong to $\mathcal{A}_0(\mathbf{Q}_{\text{ab}})$, and $U\|_\omega J = (T\|_\omega J)(S \circ J)$. As remarked at the end of §27 (or by Theorem 26.8), the entries of $S \circ J$ belong to $\mathcal{A}_0(\mathbf{Q}_{\text{ab}})$. Thus our task is to show that $T\|_\omega J \in \mathcal{A}_\omega(\mathbf{Q}_{\text{ab}})$. But this follows immediately from (28.1e) and (30.2c) with $\gamma = J$.

30.3. Theorem. *Let $\mathcal{P} = (A, \mathcal{C})$ and k be as in §30.1. Suppose that A is isomorphic to $\mathbf{C}^n/\Lambda(w, \delta)$ with $w \in \mathfrak{H}_n$ and the matrix of elementary divisors δ of the Riemann form of a divisor in $\mathcal{C}$ (which may or may not satisfy (27.12a)). Then there exists an element S of $\mathcal{A}_\omega(\mathbf{Q})$ such that* $\det[S(w)] \neq 0$. *Moreover, with any such S define 1-forms $\xi_1, \ldots, \xi_n$ on $\mathbf{C}^n/\Lambda(w, \delta)$ by*

$$(30.3\text{a}) \qquad \begin{bmatrix} \xi_1 \\ \vdots \\ \xi_n \end{bmatrix} = 2\pi i \cdot {}^t S(w) \begin{bmatrix} du_1 \\ \vdots \\ du_n \end{bmatrix},$$

where $u_1, \ldots, u_n$ are the coordinate functions on $\mathbf{C}^n$. Then $\xi_1, \ldots, \xi_n$, viewed as 1-forms on A, form a basis of holomorphic 1-forms rational over $\overline{k}$. In particular, if $\mathcal{P} = A(w, \delta)$ with δ satisfying (27.12a), we can take S in $\mathcal{A}_\omega(\Gamma_\delta, \mathbf{Q})$. The ξ_k defined with such an S are rational over $\mathfrak{K}_\delta[w]$.

PROOF. We first assume that $\mathcal{P} = A(w, \delta)$ with δ satisfying (27.12a). The existence of $S \in \mathcal{A}_\omega(\Gamma_\delta, \mathbf{Q})$ is guaranteed by Lemma 30.2(1). Fixing an index j such that $\varphi_j(0, w) \neq 0$, put

$$(30.3\text{b}) \qquad q_k(u) = \varphi_k(u, w)/\varphi_j(u, w) \quad (1 \leq k \leq m)$$

and ${}^t\xi = (\xi_1 \ \cdots \ \xi_n)$ with ξ_ν as above. Let $\mathcal{F}$ denote the field of $\mathfrak{K}_\delta[w]$-rational functions on the variety $A(w, \delta)$ (cf. Theorem 27.13). Then the q_k are the coordinate functions of an affine Zariski open subset of $A(w, \delta)$ containing the origin, and $\mathcal{F}$ is generated by the q_k over $\mathfrak{K}_\delta[w]$. Observe that

$$(30.3\text{c}) \qquad dq_k = {}^t\sigma_k(u, w)\xi$$

with a vector-valued function $\sigma_k(u, z)$ defined by

$$\sigma_k(u, z) = \varphi_j(u, z)^{-2}\big[\varphi_j(u, z)S(z)^{-1}\psi_k(u, z) - \varphi_k(u, z)S(z)^{-1}\psi_j(u, z)\big].$$

Evaluating this at $u = 0$, we see from (30.2d) that the entries of $\sigma_k(0, z)$ belong to $\mathcal{A}_0(\Gamma_\delta, \mathbf{Q}) = \mathfrak{K}_\delta$, and hence their values at w, $\sigma_k(0, w)$, belong to $\mathfrak{K}_\delta[w]$. Let $\{\omega_1, \ldots, \omega_n\}$ be a basis of $\mathfrak{K}_\delta[w]$-rational holomorphic 1-forms on $A(w, \delta)$. Since we can choose a set of local parameters at the origin from the q_k (see the paragraph preceding Proposition 5 in §10.3), we have $\omega_i = \sum_{k=1}^m f_{ik} dq_k$ with $f_{ik} \in \mathcal{F}$ finite at the origin. Let σ_{kj} denote the j-th component of σ_k. Then $\omega_i = \sum_{j=1}^n g_{ij}\xi_j$ with $g_{ij} = \sum_{k=1}^m f_{ik}\sigma_{kj}(u, w)$. Since both ω_i and ξ_j are invariant 1-forms, the g_{ij} must be constants. Therefore $g_{ij} = \sum_{k=1}^m f_{ik}(0)\sigma_{kj}(0, w) \in \mathfrak{K}_\delta[w]$. Since the ξ_j are linearly independent, this proves that they are $\mathfrak{K}_\delta[w]$-rational. Now in the general case take any integer $\mu > 2$. Then $\mathcal{P}$ is isomorphic to $A(\mu w, \mu\delta)$. Take S as above for $A(\mu w, \mu\delta)$, and put $S'(z) = S(\mu z)$. Then $\det[S'(w)] \neq 0$, and $S'(w)^{-1}S''(w)$ has entries in $\mathfrak{K}[w]$ for every $S'' \in \mathcal{A}_\omega(\mathbf{Q}_{ab})$. Since $\mathfrak{K}[w] \subset \overline{k}$, we obtain the desired result in the general case.

We are now ready to answer question (Q) of §30.1.

30.4. Theorem. *Let $\mathcal{P}$, w, δ, and S be as in the first part of Theorem 30.3. Then the period coset of $\mathcal{P}$ is represented by the matrix $2\pi i \cdot {}^tS(w)[w \quad \delta]$.*

Proof. Define ξ_k by (30.3a). Let $\{c_1, \ldots, c_{2n}\}$ be the basis of $H_1(A, \mathbf{Z})$ corresponding to the standard basis of $\mathbf{Z}^{2n}$ through the map $\mathbf{R}^{2n} \to \mathbf{C}^n/\Lambda(w, \delta)$ given by $x \to [w \quad \delta]x$ for $x \in \mathbf{R}^{2n}$. Then clearly

$$\text{(30.4a)} \qquad \begin{bmatrix} \int_{c_1}\xi_1 & \cdots & \int_{c_{2n}}\xi_1 \\ \cdots & \cdots & \cdots \\ \int_{c_1}\xi_n & \cdots & \int_{c_{2n}}\xi_n \end{bmatrix} = 2\pi i \cdot {}^tS(w)[w \quad \delta],$$

which proves our theorem.

It should be noted that we can take S to be T of (30.2f). Then $S(w)$ can be given in terms of the values of classical theta functions and their derivatives at $u = 0$.

31. Periods in the Hilbert Modular Case

31.1. Let F be a totally real algebraic number field, and $\mathbf{a}$ the set of all archimedean primes of F as in Chapter VI. Putting $[F : \mathbf{Q}] = r$ and fixing an identification of $\mathbf{a}$ with $\{1, \ldots, r\}$, we hereafter identify $X^{\mathbf{a}}$ with X^r for any set X, and write $\{a_1, \ldots, a_r\}$ for $\{a_v\}_{v\in\mathbf{a}}$. (We do not discard the symbols $\mathbf{a}$ and $X^{\mathbf{a}}$; we shall use them whenever necessary.) Take a basis $\{\beta^1, \ldots, \beta^r\}$ of

F over $\mathbf{Q}$; define ε: $\mathfrak{H}_1^r \to \mathfrak{H}_r$ and τ: $SL_2(\mathbf{R})^r \to Sp(r, \mathbf{R})$ by

$$\varepsilon(z) = {}^tB \cdot \mathrm{diag}[z_1, \ldots, z_r]B \qquad (z \in \mathfrak{H}_1^r),$$

$$\tau(\alpha) = \begin{bmatrix} {}^tB & 0 \\ 0 & B^{-1} \end{bmatrix} \begin{bmatrix} \psi(a_\alpha) & \psi(b_\alpha) \\ \psi(c_\alpha) & \psi(d_\alpha) \end{bmatrix} \begin{bmatrix} {}^tB^{-1} & 0 \\ 0 & B \end{bmatrix} \qquad (\alpha \in SL_2(\mathbf{R})^r),$$

$$B = \begin{bmatrix} \beta_1^1 & \cdots & \beta_1^r \\ \cdots & \cdots & \cdots \\ \beta_r^1 & \cdots & \beta_r^r \end{bmatrix}, \qquad \psi(a) = \mathrm{diag}[a_1, \ldots, a_r].$$

It can easily be seen that $\tau\big(SL_2(F)\big) \subset Sp(r, \mathbf{Q})$, $\tau(\alpha)\varepsilon(z) = \tau(\alpha z)$, and

$$\mu\big(\tau(\alpha), \varepsilon(z)\big) = B^{-1}\,\mathrm{diag}[\mu(\alpha_1, z_1), \ldots, \mu(\alpha_r, z_r)]B,$$

and hence $j\big(\tau(\alpha), \varepsilon(z)\big) = j(\alpha, z)^u$, where u denotes the element of $\mathbf{Z}^r$ whose components are all equal to 1.

We now consider modular forms on $\mathfrak{H}_1^{\mathbf{a}}$ with respect to congruence subgroups of $SL_2(F)$. We can view each $v \in \mathbf{a}$ as the element of $\mathbf{Z}^{\mathbf{a}}$ whose v-component is 1 and other components are 0. Then $\mathcal{A}_v(D)$ for a subfield D of $\mathbf{C}$ can be defined as in §25.5. This consists of all quotients g_1/g_2 with $g_1 \in \mathcal{A}_{k+v}(D)$ and $0 \neq g_2 \in \mathcal{A}_k(D), k \in \mathbf{Z}^{\mathbf{a}}$. In general, if a function f on $\mathfrak{H}_r$ has an expansion $f(Z) = \sum_h c_h \mathbf{e}\big(\mathrm{tr}(hZ)\big)$ for $Z \in \mathfrak{H}_r$ as in (25.5a), then $f\big(\varepsilon(z)\big) = \sum_h c_h \mathbf{e}_{\mathbf{a}}(h_B z)$, where $h_B = \sum_{j,k} h_{jk}\beta_j\beta_k$, which belongs to F.

31.2. Proposition. *Given $z_0 \in \mathfrak{H}_1^{\mathbf{a}}$, there exists a set of functions $\{h_v\}_{v\in\mathbf{a}}$ with the following properties:*

(1) *$h_v \in \mathcal{A}_v(F^v)$, where F^v is the image of F under v.*
(2) *$h_v^\sigma = h_{v\sigma}$, where $v\sigma$ is defined for $\sigma \in \mathrm{Gal}(\overline{\mathbf{Q}}/\mathbf{Q})$ as in Proposition 25.8.*
(3) *h_v is holomorphic at z_0 and $h_v(z_0) \neq 0$ for every $v \in \mathbf{a}$.*

PROOF. Take T on $\mathfrak{H}_r$ as in Lemma 30.2(1) with $n = r$ and $w = \varepsilon(z_0)$, and put

$$\begin{bmatrix} h_1(z) \\ \vdots \\ h_r(z) \end{bmatrix} = BT\big(\varepsilon(z)\big)q \qquad (z \in \mathfrak{H}_1^{\mathbf{a}})$$

with $q \in \mathbf{Q}^r$. We can put $T = f^{-1}U$ with $0 \neq f \in \mathcal{M}_\kappa(\mathbf{Q}), 0 < \kappa \in \mathbf{Z}$, so that f and U have $\mathbf{Q}$-rational Fourier coefficients. Therefore, writing $\{h_v\}_{v\in\mathbf{a}}$ for $\{h_i\}$, we can easily verify that the h_v satisfy (1) and (2), and also (3) with a suitable choice of q.

31.3. Let $\mathcal{P} = (A, \mathcal{C}, \iota)$ be an r-dimensional polarized abelian variety such that ι is a ring-injection of F into $\mathrm{End}_{\mathbf{Q}}(A)$. As in §30.1 we denote by $\overline{k}$ the algebraic closure of the field of moduli of $\mathcal{P}$, and assume $\mathcal{P}$ to be rational over $\overline{k}$. Then we can find a $\overline{k}$-basis $\{\eta_v\}_{v\in\mathbf{a}}$ of invariant 1-forms on A such that $\delta\iota(a)\eta_v = a_v\eta_v$ for every $a \in F$ such that $\iota(a) \in \mathrm{End}(A)$. Now $\iota(F)$ acts on $H_1(A, \mathbf{Q})$. Taking an F-basis $\{c_1, c_2\}$ of $H_1(A, \mathbf{Q})$ in this sense, we consider an element p_v of $\mathbf{C}_2^1$ defined by

$$p_v = \left(\int_{c_1} \eta_v, \int_{c_2} \eta_v\right). \tag{31.3a}$$

Then the coset $\overline{k}^\times\backslash\mathbf{C}_2^1/GL_2(F^v)$ represented by p_v, where F^v is the image of F under v, is completely determined by the isogeny class of $\mathcal{P}$ independently of the choice of η_v and c_i. Therefore, as an analogue of question (Q) in §30.1, we can pose a question about the nature of this coset.

31.4. To answer this question, we consider families

$$\mathcal{F}(\Omega') = \{(A_z, \mathcal{C}_z, \iota_z) \mid z \in \mathfrak{H}_1^{\mathbf{a}}\}, \quad \Omega' = \{F, \psi, L, J_1\},$$

$$\mathcal{F}(\Omega) = \{(A_Z, \mathcal{C}_Z) \mid Z \in \mathfrak{H}_r\}, \quad \Omega = \{\mathbf{Q}, \Psi, \mathbf{Z}^{2r}, J_r\}$$

in the sense of (24.5e), where Ψ and ψ are "obvious" representations of $\mathbf{Q}$ and F determined by the dimensionality; L is a $\mathbf{Z}$-lattice in F_2^1. Recall that A_Z is isomorphic to $\mathbf{C}^r/[Z \quad 1]\mathbf{Z}^{2n}$ and A_z to $\mathbf{C}^r/p_z(L)$ with $p_z(x) = \left([z_v \quad 1]\cdot{}^t x_v\right)_{v\in\mathbf{a}}$ for $x \in F_2^1$ (see (24.5c)).

Let $L = \left\{(b, c) \in F_2^1 \mid b \in \mathfrak{b}, c \in \mathfrak{c}\right\}$ with $\mathfrak{b} = \sum_{k=1}^r \mathbf{Z}\beta^k$ and $\mathfrak{c} = \left\{c \in F \mid \mathrm{Tr}_{F/\mathbf{Q}}(c\mathfrak{b}) \subset \mathbf{Z}\right\}$. With this L we easily see that the matrix ${}^tB$ gives an isomorphism of $\mathbf{C}^r/p_z(L)$ onto $\mathbf{C}^r/[\varepsilon(z) \quad 1]\mathbf{Z}^{2n}$. Thus A_z is isomorphic to $A_{\varepsilon(z)}$. Let (V, φ) be a model of $\Gamma\backslash\mathfrak{H}_1^{\mathbf{a}}$ as in Theorem 24.9 with Γ of (24.6a). If $f \in \mathcal{A}_0(\Gamma, \overline{\mathbf{Q}})$, then $f = g \circ \varphi$ with a $\overline{\mathbf{Q}}$-rational function g on V by Theorem 26.4. Then, for $w \in \mathfrak{H}_1^{\mathbf{a}}$ the value $f(w)$, when finite, belongs to $\overline{\mathbf{Q}}(\varphi(w))$. By property (3) of Theorem 24.9 we see that $\overline{\mathbf{Q}}(\varphi(w)) \subset \overline{k}$. This shows that $f(w) \in \overline{k}$ for every $f \in \mathcal{A}_0(\Gamma, \overline{\mathbf{Q}})$ finite at w.

We now determine the quantity p_v of (31.3a).

31.5. Theorem. *Given (A, ι) and $\overline{k}$ as in §31.3, let $\mathcal{P}_w = (A_w, \mathcal{C}_w, \iota_w) \in \mathcal{F}(\Omega')$ with (A_w, ι_w) isogenous to (A, ι). (Clearly we can always find such a $w \in \mathfrak{H}_1^{\mathbf{a}}$.) Take for each $v \in \mathbf{a}$ an element $h \in \mathcal{A}_v(\overline{\mathbf{Q}})$ so that $h(w) \neq 0$, as guaranteed by Proposition 31.2. Then the coset of p_v of (31.3a) is represented by $\pi \cdot h(w)[w_v \quad 1]$.*

PROOF. Represent A_w as $\mathbf{C}^r/p_w(L)$ as above. Let $u_1, \ldots, u_r$ be the standard coordinate functions on $\mathbf{C}^r$. Take S as in Theorem 30.3 for the point $\varepsilon(w)$ on $\mathfrak{H}_r$. Put $Q(z) = \mathrm{diag}[h_1(z), \ldots, h_r(z)]^{-1}BS(\varepsilon(z))$ with $h_v \in \mathcal{A}_v(\overline{\mathbf{Q}})$ such that $h_v(w) \neq 0$. Putting $u' = {}^tBu$, we obtain

$$2\pi i \begin{bmatrix} h_1(z)du_1 \\ \vdots \\ h_r(z)du_r \end{bmatrix} = {}^tQ(z)^{-1}2\pi i \cdot {}^tS(\varepsilon(z)) \begin{bmatrix} du'_1 \\ \vdots \\ du'_r \end{bmatrix} = {}^tQ(z)^{-1} \begin{bmatrix} \xi_1 \\ \vdots \\ \xi_r \end{bmatrix}$$

with 1-forms ξ_i on $A_{\varepsilon(z)}$ as in Theorem 30.3. Now we see that the entries of Q as functions on $\mathfrak{H}_1^r$ belong to $\mathcal{A}_0(\overline{\mathbf{Q}})$, and their values at w belong to $\overline{k}$ as remarked above. Therefore $\pi h_v(z)du_v$ defines a $\overline{k}$-rational form on A_w. Since $\int_c du_v \in F^v w_v + F^v$ for every $c \in H_1(A_w, \mathbf{Z})$, we obtain our theorem.

In the one-dimensional case the above theorem can be stated as follows:

31.6. Corollary. *Let E be an elliptic curve with invariant j isomorphic to $\mathbf{C}/[\mathbf{Z}\tau + \mathbf{Z}]$ with $\tau \in \mathfrak{H}_1$, and h an element of $\mathcal{M}_1(\mathbf{Q})$ such that $h(\tau) \neq 0$; denote by $\overline{k}$ the algebraic closure of $\mathbf{Q}(j)$ in $\mathbf{C}$. If E is $\overline{k}$-rational and η is a $\overline{k}$-rational holomorphic 1-form on E, then $\int_c \eta \in \pi h(\tau)\overline{k}(\mathbf{Z}\tau + \mathbf{Z})$ for every $c \in H_1(E, \mathbf{Z})$.*

We conclude this section by noting that *if A is simple, then $\pi h(w)$ of Theorem 31.5 is transcendental.* This follows from Lang [L, Theorem 1], which generalizes earlier results of Siegel and Schneider.

32. Periods on Abelian Varieties with Complex Multiplication and Their Algebraic Relations

32.1. For $a, b \in \mathbf{C}$ we write $a \sim b$ if $b \neq 0$ and $a/b \in \overline{\mathbf{Q}}^\times$. If b belongs to a coset c in $\mathbf{C}^\times/\overline{\mathbf{Q}}^\times$, we also write $a \sim c$ if $a \sim b$. Thus $a \sim 1$ if $a \in \overline{\mathbf{Q}}$.

Given a CM-field K, we denote by J_K the set of all embeddings of K into $\mathbf{C}$, and by id_K the identity embedding of K into $\mathbf{C}$. If (K, Φ) is a CM-type, then we naturally view Φ as a subset of J_K. To make our exposition easier, we assume that every CM-field in this section is a subfield of $\mathbf{C}$. We always denote complex conjugation by ρ. We also remind the reader of the fact that the composite of finitely many CM-fields is a CM-field, as noted in Lemma 18.2.

32.2. Theorem. *To each CM-type (K, Φ) and $\varphi \in \Phi$ we can associate an element $p_K(\varphi, \Phi)$ of $\mathbf{C}^\times/\overline{\mathbf{Q}}^\times$ with the following properties:*

(1) *If (A, ι) is of type (K, Φ) and rational over $\overline{\mathbf{Q}}$, and η is a nonzero $\overline{\mathbf{Q}}$-rational invariant 1-form on A such that $\delta\iota(a)\eta = a^\varphi\eta$ for every $a \in K$ with $\iota(a) \in \mathrm{End}(A)$, then $\int_c \eta \sim \pi \cdot p_K(\varphi, \Phi)$ for every $c \in H_1(A, \mathbf{Z})$.*

(2) *Suppose K has F of §31.1 as the maximal real subfield; write $\Phi = \{\varphi_v\}_{v\in\mathbf{a}}$ with $\varphi_v \in J_K$ which coincides with v on F. Let $w = \left(w_0^{\varphi_v}\right)_{v\in\mathbf{a}}$ with $w_0 \in K$ such that $\mathrm{Im}\left(w_0^{\varphi_v}\right) > 0$ for every $v \in \mathbf{a}$. Then $f(w) \sim p_K(\varphi_v, \Phi)$ for every $f \in \mathcal{A}_v(\overline{\mathbf{Q}})$ finite at w.*

PROOF. We let $\iota(K)$ act on $H_1(A, \mathbf{Q})$ and observe that $H_1(A, \mathbf{Q})$ is one-dimensional over K. Taking any nonzero element c of $H_1(A, \mathbf{Q})$, we put $q = \int_c \eta$ and observe that the coset of q in $\mathbf{C}^\times/\overline{\mathbf{Q}}^\times$ is determined by the isogeny class of (K, Φ) and φ. (Recall that any two structures of the same CM-type are isogenous.) Notice that $q \neq 0$ since all the transforms of c under $\mathrm{End}(A) \cap \iota(K)$ span $H_1(A, \mathbf{R})$ over $\mathbf{R}$. Thus the coset of $\pi^{-1}q$ in $\mathbf{C}^\times/\overline{\mathbf{Q}}^\times$ can be taken as $p_K(\varphi, \Phi)$ with the property of (1). Next, if w is as in (2), (A_w, ι_w) is of type (K, Φ) by Proposition 24.13. Take $\varphi = \varphi_v$ with a fixed $v \in \mathbf{a}$ and let $f \in \mathcal{A}_v(\overline{\mathbf{Q}})$. Since $w_v \in \overline{\mathbf{Q}}$, Theorem 31.5 shows that $q \sim \pi \cdot h(w)$ with h in that theorem. Hence $\pi q^{-1} f(w) \sim (f/h)(w) \sim 1$, since $f/h \in \mathcal{A}_0(\overline{\mathbf{Q}})$ and its value at a CM-point is algebraic by virtue of Theorem 26.5. This completes the proof.

32.3. With J_K as in §32.1, we denote by I_K the free $\mathbf{Z}$-module generated by the elements of J_K. Let L be a CM-field containing K. For $\alpha \in J_K$ and $\beta \in J_L$ we denote by $\mathrm{Inf}_{L/K}(\alpha)$ the sum of all the elements of J_L which coincide with α on K, and by $\mathrm{Res}_{L/K}(\beta)$ the restriction of β to K. We then extend these to additive maps

(32.3a) $$\mathrm{Inf}_{L/K}\colon I_K \to I_L, \qquad \mathrm{Res}_{L/K}\colon I_L \to I_K.$$

The above theorem establishes the invariant $p_K(\tau, \Phi)$ for every $\tau \in \Phi$. Define now $p_K(\tau, \Phi) \in \mathbf{C}^\times/\overline{\mathbf{Q}}^\times$ for $\tau \notin \Phi$ by putting $p_K(\tau, \Phi) = p_K(\tau\rho, \Phi)^{-1}$. We can thus speak of $p_K(\tau, \Phi)$ for every $\tau \in J_K$. We then define $p_K(\xi, \Phi)$ for every $\xi = \sum_{\tau\in J_K} c_\tau \tau \in I_K$ by $p_K(\xi, \Phi) = \prod_{\tau\in J_K} p_K(\tau, \Phi)^{c_\tau}$. We identify Φ with the element of I_K which is the sum of the members of Φ. From this definition and Theorem 32.2(2) we obtain the following result: *If (K, Φ) and w are as in Theorem 32.2(2) and $k = \sum_{v\in\mathbf{a}} k_v v \in \mathbf{Z}^{\mathbf{a}}$, then*

(32.3b) $$f(w) \sim p_K\left(\sum_{v\in\mathbf{a}} k_v\varphi_v, \Phi\right) \quad \textit{for every } f \in \mathcal{A}_k(\overline{\mathbf{Q}}) \textit{ finite at } w.$$

In fact, take $h_v \in \mathcal{A}_v(\overline{\mathbf{Q}})$ so that $h_v(w) \neq 0$, as guaranteed by Proposition 31.2. Then $f\prod_{v\in\mathbf{a}} h_v^{-k_v} \in \mathcal{A}_0(\overline{\mathbf{Q}})$, and its value at w is algebraic. Therefore we obtain (32.3b).

We are going to extend p_K to a map of $I_K \times I_K$ into $\mathbf{C}^\times/\overline{\mathbf{Q}}^\times$, which we call *the period symbol*. To that end, we first state a theorem which is fundamental in our theory:

32.4. Theorem. (1) *Let* (K, Φ_i) *for* $1 \le i \le m$ *be CM-types with the same* K*, and let* $s_1, \ldots, s_m$ *be integers. Then for each fixed* $\tau \in J_K$ *the product* $\prod_{i=1}^{n} p_K(\tau, \Phi_i)^{s_i}$ *depends only on* τ *and the element* $\sum_{i=1}^{m} s_i\Phi_i$ *of* I_K.

(2) *The notation being as in (1), let* (Y, ψ) *be a CM-type such that* $K \subset Y$ *and* $\mathrm{Res}_{Y/K}(\psi) = \sum_{i=1}^{m} s_i\Phi_i$. *Then* $p_Y(\mathrm{Inf}_{Y/K}(\tau), \psi) = \prod_{i=1}^{n} p_K(\tau, \Phi_i)^{s_i}$ *for every* $\tau \in J_K$.

(3) *Let* (K, φ) *and* (L, ψ) *be CM-types such that* $K \subset L$ *and* $Res_{L/K}(\psi) = [L : K]\varphi$. *Then* $p_L(\beta, \psi) = p_K(Res_{L/K}(\beta), \varphi)$ *for every* $\beta \in J_L$.

(4) *If* γ *is an isomorphism of* K' *onto* K *and* (K, φ) *is a CM-type, then* $p_{K'}(\gamma\tau, \gamma\varphi) = p_K(\tau, \varphi)$ *for every* $\tau \in J_K$.

PROOF. The last assertion is obvious. To prove (3), take $\overline{\mathbf{Q}}$-rational (A, ι) and (A', ι') of type (K, φ) and (L, ψ), respectively. By Theorem 24.15 (or by Theorem 3 of §6.2 and its proof) there is an isogeny λ of A^m onto A', where $m = [L : K]$. Given $\beta \in \psi$, let η be a $\overline{\mathbf{Q}}$-rational 1-form on A' such that $\delta\iota(a)\eta = a^{\beta}\eta$ for every $a \in L$. Put $\alpha = \mathrm{Res}_{L/K}(\beta)$ and $\delta\lambda\eta = \xi_1 + \cdots + \xi_m$ with a 1-form ξ_i on the i-th factor A of A^m for each i. Clearly $\delta\iota(a)\xi_i = a^{\alpha}\xi_i$ for every i and every $a \in K$. Now assertion (3) follows easily from Theorem 32.2(1).

Assertions (1) and (2) will be proved in Section 33. Now the principal result of this section concerns various properties of the period symbol p_K:

32.5. Theorem. *There exists a bilinear map* $p_K\colon I_K \times I_K \to \mathbf{C}^\times/\overline{\mathbf{Q}}^\times$ *defined for every CM-field* K *with the following properties:*

(1) $p_K(\tau, \varphi)$ *is defined as above if* (K, φ) *is a CM-type.*
(2) $p_K(\xi\rho, \eta) = p_K(\xi, \eta\rho) = p_K(\xi, \eta)^{-1}$ *for every* $\xi, \eta \in I_K$.
(3) $p_K(\xi, \mathrm{Res}_{L/K}(\zeta)) = p_L(\mathrm{Inf}_{L/K}(\xi), \zeta)$ *if* $\xi \in I_K$, $\zeta \in I_L$, *and* $K \subset L$.
(4) $p_K(\mathrm{Res}_{L/K}(\zeta), \xi) = p_L(\zeta, \mathrm{Inf}_{L/K}(\xi))$ *if* $\xi \in I_K$, $\zeta \in I_L$, *and* $K \subset L$.
(5) $p_{K'}(\gamma\xi, \gamma\eta) = p_K(\xi, \eta)$ *if* γ *is an isomorphism of* K' *onto* K.

PROOF. Let I_K^0 denote the submodule of I_K consisting of the elements of the form $\sum_i s_i\Phi_i$ as in Theorem 32.4(1). An element $\sum_{\tau \in J_K} c_\tau \tau$ belongs to I_K^0 if and only if $c_\tau + c_{\tau\rho}$ does not depend on τ. Then we define $p_K(\xi, \eta)$ for $\xi \in I_K$ and $\eta \in I_K^0$ by $p_K(\xi, \sum_i s_i\Phi_i) = \prod_i p_K(\xi, \Phi_i)^{s_i}$, which is well defined by virtue of Theorem 32.4(1). Then from Theorem 32.4 we can easily derive properties (3), (4), and (5) for $p_K(\xi, \eta)$ when $\eta \in I_K^0$. Observing that $\zeta - \zeta\rho \in I_K^0$ for every $\zeta \in I_K$, we put $p_K(\xi, \zeta) = p_K(\xi, \zeta - \zeta\rho)^{1/2}$. Then all the properties of p_K on $I_K \times I_K$ can be verified in a straightforward way.

32.6. Example. The above two theorems show that there are many nontrivial algebraic relations among the periods on abelian varieties which are not isogenous. For example, let K be an imaginary cyclic extension of $\mathbf{Q}$ of degree 10 and M the imaginary quadratic field contained in K. Define three CM-types (K, φ_i) for $i = 1, 2, 3$ by

$$\varphi_1 = \{1, \sigma, \sigma^2, \sigma^3, \sigma^4\},$$
$$\varphi_2 = \{1, \sigma, \sigma^2, \sigma^8, \sigma^4\},$$
$$\varphi_3 = \{1, \sigma^2, \sigma^4, \sigma^6, \sigma^8\},$$

where σ is a generator of $\mathrm{Gal}(K/\mathbf{Q})$. Clearly (K, φ_1) and (K, φ_2) are primitive but (K, φ_3) is not. Also, two abelian varieties belonging to (K, φ_1) and (K, φ_2) are not isogenous, since there is no automorphism of K which sends φ_1 to φ_2. Since $\varphi_1 + \varphi_3 = \varphi_2 + \sigma^2\varphi_2$, we have

$$p_K(\tau, \varphi_1)p_K(\tau, \varphi_3) = p_K(\tau, \varphi_2)p_K(\sigma^{-2}\tau, \varphi_2)$$

and $p_K(\tau, \varphi_3) = p_M\big(\mathrm{Res}_{K/M}(\tau), \mathrm{id}_M\big)$ for every $\tau \in J_K$. One can of course produce many more examples in a similar fashion. Theorem 32.8 below gives another type of example.

32.7. Let M be a CM-field contained in $\mathbf{C}$ which is normal over $\mathbf{Q}$, and let $G = \mathrm{Gal}(M/\mathbf{Q})$. Every $\delta \in I_M$ can be written $\delta = \sum_{\gamma\in G} c_\gamma \gamma$ with $c_\gamma \in \mathbf{Z}$. We then put $\delta^* = \sum_{\gamma\in G} c_\gamma \gamma^{-1}$. Now let K and L be CM-fields contained in $\mathbf{C}$, and let $\xi \in I_K$ and $\eta \in I_L$. We call (K, ξ) and (L, η) a *reflexive pair* if $\mathrm{Inf}_{M/K}(\xi)^* = \mathrm{Inf}_{M/L}(\eta)$ for a CM-field M which is normal over $\mathbf{Q}$ and contains both K and L. It can easily be seen that if this is so for *one* M, then it holds for *every* M. Let $\mathrm{Gal}(M/\mathbf{Q})$ act on I_K by right multiplication. If (K, ξ) and (L, η) form a reflexive pair, we see that $\eta\beta$ for $\beta \in \mathrm{Gal}(M/\mathbf{Q})$ depends only on η and $\mathrm{Res}_{M/K}(\beta)$. Therefore we write $\eta\alpha$ for $\eta\beta$ if $\alpha = \mathrm{Res}_{M/K}(\beta)$. Given $\xi \in I_K$, there is a unique CM-field P and a unique $\eta_0 \in I_P$ with the property that (K, ξ) and (L, η) form a reflexive pair if and only if $P \subset L$ and $\eta = \mathrm{Inf}_{L/P}(\eta_0)$. We call then (P, η_0) *the reflex of* (K, ξ). If (K, ξ) is a CM-type, this coincides with the reflex of (K, ξ) defined in §8.3. For the treatment of this topic for arbitrary number fields, we refer the reader to [S70, §1].

Given a CM-field K and $\xi \in I_K$, we denote by $T(\xi)$ the $\mathbf{Z}$-submodule of I_K generated by $\xi\gamma$ for all $\gamma \in \mathrm{Gal}(M/\mathbf{Q})$, and by $r(\xi)$ its rank, where M is any CM-field which is normal over $\mathbf{Q}$ and contains K. Clearly $r(\xi)$ is determined independently of the choice of M.

32.8. Theorem. *If (K, ξ) and (L, η) form a reflexive pair, then $p_K(\alpha, \xi) = p_L(\eta\alpha, \mathrm{id}_L)$ for every $\alpha \in J_K$ and $p_K(\mathrm{id}_K, \xi\beta) = p_L(\eta, \beta)$ for every $\beta \in J_L$.*

PROOF. Take M as above. By Theorem 32.5(5) we have $p_M(\beta, \gamma) = p_M(\gamma^{-1}\beta, \mathrm{id}_M)$ for $\beta, \gamma \in \mathrm{Gal}(M/\mathbf{Q})$, so that $p_M(\beta, \zeta) = p_M(\zeta^*\beta, \mathrm{id}_M)$

for every $\zeta \in I_M$. If $\alpha = \mathrm{Res}_{M/K}(\beta)$, then from Theorem 32.5(3, 4) we obtain

$$\begin{aligned} p_K(\alpha, \xi) &= p_M\big(\beta, \mathrm{Inf}_{M/K}(\xi)\big) = p_M\big(\mathrm{Inf}_{M/K}(\xi)^*\beta, \mathrm{id}_M\big) \\ &= p_M\big(\mathrm{Inf}_{M/L}(\eta)\beta, \mathrm{id}_M\big) = p_L(\eta\alpha, \mathrm{id}_L). \end{aligned}$$

The second formula can be proved in the same manner.

32.9. Proposition. *If (K, ξ) and (L, η) form a reflexive pair, then $r(\xi) = r(\eta)$.*

PROOF. Take M containing both K and L. Since $r(\xi) = r\big(\mathrm{Inf}_{M/K}(\xi)\big)$, it is sufficient to prove $r(\zeta) = r(\zeta^*)$ for $\zeta \in I_M$. Put $G = \mathrm{Gal}(M/\mathbf{Q})$, $\zeta = \sum_{\gamma\in G} c(\gamma)\gamma$, $\zeta\alpha = \sum_{\gamma\in G} m(\alpha, \gamma)\gamma$, and $\zeta^*\beta = \sum_{\gamma\in G} m^*(\gamma, \beta)\gamma$ for $\alpha, \beta \in G$ with a fixed ζ. Then $m(\alpha, \beta) = c(\beta\alpha^{-1}) = m^*(\alpha, \beta)$, so that $r(\zeta) = \mathrm{rank}\,\big(m(\alpha, \beta)\big)_{\alpha,\beta} = \mathrm{rank}\,\big(m^*(\alpha, \beta)\big)_{\alpha,\beta} = r(\zeta^*)$ as expected.

32.10. Proposition. *Let (K, φ) be a CM-type and (L, ψ) its reflex. Then $r(\varphi - \varphi\rho) = r(\varphi) - 1 \le (1/2)\mathrm{Min}\big([K : \mathbf{Q}], [L : \mathbf{Q}]\big)$.*

PROOF. Clearly we may assume that $\mathrm{id}_K \in \varphi$. Take M and G as above; put $H = \mathrm{Gal}(M/K)$ and $H' = \mathrm{Gal}(M/L)$. Then $\varphi\gamma = \varphi$ for $\gamma \in H'$ and $\psi\delta = \psi$ for $\delta \in H$. Put $\omega = \sum_{\tau\in J_K} \tau$. Since $\omega = \varphi\gamma + \varphi\gamma\rho$ for every $\gamma \in G$, we see that $T(\varphi)$ is spanned by ω and $\varphi\gamma$ with $\gamma \in (H' \cup H'\rho)\backslash G$. Therefore $r(\varphi) \le 1 + [L : \mathbf{Q}]/2$. Similarly $r(\psi) \le 1 + [K : \mathbf{Q}]/2$. Since $r(\varphi) = r(\psi)$ by Proposition 32.9, we obtain the last inequality of our proposition. To prove the remaining part, define $f\colon T(\varphi) \to T(\varphi - \varphi\rho)$ by $f(\zeta) = \zeta - \zeta\rho$ for $\zeta \in T(\varphi)$. Clearly f is surjective and $\omega \in \mathrm{Ker}(f)$. For every $\zeta \in T(\varphi)$ we can put $\zeta + \zeta\rho = m\omega$ with $m \in \mathbf{Z}$. If $f(\zeta) = 0$, then $2\zeta = m\omega$ and hence $\mathrm{Ker}(f) = \mathbf{Z}\omega$. Therefore we obtain $r(\varphi - \varphi\rho) = r(\varphi) - 1$.

32.11. Proposition. *Let $\big(K\colon \{\tau_i\}_{i=1}^n\big)$ be a CM-type with $[K : \mathbf{Q}] = 2n$. Then, for every $\xi \in I_K$, the module*

$$\left\{ (x_1, \ldots, x_n) \in \mathbf{Z}^n \;\middle|\; \prod_{i=1}^n p_K(\tau_i, \xi)^{x_i} = 1 \right\}$$

has rank at least $n - r(\xi - \xi\rho)$. In particular, it has rank at least $n + 1 - r(\xi)$ if (K, ξ) is a CM-type.

PROOF. Let (L, η) be the reflex of (K, ξ). Then $(L, \eta - \eta\rho)$ and $(K, \xi - \xi\rho)$ form a reflexive pair. Observe that $T(\eta - \eta\rho)$ is generated by $(\eta - \eta\rho)\tau_i$ for $i = 1, \ldots, n$. Hence $r(\eta - \eta\rho) \le n$. By Theorem 32.8 we have $p_K(\tau_i, \xi)^2 =$

$p_K(\tau_i, \xi-\xi\rho) = p_L((\eta-\eta\rho)\tau_i, \mathrm{id}_L)$. Therefore if $\sum_{i=1}^n x_i(\eta-\eta\rho)\tau_i = 0$, then $\prod_{i=1}^n p_K(\tau_i, \xi)^{x_i} = 1$. This proves the first assertion, since $r(\xi - \xi\rho) = r(\eta - \eta\rho)$ by Proposition 32.9. The last assertion then follows from Proposition 32.10.

If (K, ξ) and (L, η) are CM-types which form a reflexive pair and $2n = [K : \mathbf{Q}] \geq [L : \mathbf{Q}] = 2m$, then $n + 1 - r(\xi) \geq n - m$ by Propositions 32.9 and 32.10. It can often happen that $n > m$. In fact, for every positive integer m we can find an example of a reflexive pair of CM-types such that $n = 2^{m-1}$ (see [S70, (1.10.1)]). Anyway, if $n > m$, we have a nontrivial relation among the $p_K(\tau_i, \xi)$.

Sections 30 through 33 are reorganized and simplified versions of relevant portions of [S77a], [S79], and [S80], though other related topics are investigated in those papers. The period symbol p_K is closely connected with the critical values of certain L-functions. We give here a statement in the easiest case:

32.12. Theorem. *Let $(K : \{\varphi_v\}_{v\in\mathbf{a}})$ be a CM-type as in Theorem 32.2(2) and χ a Hecke character of K such that $\chi(x) = \prod_{v\in\mathbf{a}}(x_v/|x_v|)^{t_v}$ for every $x \in K_\mathbf{a}^\times$ with $t = (t_v)_{v\in\mathbf{a}} \in \mathbf{Z}^\mathbf{a}$, $t_v \geq 0$ for every $v \in \mathbf{a}$, where the v-component x_v of x is defined so that $x_v = x^{\varphi_v}$ if $x \in K$, and $|x_v|$ is the standard absolute value of x_v in $\mathbf{C}$. Then*

$$L(m/2, \chi) \sim \pi^{e/2} p_K\left(\sum_{v\in\mathbf{a}} t_v\varphi_v, \Phi\right),$$

for every integer m such that $m - t_v \in 2\mathbf{Z}$ and $-t_v < m \leq t_v$ for every $v \in \mathbf{a}$, where $e = m[F : \mathbf{Q}] + \sum_{v\in\mathbf{a}} t_v$.

The reader is referred to [S78a, §5] for the proof and other related facts, and to [S75a], [S80], [S88], and [Bl], as well as other articles cited in those papers, for further ramifications of this subject.

33. *Proof of Theorem 32.4*

33.1. We consider $\Omega = \{K, \Psi, L, T, \{u_i\}_{i=1}^r\}$ in Case U as in (24.1a, b, c) under (24.5f). Thus we have a CM-type $(K, \{\tau_v\}_{v\in\mathbf{a}})$ as in §23.4 such that $2d = m[K : \mathbf{Q}]$, and $\mathcal{H} = \prod_{v\in\mathbf{a}} \mathfrak{B}(r_v, s_v)$ with $r_v + s_v = m$ for all $v \in \mathbf{a}$; also $\Psi(a) = \mathrm{diag}\left[\Psi_v(a)\right]_{v\in\mathbf{a}}$ and $\Psi_v(a) = \mathrm{diag}[a_v 1_{r_v}, \bar{a}_v 1_{s_v}]$ for $a \in K$. We are going to define an embedding ε: $\mathcal{H} \to \mathfrak{H}_d$. Since $(x, y) \mapsto \mathrm{Tr}_{K/\mathbf{Q}}(xTy^*)$ with T of (24.1f) is a nondegenerate alternating form, we can find a $\mathbf{Q}$-linear map g: $K_m^1 \to \mathbf{Q}_{2d}^1$ so that

(33.1a) $$\mathrm{Tr}_{K/\mathbf{Q}}(xTy^*) = g(x)J_d \cdot {}^t g(y) \qquad (x, y \in K_m^1).$$

Let $\{ e_k \}_{k=1}^{2d}$ be the $\mathbf{Q}$-basis of K_m^1 such that $\{ g(e_k) \}_{k=1}^{2d}$ is the standard basis of $\mathbf{Q}_{2d}^1$. Given $z \in \mathcal{H}$, we consider the map p_z: $(K_{\mathbf{a}})_m^1 \to \mathbf{C}^d$ of (24.5a, b) and put $x_k = p_z(e_k)$. Now we can take $(\mathbf{Q}, p_z \circ g^{-1}, J_d)$ as (W, q, T) in §§24.1 and 24.2, and we can write (24.2a) and (24.4c) in the form

$$(33.1\text{b}) \qquad X\big(p_z \circ g^{-1}\big) = \begin{bmatrix} x_1 & \cdots & x_{2d} \\ \overline{x}_1 & \cdots & \overline{x}_{2d} \end{bmatrix} = \begin{bmatrix} \kappa & 0 \\ 0 & \overline{\kappa} \end{bmatrix} \begin{bmatrix} Z & 1_d \\ \overline{Z} & 1_d \end{bmatrix}$$

with $\kappa \in GL_d(\mathbf{C})$ and $Z \in \mathfrak{H}_d$. Put $\kappa = \kappa(z)$ and $Z = \varepsilon(z)$. Then

$$(33.1\text{c}) \quad \varepsilon(z) = \kappa(z)^{-1} \begin{bmatrix} x_1 & \cdots & x_d \end{bmatrix}, \quad \kappa(z) = \begin{bmatrix} x_{d+1} & \cdots & x_{2d} \end{bmatrix}.$$

Since p_z is holomorphic in z, as noted in §24.5, we see that both $\kappa(z)$ and $\varepsilon(z)$ are holomorphic in z. It should be remembered that $\kappa(z)$ is invertible for every $z \in \mathcal{H}$. Now $p_z\big(\sum_{k=1}^m c_k e_k\big) = \kappa(z)[\varepsilon(z) \quad 1]c$ for every $c \in \mathbf{Q}^{2d}$, that is,

$$(33.1\text{d}) \qquad p_z(x) = \kappa(z)[\varepsilon(z) \quad 1] \cdot {}^t g(x) \qquad (x \in K_m^1).$$

Denote by $U(T)$ the group of (23.4c). For $\alpha \in K_m^m$ define $\tilde{\alpha} \in \mathbf{Q}_{2d}^{2d}$ by $g(x\alpha) = g(x)\tilde{\alpha}$. From (33.1a) we see that $\tilde{\alpha} \in Sp(d, \mathbf{Q})$ if $\alpha \in U(T)$. Also, from (24.7c) and (33.1c) we obtain

$$\begin{aligned} {}^t M(\alpha, z)\kappa(\alpha z)[\varepsilon(\alpha z) \quad 1] \cdot {}^t g(x) &= {}^t M(\alpha, z) p_{\alpha z}(x) = p_z(x\alpha) \\ &= \kappa(z)[\varepsilon(z) \quad 1] \cdot {}^t \tilde{\alpha} \cdot {}^t g(x) \\ &= \kappa(z) \cdot {}^t \mu\big(\tilde{\alpha}, \varepsilon(z)\big)[\tilde{\alpha}(\varepsilon(z)) \quad 1] \cdot {}^t g(x). \end{aligned}$$

This proves that $\varepsilon(\alpha z) = \tilde{\alpha}\big(\varepsilon(z)\big)$ and

$$(33.1\text{e}) \quad \mu\big(\tilde{\alpha}, \varepsilon(z)\big) = {}^t\kappa(\alpha z) M(\alpha, z) \cdot {}^t\kappa(z)^{-1} \qquad (\alpha \in U(T), z \in \mathcal{H}).$$

33.2. Lemma. *If w is a CM-point on $\mathcal{H}$, then the entries of $X_v(p_w)$, w_v, $\kappa(w)_v$ for every $v \in \mathbf{a}$ and $\varepsilon(w)$ are all algebraic.*

PROOF. The point w can be obtained as the unique fixed point of $h(\alpha)$ with some $\alpha \in Y^u$ as shown in §24.14, where Y^u is as in §24.10. Since we took Q_v of (23.4h) to be algebraic, the action of $h(\alpha)$ on $\mathcal{H}$ is $\overline{\mathbf{Q}}$-rational. (Notice that $\mathcal{H}$ has an obvious $\overline{\mathbf{Q}}$-rational structure.) Therefore the fixed point w_v of $Q_v^{-1} h(\alpha)_v^\rho Q_v$ has algebraic entries. Then from (24.5a) and the formulas in §33.1 we immediately see the algebraicity of the other quantities in question.

33.3. With Γ as in (24.6a) in Case U, let (V, φ) be a model of $\Gamma \backslash \mathcal{H}$ as in Theorem 24.9. As explained in §25.4, we can identify $\mathcal{A}_0(\Gamma)$ with the function field of V. Let $\mathcal{A}_0(\Gamma, \overline{\mathbf{Q}})$ denote the set of all functions of the form $f \circ \varphi$ with a $\overline{\mathbf{Q}}$-rational function f on V, which is meaningful since V is defined over $\overline{\mathbf{Q}}$. Then let $\mathcal{A}_0(\overline{\mathbf{Q}})$ denote the union of the $\mathcal{A}_0(\Gamma, \overline{\mathbf{Q}})$ for all such Γ. To avoid confusion, we hereafter denote by $\mathfrak{A}_0(\overline{\mathbf{Q}})$ the field $\mathcal{A}_0(\overline{\mathbf{Q}})$ defined on $\mathfrak{H}_d$ with respect to $Sp(d, \mathbf{Q})$.

Before stating the next proposition, we recall the simple fact that if w is a CM-point, then $\alpha(w)$ is also a CM-point for every $\alpha \in U(T)$, as remarked in §24.10.

33.4. Proposition. *The field $\mathcal{A}_0(\overline{\mathbf{Q}})$ has the following properties:*

(1) *If $f \in \mathcal{A}_0(\overline{\mathbf{Q}})$ and $\alpha \in U(T)$, then $f \circ \alpha \in \mathcal{A}_0(\overline{\mathbf{Q}})$.*
(2) *Let $\mathcal{W}$ be a set of CM-points on $\mathcal{H}$ which is dense in $\mathcal{H}$. If $f \in \mathcal{A}_0(\mathbf{C})$, then $f \in \mathcal{A}_0(\overline{\mathbf{Q}})$ if and only if $f(w) \in \overline{\mathbf{Q}}$ for every CM-point $w \in \mathcal{W}$ where f is finite.*
(3) *If $g \in \mathfrak{A}_0(\overline{\mathbf{Q}})$ and $g \circ \varepsilon$ is finite, then $g \circ \varepsilon \in \mathcal{A}_0(\overline{\mathbf{Q}})$.*

PROOF. With (V, φ) as in §33.3 and $\mathcal{W}$ as in (2), let $f = g \circ \varphi$ with an element g of the function field of V over $\mathbf{C}$, and let $\mathcal{W}'$ be the subset of $\mathcal{W}$ consisting of the points where f is defined. Take a field of rationality k for g containing $\overline{\mathbf{Q}}$ and take also an isomorphism σ of k onto a subfield of $\mathbf{C}$ over $\overline{\mathbf{Q}}$. Suppose $f(w) \in \overline{\mathbf{Q}}$ for every $w \in \mathcal{W}'$. Now $\varphi(w)$ is $\overline{\mathbf{Q}}$-rational by Theorem 24.11, and hence $g^\sigma\big(\varphi(w)\big) = g\big(\varphi(w)\big)^\sigma = f(w)^\sigma = f(w) = g\big(\varphi(w)\big)$. Since $\varphi(\mathcal{W}')$ is dense in V, we obtain $g^\sigma = g$, that is, g is $\overline{\mathbf{Q}}$-rational, and so $f \in \mathcal{A}_0(\overline{\mathbf{Q}})$. Conversely, if $f \in \mathcal{A}_0(\overline{\mathbf{Q}})$, then g is $\overline{\mathbf{Q}}$-rational, and $f(w) = g\big(\varphi(w)\big) \in \overline{\mathbf{Q}}$ for $w \in \mathcal{W}'$ by Theorem 24.11. This proves (2). Clearly (1) follows from (2) because of the fact mentioned at the end of §33.3. To prove (3), take a CM-point w on $\mathcal{H}$ fixed by $h(Y^u)$ with $h\colon Y \to K_m^m$ satisfying (24.10a). Define $\tilde{h}\colon Y \to \mathbf{Q}_{2d}^{2d}$ by $\tilde{h}(a) = \widetilde{h(a)}$. Then (24.10a) together with (33.1a) shows that $\tilde{h}(a^\rho) = J_d \cdot {}^t\tilde{h}(a) J_d^{-1}$, and $\varepsilon(w)$ is the fixed point of $\tilde{h}(Y^u)$. Thus $\varepsilon(w)$ is a CM-point on $\mathfrak{H}_d$. Since the statement corresponding to (2) is true for $\mathfrak{A}_0(\overline{\mathbf{Q}})$ and $\mathfrak{H}_d$, we obtain (3) from (2).

It should of course be remarked that $\mathcal{W}$ as in (2) exists. In fact, we can easily find m CM-types (K, Φ_i) such that Ψ is equivalent to $\sum_{i=1}^m \Phi_i$. Then Lemma 24.16 produces at least one CM-point w_0 on $\mathcal{H}$. By Proposition 23.5(3) we can take the set of points $\alpha(w)$ for all $\alpha \in U(T)$ as $\mathcal{W}$. We can also find a CM-point on $\mathcal{H}$ fixed by $h(Y^u)$ with a CM-field Y, as shown in [S64, Proposition 4.10].

33.5. Take $b \in K$ so that $K = \mathbf{Q}(b)$ and $bb^\rho = 1$; put $\alpha = b1_m$ and

$$U(z) = V\big(\varepsilon(z)\big), \quad V(Z) = P(Z)^{-1}(P\|_\omega \tilde{\alpha})(Z) \quad (z \in \mathcal{H}, Z \in \mathfrak{H}_d)$$

with ω of (30.2a) with d as n and $P \in \mathcal{A}_\omega(\mathbf{Q}_{ab})$ such that $\det[P \circ \varepsilon] \neq 0$. Lemma 30.2 guarantees such a P. By (2) of the same lemma, the entries of V belong to $\mathfrak{A}_0(\mathbf{Q}_{ab})$, and hence the entries of U belong to $\mathcal{A}_0(\overline{\mathbf{Q}})$ by Proposition 33.4(3). From (23.3d), (23.4k), (24.1a, b), and (24.7a) we see that $M(\alpha, z) = \Psi(b)$, which combined with (33.1e) shows that $\mu\big(\tilde{\alpha}, \varepsilon(z)\big) = {}^t\kappa(z)\Psi(b) \cdot {}^t\kappa(z)^{-1}$, and hence $U(z)^{-1} = X(z)^{-1}\Psi(b)X(z)$ with $X(z) = {}^t\kappa(z)^{-1}P\big(\varepsilon(z)\big)$. Since $K = \mathbf{Q}[b]$, we see that $a \mapsto X(z)^{-1}\Psi(a)X(z)$ is a ring-injection of K into $\mathcal{A}_0(\overline{\mathbf{Q}})^d_d$. Therefore we can find an element W of $GL_d\big(\mathcal{A}_0(\overline{\mathbf{Q}})\big)$ such that $X(z)^{-1}\Psi(a)X(z) = W^{-1}\Psi(a)W$ for every $a \in K$. In view of (24.1b) we have

$$(33.5a) \qquad {}^t\kappa(z)^{-1}P\big(\varepsilon(z)\big) = X(z) = \mathrm{diag}[R_v, S_v]_{v\in\mathbf{a}} \cdot W(z)$$

with square matrices R_v and S_v of size r_v and s_v, whose entries are meromorphic functions on $\mathcal{H}$. Employing (33.1e) and (24.7a), we easily see that

$$(33.5b) \quad R_v\big(\gamma(z)\big) = \lambda_v(\gamma, z)R_v(z), \quad S_v\big(\gamma(z)\big) = \mu_v(\gamma, z)S_v(z) \quad (v \in \mathbf{a})$$

if γ belongs to a sufficiently small congruence subgroup Γ of $U(T)$.

33.6. Let us now consider the family $\mathcal{F}(\Omega)$ of (24.5e) with $\Omega = \{K, \Psi, L, T\}$, where $L = g^{-1}(\mathbf{Z}^1_{2d})$ with g: $K^1_{2m} \to \mathbf{Q}^1_{2d}$ as in §33.1. Take a CM-algebra Y, a map h, and the fixed point w of $h(Y^u)$ as in §24.10. Changing w for $\alpha(w)$ with a suitable $\alpha \in U(T)$, we may assume that both $P\big(\varepsilon(z)\big)$ and $W(z)$ are finite and invertible at $z = w$. Then, so are R_v and S_v by (33.5a).

We first assume that Y itself is a CM-field. Thus $[Y : K] = m$, and if $\mathcal{P}_w = (A_w, \mathcal{C}_w, \iota_w)$ is the member of our family at w, then (A_w, ι_w) is of type (Y, Φ) with some Φ whose restriction to K is equivalent to Ψ. We take here a $\overline{\mathbf{Q}}$-rational model of $\mathcal{P}_w$. Now A_w is isomorphic to $\mathbf{C}^d/p_w(L)$. Also, we see that $u \mapsto \kappa(w)^{-1}u$ for $u \in \mathbf{C}^d$ gives an isomorphism of $\mathbf{C}^d/p_w(L)$ onto $\mathbf{C}^d/[\varepsilon(w) \quad 1_d]\mathbf{Z}^{2d}$. By Theorem 30.3 we obtain $\overline{\mathbf{Q}}$-rational 1-forms ξ_k on A_w by putting

$$\begin{bmatrix} \xi_1 \\ \vdots \\ \xi_d \end{bmatrix} = \pi \cdot {}^tP(\varepsilon(w))\kappa(w)^{-1} \begin{bmatrix} du_1 \\ \vdots \\ du_d \end{bmatrix},$$

where $u_1, \ldots, u_d$ are the standard coordinate functions on $\mathbf{C}^d$.

We have a decomposition $\mathbf{C}^d = \bigoplus_{v\in\mathbf{a}} V_v$ with V_v, isomorphic to $\mathbf{C}^m$, on which Ψ_v of (24.1b) acts. Let $x^v_1, \ldots, x^v_{r_v}, y^v_1, \ldots, y^v_{s_v}$ be the coordinate functions on V_v. Then these for all v are renamings of the u_k. Since $W(w)$ is

$\overline{\mathbf{Q}}$-rational, we find that the components of

$$(*) \quad \pi \cdot {}^t R_v(w) \begin{bmatrix} dx_1^v \\ \vdots \\ dx_r^v \end{bmatrix} \quad \text{and} \quad \pi \cdot {}^t S_v(w) \begin{bmatrix} dy_1^v \\ \vdots \\ dy_r^v \end{bmatrix} \qquad (r = r_v, s = s_v)$$

correspond to $\overline{\mathbf{Q}}$-rational 1-forms on A_w.

Let Φ_v, ψ_v, and φ_v be as in (24.10f) for the present Y. For $\alpha \in Y^u$ we have $\psi_v(\alpha) = \lambda_v(h(\alpha), w)$ and $\varphi_v(\alpha) = \mu_v(h(\alpha), w)$. Moreover $\operatorname{diag}[{}^t\psi_v(\alpha), {}^t\varphi_v(\alpha)]$, being the v-component of ${}^t M(h(\alpha), w)$, represents $\iota_w(\alpha)$ on V_v. Since the matrices $X_v(p_w)$ have algebraic entries by Lemma 33.2, (24.10d) shows that the same is true for $\psi_v(\alpha)$ and $\varphi_v(\alpha)$. Take $B_v \in GL_{r_v}(\overline{\mathbf{Q}})$ and $C_v \in GL_{s_v}(\overline{\mathbf{Q}})$ so that

$$B_v \psi_v(a) B_v^{-1} = \operatorname{diag}[\psi_{v1}(a), \ldots, \psi_{vr}(a)] \qquad (r = r_v),$$

$$C_v \varphi_v(a) C_v^{-1} = \operatorname{diag}[\varphi_{v1}(a), \ldots, \varphi_{vs}(a)] \qquad (s = s_v)$$

for every $a \in Y$ with $\psi_{vi}, \varphi_{vj} \in J_Y$. Putting

$$\begin{bmatrix} du_1^v \\ \vdots \\ du_r^v \end{bmatrix} = {}^t B_v^{-1} \begin{bmatrix} dx_1^v \\ \vdots \\ dx_r^v \end{bmatrix} \qquad (r = r_v),$$

we find that $\delta\iota(a) du_i^v = \psi_{vi}(a) du_i^v$. Now the periods of dx_i^v are entries of $p_w(L)$, which are algebraic, and hence so are the periods of du_i^v. Therefore by Theorem 32.2(1) we see that $\pi p_Y(\psi_{vi}, \Phi) du_i^v$, viewed as a 1-form on A_w, must be $\overline{\mathbf{Q}}$-rational. Comparing this result with $(*)$, we find that

$$\operatorname{diag}[p_Y(\psi_{v1}, \Phi), \ldots, p_Y(\psi_{vr}, \Phi)]^{-1} B_v R_v(w) \in GL_r(\overline{\mathbf{Q}}),$$

and similarly

$$\operatorname{diag}[p_Y(\varphi_{v1}, \Phi), \ldots, p_Y(\varphi_{vs}, \Phi)]^{-1} C_v S_v(w) \in GL_s(\overline{\mathbf{Q}}).$$

Put $f_v(z) = \det[R_v(z)]$ and $g_v(z) = \det[S_v(z)]$. Then

$$f_v(w) \sim \prod_{i=1}^{r} p_Y(\psi_{vi}, \Phi), \quad g_v(w) \sim \prod_{i=1}^{s} p_Y(\varphi_{vi}, \Phi).$$

Since $\psi_{vi} = \rho\varphi_{vi} = \tau_v$ on K, this shows that

$$(33.6a) \qquad (f_v/g_v)(w) \sim p_Y(\operatorname{Inf}_{Y/K}(\tau_v), \Phi).$$

33.7. We can apply the whole procedure to the case where Y is the direct sum of m copies of K. Let (K, Φ_i) for $1 \leq i \leq m$ be m CM-types such that Ψ is equivalent to the direct sum of the Φ_i. By Lemma 24.16 there is a CM-point w' on $\mathcal{H}$ such that $A_{w'}$ is isogenous to $\prod_{i=1}^{m} A_i$ with A_i of type (K, Φ_i). Then, changing w' for $\gamma(w')$ with a suitable $\gamma \in U(T)$ if necessary, we find for every fixed $v \in \mathbf{a}$ that

$$f_v(w') \sim \prod_{\tau_v \in \Phi_i} p_K(\tau_v, \Phi_i), \quad g_v(w') \sim \prod_{\tau_v \notin \Phi_i} p_K(\tau_v \rho, \Phi_i),$$

where the products are taken over all i as specified. Thus

$$(f_v/g_v)(w') \sim \prod_{i=1}^{m} p_K(\tau_v, \Phi_i)$$

because of our definition of $p_K(\tau\rho, \ldots)$ in §32.3. Call the last product $q(\tau_v)$.

Now if γ belongs to a sufficiently small congruence subgroup of $U(T)$, then $\det(\gamma) = \nu(\gamma) = 1$, and hence $\det[\lambda_v(\gamma, z)] = \det[\mu_v(\gamma, z)]$ for every $v \in \mathbf{a}$ by (23.3g). Therefore, by (33.5b) we see that $f_v/g_v \in \mathcal{A}_0$, and moreover we have seen that $q(\tau_v)^{-1}(f_v/g_v)(w') \sim 1$. Replace w' and $\alpha(w')$ with any $\alpha \in U(T)$ such that the functions W and $(P \circ \varepsilon)^{-1}$ are finite at $\alpha(w')$. Since such points form a dense subset of $\mathcal{H}$, we see that $q(\tau_v)^{-1} f_v/g_v \in \mathcal{A}_0(\overline{\mathbf{Q}})$ by virtue of Proposition 33.4(2).

33.8. We are now ready to complete the proof of Theorem 32.4. The quotient f_v/g_v depends only on v, Ω, and the choice of P and W, while $q(\tau_v)$ depends only on v and the (K, Φ_i). Therefore we obtain Theorem 32.4(1) when all s_i are nonnegative. In the general case we have $\sum_{i=1}^{m} s_i\Phi_i = \sum_{s_i \geq 0} s_i\Phi_i - \sum_{s_i<0} |s_i|\Phi_i$, and $\prod_{i=1}^{m} p_K(\tau, \Phi_i)^{s_i} = q_+(\tau)/q_-(\tau)$ with $q_+(\tau) = \prod_{s_i \geq 0} p_K(\tau, \Phi_i)^{s_i}$ and $q_-(\tau) = \prod_{s_i<0} p_K(\tau, \Phi_i)^{-s_i}$. Now suppose $\sum_{i=1}^{m} s_i\Phi_i = \sum_{i=1}^{m'} s_i'\Phi_i'$ with some s_i' and Φ_i'. Then $\prod_{i=1}^{m'} p_K(\tau, \Phi_i')^{s_i} = q_+'(\tau)/q_-'(\tau)$ with $q_\pm'(\tau)$ defined similarly with s_i' and Φ_i'. By virtue of the above result, we find that $q_+(\tau)q_-'(\tau) = q_-(\tau)q_+'(\tau)$, and so $q_+(\tau)/q_-(\tau) = q_+'(\tau)/q_-'(\tau)$, which proves the general case.

Next, the notation being as in (2) of Theorem 32.4, let Ψ be the restriction of ψ to K and let $m = [Y : K]$. Define r_v and s_v for this Ψ as in §24.1. Take $\zeta \in Y^\times$ so that $\zeta^\rho = -\zeta$ and $\mathrm{Im}(\zeta^\gamma) > 0$ for every $\gamma \in \psi$. Then $(x, y) \mapsto \mathrm{Tr}_{Y/K}(\zeta x y^\rho)$ is a K-valued nondegenerate skewhermitian form on $Y \times Y$. Take a K-linear bijection r: $Y \to K_m^1$. Then we find $T \in GL_m(K)$ such that $\mathrm{Tr}_{Y/K}(\zeta x y^\rho) = r(x)Tr(y)^*$ and $T^* = -T$. We then consider $\Omega = (K, \Psi, L, T)$ with these Ψ and T. Our choice of ζ implies condition (24.5f) in the present case. Defining a K-linear map h: $Y \to K_m^m$ by $r(ax) = r(x)h(a)$ for every $a, x \in Y$, we see that (24.10a) is satisfied. Thus we obtain a CM-point

w on $\mathcal{H}$ fixed by $h(Y^u)$, and the member (A_w, ι_w) is of type (Y, Φ) with Φ as in (24.10f). By (24.1f) we have

$$E_w\big(p_w(r(x)), p_w(r(y))\big) = \mathrm{Tr}_{K/\mathbf{Q}}\big(r(x)Tr(y)^*\big) = \mathrm{Tr}_{Y/\mathbf{Q}}(\zeta x y^\rho).$$

This shows that $(A_w, \mathcal{C}_w, \iota_w)$ is of type (Y, ζ, Φ), and so $\mathrm{Im}(\zeta^\gamma) > 0$ for every $\gamma \in \Phi$. This proves that Φ is equivalent to ψ. Taking this (Y, ψ) as (Y, Φ) in §33.6, we have (33.6a) with $p_Y\big(\mathrm{Inf}_{Y/K}(\tau_v), \psi\big)$ on the right-hand side.

Now we can easily find m CM-types (K, φ_i) such that $\mathrm{Res}_{Y/K}(\psi) = \sum_{i=1}^m \varphi_i$. Put $r_v = \prod_{i=1}^m p_K(\tau_v, \varphi_i)$. In §33.7 we showed that $r_v^{-1} f_v/g_v \in \mathcal{A}_0(\overline{\mathbf{Q}})$, and hence $r_v^{-1}(f_v/g_v)(w) \sim 1$ for the present w. This shows that $p_Y\big(\mathrm{Inf}_{Y/K}(\tau_v), \psi\big) \sim r_v$. Now, if $\mathrm{Res}_{Y/K}(\psi) = \sum_{i=1}^n s_i\Phi_i$ with $s_i \in \mathbf{Z}$ and CM-types Φ_i of K, then $r_v = \prod_{i=1}^n p_K(\tau_v, \Phi_i)^{s_i}$ by virtue of assertion (1), which we already proved. This proves (2), and our proof of Theorem 32.4 is complete. It should be remarked that our reasoning is valid even if $r_v s_v = 0$ for some or all v by virtue of our convention (23.3h, i). In fact, if $r_v = 0$ (resp. $s_v = 0$), we ignore R_v (resp. S_v) and put $f_v = 1$ (resp. $g_v = 1$).

Bibliography

[1] A. A. Albert, On the construction of Riemann matrices, I, II, Ann. of Math., **35** (1934), 1–28, **36** (1935), 376–394.

[2] A. A. Albert, A solution of the principal problem in the theory of Riemann matrices, Ann. of Math., **35** (1934), 500–515.

[3] I. Barsotti, Abelian varieties over fields of positive characteristic, Rendiconti del circolo di Palermo, **5** (1956), 1–25.

[4] O. Blumenthal, Über Modulfunktionen von mehreren Veränderlichen, I, II, Math. Ann., **56** (1903), 509–548, **58** (1904), 497–527.

[5] W. L. Chow, The Jacobian variety of an algebraic curve, Amer. J. Math., **76** (1954), 453–476.

[6] H. Davenport und H. Hasse, Die Nullstellen der Kongruenzzetafunktionen im gewissen zyklischen Fällen, Journ. Reine Angew, Math., **172** (1935), 151–182.

[7] M. Deuring, Algebren, Ergebnisse der Math., Berlin, 1935.

[8] M. Deuring, Die Typen der Multiplikatorenringe elliptischer Funktionenkörper, Abh. Math. Sem. Univ. Hamburg, **14** (1941), 197–272.

[9] M. Deuring, Reduktion algebraischer Funktionenkörper nach Primdivisoren des Konstantenkörpers, Math. Zeitschr., **47** (1942), 643–654.

[10] M. Deuring, Algebraische Begründung der komplexen Multiplikation, Abh. Math. Sem. Univ. Hamburg, **16** (1949), 32–47.

[11] M. Deuring, Die Struktur der elliptischen Funktionenkörper und Klassenkörper der imaginären quadratischen Zahlkörper, Math. Ann., **124** (1952), 393–426.

[12] M. Deuring, Die Zetafunktion einer algebraischen Kurve vom Geschlechte Eins, I, II, III, IV, Nachr. Akad. Wiss. Göttingen, (1953) 85–94, (1955) 13–42, (1956) 37–56, (1957) 55–80.

[13] M. Eichler, Quaternäre quadratische Formen und die Riemannsche Vermutung für die Kongruenzzetafunktion, Arch. Math., **5** (1954), 355–366.

[14] M. Eichler, Der Hilbertsche Klassenkörper eines imaginärquadratischen Zahlkörpers, Math. Zeitschr., **64** (1956), 229–242.

[15] R. Fricke, Lehrbuch der Algebra III, Braunschweig, 1928.

[16] R. Fueter, Vorlesungen über die singulären Moduln und die komplexe Multiplikation der elliptischen Funktionen, I, II, 1924, 1927.

[17] H. Hasse, Neue Begründung der komplexen Multiplikation, I, II, Journ. Reine Angew. Math., **157** (1927), 115–139, **165** (1931), 64–88.

[18] H. Hasse, Abstrakte Begründung der komplexen Multiplikation und Riemannsche Vermutung in Funktionenkörpern, Abh. Math. Sem. Univ. Hamburg, **10** (1934), 325–348.

[19] H. Hasse, Zetafunktion und L-Funktionen zu einem arithmetischen Funktionenkörper vom Fermatschen Typus, Abh. Deutscher Akad. Wiss., 1955.

[20] E. Hecke, Höhere Modulfunktionen und ihre Anwendung auf die Zahlentheorie, Math. Ann., **71** (1912), 1–37.

[21] E. Hecke, Über die Konstruktion relativ-Abelscher Zahlkörper durch Modulfunktionen von zwei Variablen, Math. Ann., **74** (1913), 465–510.

[22] E. Hecke, Eine neue Art von Zetafunktionen und ihre Beziehungen zur Verteilung der Primzahlen, I, II, Math. Zeitschr., **1** (1918), 357–376, **6** (1920), 11–51.

[23] S. Koizumi, On the differential forms of the first kind on algebraic varieties, J. Math. Soc. Japan, **1** (1949), 273–280.

[24] S. Koizumi and G. Shimura, On specializations of abelian varieties, Scientific Papers of the College of General Education, University of Tokyo, **9** (1959), 187–211.

[25] L. Kronecker, Zur Theorie der elliptischen Funktionen, 1883–1889, Werke IV.

[26] S. Lang, Abelian varieties, Interscience Tracts, New York, 1959.

[27] S. Lefschetz, On certain numerical invariants of algebraic varieties with applications to abelian varieties, Trans. Amer. Math. Soc., **22** (1921), 327–482.

[28] T. Matsusaka, Polarized varieties, fields of moduli and generalized Kummer varieties of polarized abelian varieties, Amer. J. Math., **80** (1958), 45–82.

[29] Y. Nakai, On the divisors of differential forms on algebraic varieties, J. Math. Soc. Japan, **5** (1953), 184–199.

[30] Séminaire H. Cartan, E.N.S., 1957/1958, Fonctions automorphes.

[31] J.-P. Serre, Quelques propriétés des variétés abéliennes en caractéristique p. Amer. J. Math., **80** (1958), 715–739.

[32] J.-P. Serre, Groupes algébriques et corps de classes, Hermann, Paris, 1959.

[33] G. Shimura, Reduction of algebraic varieties with respect to a discrete valuation of the basic field, Amer. J. Math., **77** (1955), 134–176.

[34] G. Shimura, On complex multiplications, Proceedings of the International Symposium on Algebraic Number Theory, Tokyo-Nikko, 1955, 23–30.

[35] G. Shimura, Correspondances modulaires et les fonctions ζ de courbes algébriques, J. Math. Soc. Japan, **10** (1958), 1–28.

[36] G. Shimura, On the theory of automorphic functions, Ann. of Math., **70** (1959), 101–144.

[37] G. Shimura, Fonctions automorphes et correspondances modulaires, Proceedings of the International Congress of Mathematicians, 1958, 330–338.

[38] C. L. Siegel, Einführung in die Theorie der Modulfunktionen n-ten Grades, Math. Ann., **116** (1939), 617–657.

[39] L. Stickelberger, Über eine Verallgemeinerung der Kreisteilung, Math. Ann., **37** (1890), 321–367.

[40] T. Takagi, Über eine Theorie des relativ-Abelschen Zahlkörpers, J. Coll. Science, Tokyo, **41** (1920), 1–132.

[41] Y. Taniyama, Jacobian varieties and number fields, Proceedings of the International Symposium on Algebraic Number Theory, Tokyo-Nikko, 1955, 31–45.

[42] Y. Taniyama, L-functions of number fields and zeta functions of abelian varieties, J. Math. Soc. Japan, **9** (1957), 330–366.

[43] H. Weber, Lehrbuch der Algebra, III, Braunschweig, 2 Auflage, 1908.

[44] A. Weil, Foundations of algebraic geometry, New York, 1946.

[45] A. Weil, Sur les courbes algébriques et les variétés qui s'en déduisent, Hermann, Paris, 1948.

[46] A. Weil, Variétés abéliennes et courbes algébriques, Hermann, Paris, 1948.

[47] A. Weil, Number of solutions of equations in finite fields, Bull. Amer. Math. Soc., **55** (1949), 497–508.

[48] A. Weil, Number-theory and algebraic geometry, Proceedings of the International Congress of Mathematicians, 1950, 90–100.

[49] A. Weil, Arithmetic on algebraic varieties, Ann. of Math., **53** (1951), 412–444.

[50] A. Weil, Jacobi sums as "Grössencharactere", Trans. Amer. Math. Soc., **73** (1952), 487–495.

[51] A. Weil, On algebraic groups of transformations, Amer. J. Math., **77** (1955), 355–391.

[52] A. Weil, On algebraic groups and homogeneous spaces, Amer. J. Math., **77** (1955), 493–512.

[53] A. Weil, On a certain type of characters of the idèle-class group of an algebraic number-field, Proceedings of the International Symposium on Algebraic Number Theory, Tokyo-Nikko, 1955, 1–7.

[54] A. Weil, On the theory of complex multiplication, ibid., 9–22.

[55] A. Weil, The field of definition of a variety, Amer. J. Math., **78** (1956), 509–524.

[56] A. Weil, On the projective embedding of abelian varieties, Algebraic geometry and topology, a symposium in honor of S. Lefschetz, Princeton, 1957.

[57] A. Weil, Introduction à l'étude des variétés kählériennes, Hermann, Paris, 1958.

Supplementary References

[BB] W. L. Baily and A. Borel, Compactification of arithmetic quotients of bounded symmetric domains, Ann. of Math., **84** (1966), 442–528.

[Bl] D. Blasius, On the critical values of Hecke L-series, Ann. of Math., **124** (1986), 23–63.

[De] M. Deuring, Die Klassenkörper der komplexen Multiplikation, Enzyklopädie Math. Wiss. Neue Aufl. Band I-2, Heft 10-II, Stuttgart, 1958.

[Do] B. Dodson, Solvable and nonsolvable CM-fields, Amer. J. Math., **108** (1986), 75–93.

[E] M. Eichler, Einführung in die Theorie der algebraischen Zahlen und Funktionen, Birkhäuser, Basel, 1963.

[H] T. Honda, Isogeny classes of abelian varieties over finite fields, J. Math. Soc. Japan, **20** (1968), 83–95.

[KP] A. Krazer and F. Prym, Neue Grundlagen einer Theorie der Allgemeinen Thetafunktionen, Teubner, Leipzig, 1892.

[L] S. Lang, Diophantine approximation on abelian varieties with complex multiplication, Advances in Math., **17** (1975), 281–336.

[M] T. Miyake, On automorphism groups of the fields of automorphic functions, Ann. of Math., **95** (1972), 243–252.

[N] A. Néron, Models Minimaux des variétés abéliennes sur les corps locaux et globaux, Publ. Math. I. H. E. S., No. 21 (1964), 5–128.

[Ra] K. Ramachandra, Some applications of Kronecker's limit formula, Ann. of Math., **80** (1964), 104–148.

[Ru83a] R. Rumely, On the Grössen character of an abelian variety in a parametrized family, Trans. Amer. Math. Soc., **276** (1983), 213–233.

[Ru83b] R. Rumely, A formula for the Grössen character of a parametrized elliptic curve, J. Number Theory, **17** (1983), 389–402.

[ST] J.-P. Serre and J. Tate, Good reduction of abelian varieties, Ann. of Math., **88** (1968), 492–517.

[S61] G. Shimura, On the zeta functions of the algebraic curves uniformized by certain automorphic functions, J. Math. Soc. Japan, **13** (1961), 275–331.

[S63] G. Shimura, On analytic families of polarized abelian varieties and automorphic functions, Ann. of Math., **78** (1963), 149–192.

[S64] G. Shimura, On the field of definition for a field of automorphic functions, Ann. of Math., **80** (1964), 160–189.

[S65] G. Shimura, On the field of definition for a field of automorphic functions: II, Ann. of Math., **81** (1965), 124–165.

[S66] G. Shimura, Moduli and fibre systems of abelian varities, Ann. of Math., **83** (1966), 294–338.

[S67] G. Shimura, Construction of class fields and zeta functions of algebraic curves,

Ann. of Math., **85**, 58–159 (1967).

[S70] G. Shimura, On canonical models of arithmetic quotients of bounded symmetric domains, Ann. of Math., **91** (1970), 144–222; II, **92** (1970), 528–549.

[S71a] G. Shimura, Introduction to the arithmetic theory of automorphic functions, Publ. Math. Soc. Japan, No. 11, Iwanami Shoten and Princeton Univ. Press, 1971.

[S71b] G. Shimura, On the zeta-function of an abelian variety with complex multiplication, Ann. of Math., **94** (1971), 504–533.

[S71c] G. Shimura, On the field of rationality for an abelian variety, Nagoya Math. J., **45** (1971), 167–178.

[S75a] G. Shimura, On some arithmetic properties of modular forms of one and several variables, Ann. of Math., **102** (1975), 491–515.

[S75b] G. Shimura, On the Fourier coefficients of modular forms of several variables, Göttingen Nachr. Akad. Wiss., 1975, 261–268.

[S76] G. Shimura, Theta functions with complex multiplication, Duke Math. J., **43** (1976), 673–696.

[S77a] G. Shimura, On the derivatives of theta functions and modular forms, Duke Math. J., **44** (1977), 365–387.

[S77b] G. Shimura, On abelian varieties with complex multiplication, Proc. London Math. Soc., 3d ser., **34** (1977), 65–86.

[S78a] G. Shimura, The special values of the zeta functions associated with Hilbert modular forms, Duke Math. J., **45** (1978), 637–679.

[S78b] G. Shimura, On certain reciprocity-laws for theta functions and modular forms, Acta Math., **141** (1978), 35–71.

[S79] G. Shimura, Automorphic forms and the periods of abelian varieties, J. Math. Soc. Japan, **31** (1979), 561–592.

[S80] G. Shimura, The arithmetic of certain zeta functions and automorphic forms on orthogonal groups, Ann. of Math., **111** (1980), 313–375.

[S82] G. Shimura, Models of an abelian variety with complex multiplication over small fields, J. Number Theory, **15** (1982), 25–35.

[S83] G. Shimura, On Eisenstein series, Duke Math. J., **50** (1983), 417–476.

[S88] G. Shimura, On the critical values of certain Dirichlet series and the periods of automorphic forms, Inv. Math., **94** (1988), 245–305.

[S93] G. Shimura, On the transformation formulas of theta series, Amer. J. Math., **115** (1993), 1011–1052.

[S94] G. Shimura, Differential operators, holomorphic projection, and singular forms, Duke Math. J., **76** (1994), 141–173.

[Y81] H. Yoshida, Abelian varieties with complex multiplication and representations of the Weil group, Ann. of Math., **114** (1981), 87–102.

[Y82] H. Yoshida, Hecke characters and models of abelian varieties with complex multiplication, J. Fac. Sci. Univ. of Tokyo, Sec. IA, **28** (1982), 633–649.

Index

About the Author

Goro Shimura is Professor of Mathematics at Princeton University. He was awarded the Leroy P. Steele Prize in 1996 for lifetime achievement in mathematics by the American Mathematical Society. He is the author of *Introduction to Arithmetic Theory of Automorphic Functions* (Princeton).